D-Amino Acids

Tohru Yoshimura • Toru Nishikawa •
Hiroshi Homma
Editors

D-Amino Acids

Physiology, Metabolism, and Application

Springer

Editors
Tohru Yoshimura
Department of Applied Molecular Biosciences
Graduate School of Bioagricultural Sciences
Nagoya University
Nagoya
Japan

Toru Nishikawa
Department of Psychiatry and Behavioral Sciences
Graduate School of Medical and Dental Sciences
Tokyo Medical and Dental University
Tokyo
Japan

Hiroshi Homma
Laboratory of Biomolecular Science
Department of Pharmaceutical Life Sciences
Kitasato University
Tokyo
Japan

ISBN 978-4-431-56075-3 ISBN 978-4-431-56077-7 (eBook)
DOI 10.1007/978-4-431-56077-7

Library of Congress Control Number: 2016947710

Printed on acid-free paper

This Springer imprint is published by Springer Nature
The registered company is Springer Japan KK

Preface

D-Amino acids were once called unnatural amino acids and were considered to be insignificant for eukaryotes especially in mammals. D-Amino acids had been known to have only a few roles, for example, as components of the peptidoglycan layers of bacterial cell walls and as antibiotics. However, in the 1990s, D-serine was found to serve as a co-agonist of the *N*-methyl-D-aspartate (NMDA) receptor in mammalian brains and to be involved in various brain functions. These findings encouraged further studies on D-amino acids. Currently, they have been revealed to bear important physiological roles. For example, D-serine is implicated in memory formation and learning, and its abnormal concentration in tissues has been reported in various neurological diseases such as schizophrenia and amyotrophic lateral sclerosis. D-Aspartate is found in various mammalian tissues, particularly in the central nervous system and the genitals. D-Aspartate facilitates the endocrine secretion of prolactin, inhibits the secretion of melatonin, and plays a peculiar role in the control of reproductive functions in mammals, including the stimulation of testosterone synthesis.

The objective of this book is to provide an overview of the roles of D-amino acids and to introduce recent progress in studies of them. Part I reviews the indispensable analytical methods by which D-amino acids are studied. Results of studies on D-serine and D-aspartate are shown in Parts II and III, respectively. Remarkable progress in studies on D-amino acids in peptides is described in Part IV. Enzymes producing and degrading D-amino acids are reviewed in Part V, as the knowledge of these enzymes is indispensable for understanding the profound functions of D-amino acids. Part VI introduces their role in foods, especially in fermented foods, which is probably the source of exogenous D-amino acids. It will give me great pleasure if this book facilitates a better understanding of the importance and the fascinating aspects of D-amino acids.

Nagoya, Japan
January 2016

Tohru Yoshimura, on behalf of the editors

Contents

Part I
Detection of D-Amino Acids

Chapter 1
Determination of D-Amino Acids and Their Distribution in Mammals

Yurika Miyoshi, Reiko Koga, and Kenji Hamase

Abstract D-Amino acids are the enantiomers of L-amino acids. Although the L-forms are predominantly present in life systems in the free form and also in the protein-bound form, the optically inversed D-forms were hardly found especially in higher animals. This is due to the very low amounts of these D-amino acids, and the determination is frequently disturbed by large amounts of L-amino acids and uncountable numbers of peptides and amino compounds. However, due to the recent progress in analytical technologies, several D-amino acids have been found in mammals including humans. In this chapter, the analytical methods for determining these D-amino acids and their distribution in mammals are described.

Keywords D-Amino acids • Chiral separation • Analytical method • Distribution

D-Amino acids are the enantiomers of L-amino acids. Although L-amino acids are predominantly present in organisms with distinct physiological functions, the chiral counterparts, D-amino acids, were rarely found, especially in the higher animals. Therefore, these D-amino acids had been long believed that they do not have physiological meanings in mammals. However, along with the progress in analytical techniques, several free D-amino acids have been found in mammals including humans and their distributions, functions, and regulation mechanisms have gradually been clarified. In this chapter, the analytical methods of these free D-amino acids and their distributions in mammals are described. Concerning the D-amino acid residues in proteins, please refer to an excellent review (Fujii et al. 2011).

Y. Miyoshi • R. Koga • K. Hamase (✉)
Graduate School of Pharmaceutical Sciences, Kyushu University, 3-1-1 Maidashi, Higashi-ku, Fukuoka 812-8582, Japan
e-mail: hamase@phar.kyushu-u.ac.jp

T. Yoshimura et al. (eds.), *D-Amino Acids*, DOI 10.1007/978-4-431-56077-7_1

1.1 Determination of D-Amino Acids in Biological Samples

In most cases, the amounts of D-amino acids are at trace levels in the tissues and physiological fluids of higher animals, and the determination is interfered by many of the intrinsic substances (large amounts of L-amino acids, peptides, and amines). Therefore, sensitive and selective analytical methods consisting of chromatographic or electro-driven separation techniques, or using highly selective enzymes/antibodies, have been established. In this section, colorimetric methods using enzymes/antibodies and separation analyses including GC, HPCE, and HPLC are described.

1.1.1 Colorimetric Analysis Using Enzymes and Antibodies

Colorimetric analyses are useful as simple and rapid methods. Concerning the D-amino acid analysis, several methods have already been established. Historically, simple methods using D-amino acid oxidase (DAO, EC 1.4.3.3) and D-aspartic acid oxidase (DDO, EC 1.4.3.1) have frequently been used. These enzymes react with the D-amino acids to form α-keto acids, and the keto acids further react with hydrazine to form hydrazones. Therefore, D-amino acids could be determined by a simple colorimetric analysis (Nagata et al. 1985). However, the substrate specificity of these enzymes is broad, and the D-amino acids are determined as their sum in the target matrices. The combination of DAO and HPLC is also used to separately determine various D-amino acids (Kato et al. 2011). For the specific determination of D-Ser, D-Ser dehydratase (DSD, EC 4.3.1.18) could also be used (Suzuki et al. 2011). Concerning the method using an antibody, an enzyme immunoassay using a specific monoclonal antibody against D-Asp has been established (Ohgusu et al. 2006). These methods are rapid and simple, and the amounts of the D-amino acids could therefore be easily determined. However, a variety of substances is present in the tissues and physiological fluids; thus, a thorough confirmation of the determined values would be necessary.

1.1.2 GC Methods for the Determination of D-Amino Acids

For the chiral amino acid analysis, enantioselective separations should be carried out. For the chromatographic and electrophoretic separations, the use of chiral derivatizing reagents, chiral stationary phases, or chiral mobile phases is essential. Concerning the GC methods, chiral derivatization reagents and chiral stationary phases have been widely used. As the chiral derivatization reagents, historically, a chiral alcohol ((+)-2-butanol) and a chiral amino acid derivative (*N*-trifluoroacetyl-L-prolyl chloride (TPC)) were used (Hamase et al. 2002). A chiral derivatization

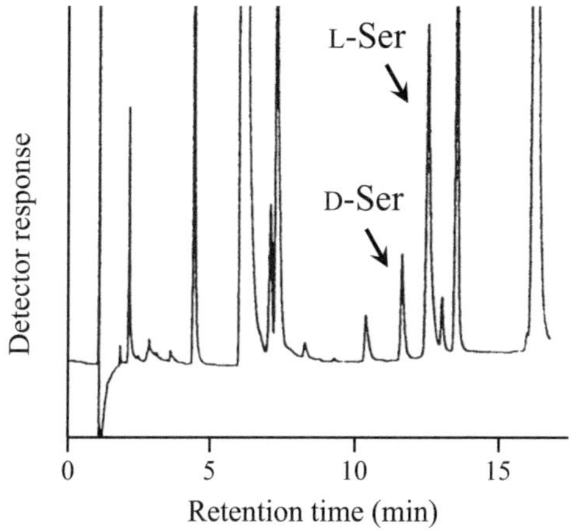

Fig. 1.1 Chiral GC determination of D-Ser in rat brain. Ser enantiomers were determined as their *N*,*O*-PFP-isopropyl derivatives using Chirasil-L-Val as a chiral stationary phase (Reproduced from a reference (Hashimoto et al. 1992a) with permission)

reagent, (*S*)-α-methoxy-α-trifluoromethylphenylacetyl chloride (MPTA-Cl, Gal and Ames 1977), was also used as a promising reagent without racemization during the derivatization procedure. As the chiral stationary phase, Chirasil-L-Val and Chirasil-γ-Dex are reported as useful columns, and most of the proteinogenic amino acids could be separated within 60 min (Schurig 2011). Using Chirasil-L-Val, a large amount of D-Ser in mammalian brain was discovered for the first time (Hashimoto et al. 1992a, Fig. 1.1.). For the enantioselective analysis of amino acids using GC, please refer to a previous review (Schurig 2011). In order to obtain a higher selectivity, two-dimensional GC methods have also been established (Junge et al. 2007; Waldhier et al. 2011). For the two-dimensional analysis, a long enantioselective column was used as the first dimension, and a short non-enantioselective column was used as the rapid second dimension. Although complicated analytical instruments are needed, the 2D-GC methods provide a selective and comprehensive determination of the D-amino acids in biological samples.

1.1.3 HPCE Methods for the Determination of D-Amino Acids

The separation of chiral amino acids using HPCE is mainly performed by chiral mobile phase methods along with adding some chiral selectors to the electrophoresis buffer system. For the sensitive determination, amino acids are normally derivatized with a fluorescence reagent and determined by a laser-induced fluorescence detector. As the derivatization reagent, naphthalene-2,3-dicarboxaldehyde (NDA) was often used, and cyclodextrin derivatives are added to the buffer system as the chiral selector. Using this method, D-Ser, D-Asp, D-Glu and D-Ala could be determined (Miao et al. 2005; Ota et al. 2014). 4-Fluoro-7-nitro-2,1,3-benzoxadiazole (NBD-F) and fluorescein isothiocyanate (FITC) are also used as the derivatization reagent and successfully applied to the cyclodextrin-based separation system. For the HPCE analysis of D-amino acids, please refer to a previous review (Kitagawa and Otsuka 2011).

1.1.4 HPLC Methods for the Determination of D-Amino Acids

Concerning HPLC, chiral derivatization reagents and chiral stationary phases are frequently used for the determination of D-amino acids in real biological samples. Historically, a chiral derivatizing reagent, 1-fluoro-2,4-dinitrophenyl-5-L-alanine amide (FDAA, Marfey's reagent), was reported in 1984 (Marfey 1984). This reagent reacts with amino acids to form peptide-type diastereomers and could be separated by a widely used reversed-phase column. The FDAA-derivatized amino acids could be monitored by the absorbance at 340 nm, and various FDAA analogues are currently used as simple analytical tools for a user-friendly HPLC system equipped with a UV detector (Bhushan and Brückner 2011). *o*-Phthalaldehyde (OPA) plus chiral thiol is also historically used for the determination of chiral amino acids. OPA was originally developed for the highly sensitive fluorescence derivatization of amino acids as a post-column derivatization reagent and also a pre-column derivatization reagent with the a chiral thiol, 2-mercaptoethanol. In 1984, Aswad established a method using OPA and a chiral thiol compound, *N*-acetyl-L-cysteine (Aswad 1984). Subsequently, various chiral thiols were designed and used for the D-amino acid analysis (Buck and Krummen 1984; Nimura and Kinoshita 1986; Brückner et al. 1994). The OPA method enables the highly sensitive fluorescence determination of chiral amino acids in combination with a reversed-phase HPLC system and widely used for biological samples. Brain D-Ser is easily determined by OPA plus a chiral thiol, *N-tert*-butyloxycarbonyl-L-cysteine (Boc-L-Cys, Hashimoto et al. 1992b). An analogue of 9-fluorenylmethyl chloroformate (Fmoc-Cl), (+)-1-(9-fluorenyl)ethyl chloroformate (FLEC), was designed (Einarsson et al. 1987). FLEC has a chiral carbon, and most of the proteinogenic amino acids were nicely separated using a reversed-phase column. For the OPA and FLEC methods, an HPLC system with a fluorescence detector was used, and the highly sensitive determination of chiral amino acids could be carried out. These days, several chiral derivatization reagents, such as *R*(−)-4-(3-isothiocyanatopyrrolidin-1-yl)-7-(*N*,*N*-dimethylaminosulfonyl)-2,1,3-benzoxadiazole (*R*(−)-DBD-PyNCS (Min et al. 2011) and (*S*)-*N*-(4-nitrophenoxycarbonyl)-L-phenylalanine-2-methoxyethyl ester (*S*)-NIFE (Visser et al. 2011), have also been developed for the HPLC-MS/MS technique, and the highly sensitive and selective determination of D-amino acids in biological matrices was successfully performed. Although the chiral derivatization methods have several fundamental problems, such as the optical purity of the reagents and a difference in the chemical properties of the diastereomers, they are frequently used as user-friendly analytical tools. For these series of reagents, please refer to the literature (Miyoshi et al. 2012a, 2013).

The non-chiral derivatization of amino acids and enantioselective separation with a chiral stationary phase are also widely used for the determination of chiral amino acids in biological matrices. Using the cyclodextrin-bonded column, amino acid enantiomers were separated as their 1-dimethylaminonaphthalene-5-sulfonyl (DNS) derivatives and also by their native (underivatized) forms (Armstrong et al. 1987). Pirkle-type chiral stationary phases are also widely used for the separation of amino acid enantiomers (Pirkle et al. 1980; Ôi et al. 1989). Especially, Pirkle-type columns having an amino acid derivative in the chiral center and also having a π-π interacting moiety are powerful tools for separating *N*-protected chiral amino acids. In combination with a fluorescence derivatization reagent, NBD-F (Imai and Watanabe 1981), and a Pirkle-type column, Sumichiral OA-2500S, a high amount of D-Asp in the pineal gland of the rat was demonstrated (Imai et al. 1995). Normally, the separation efficiency of the chiral stationary phases is not sufficient for the determination of D-amino acids in biological matrices and is only applicable to specific samples containing high amounts of the D-forms. Therefore, the combination of the chiral LC system with tandem mass spectrometry (LC-MS/MS) is frequently used these days (Reischl et al. 2011, Reischl and Lindner 2012). Cinchona alkaloid-based chiral stationary phases (Chiralpak QN-AX, QD-AX, and ZWIX) are powerful columns (Hoffmann et al. 2008), and several chiral HPLC-MS/MS methods have been established for the determination of D-amino acids in mammals.

1.1.5 Two-Dimensional HPLC Determination of D-Amino Acids

Although chiral LC methods are rapid and convenient, the determination of trace amounts of D-amino acids in mammalian tissues and physiological fluids is difficult in many cases. This is due to the presence of various interfering substances in real-world samples, and the analytical methods with a higher selectivity are needed. In this context, two-dimensional chromatographic approaches consisting of two different separation modes are effective (Miyoshi et al. 2014, Fig. 1.2.). Especially, heart-cut two-dimensional methods isolating the target amino acid fractions in the first dimension and further separating the target enantiomers in the second dimension are effective. Normally, amino acids are pre-column labeled with some non-chiral reagents and separated by a microbore reversed-phase column in the first dimension. The target amino acid fractions were online collected and transferred to the enantioselective column representing the second dimension in which the D- and L-forms were separated. Applying this two-dimensional HPLC concept, analytical methods for neuroactive D-amino acids (Miyoshi et al. 2011), acidic D-amino acids (Han et al. 2011a), branched aliphatic D-amino acids (Hamase

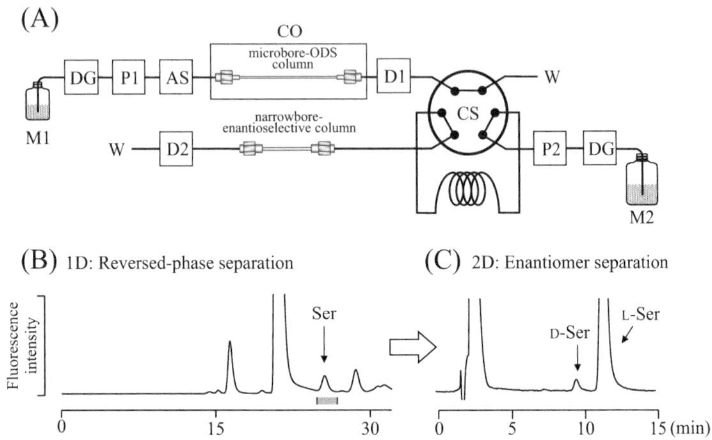

Fig. 1.2 2D-HPLC determination of D-Ser in mouse serum. (**a**) Flow diagram of the 2D-HPLC system, (**b**) reversed-phase separation of Ser in the first dimension as a D plus L mixture, (**c**) enantiomer separation of D- and L-Ser in the mouse serum in the second dimension (Reproduced from a reference (Miyoshi et al. 2009) with permission)

et al. 2007), proline analogues (Tojo et al. 2008), and hydrophilic D-amino acids (Hamase et al. 2010) have been established. The 2D-HPLC could be combined with an MS/MS detection system and the higher selective analysis of trace amounts of D-amino acids could be carried out (Koga et al. 2012). These methods are powerful tools for the determination of a variety of D-amino acids in complicated biological matrices.

1.2 Distribution of D-Amino Acids in Mammals

1.2.1 Distribution of D-Ser

In 1992, a high amount of D-Ser has been found in rat brain extracts by using GC-MS and HPLC (Hashimoto et al. 1992a, b). Until now, different research groups have confirmed the presence of D-Ser in mammalian brain using various analytical methods, and the distribution of D-Ser in various tissues and physiological fluids of mammals was also investigated (Hamase et al. 1997; Hashimoto et al. 1995; Morikawa et al. 2001; Nagata et al. 1994; Nishikawa 2011; Schell et al. 1997). The amounts of D-Ser in various tissues and physiological fluids of adult rats and mice are summarized in Table 1.1 (obtained by the 2D-HPLC methods, Hamase et al. 2010; Miyoshi et al. 2009, 2011, 2012b). The profiles of D-Ser distribution in rats and mice were almost the same. High amounts of D-Ser were found in the frontal brain areas such as the cerebrum and hippocampus (rats,

Table 1.1 Determination of D-Ser in mammalian tissues and physiological fluids using the 2D-HPLC

	Cerebrum	Hippo-campus	Hypothalamus	Olfactory bulb	Cerebellum	Medulla oblongata	Pituitary gland[a]	Spinal cord	Pancreas	Kidney	Liver	Plasma/Serum[b]	Urine	CSF
Rat	257.6–309.5	253.6–271.0	135.6–136.8	78.7–83.8	7.4–7.7	5.4	16.5–17.8	3.7–3.8	11.7–11.8	5.7–6.3	6.4–9.3	1.1–1.9	3.2–8.8	–
Mice	300.0–329.9	268.8–296.2	127.4	104.4	3.7–10.0	13.8	8.4	6.4	34.8	15.0	21.9	2.1–2.5	54.1	1.5

Values represent nmol/g or mL
[a]Values of rat were D-Ser contents in anterior pituitary
[b]Values of rat were D-Ser contents in plasma and those of mice were in serum

253.6–309.5 nmol/g wet tissue; mice, 268.8–329.7 nmol/g wet tissue) and the % D values (ratio of D-form to total Ser) of both species were 18–25 %. Relatively high amounts of D-Ser were observed in the hypothalamus and olfactory bulb (rats, 78.7–136.8 nmol/g wet tissue; mice, 104.4–127.4 nmol/g wet tissue). On the other hand, in the cerebellum, medulla oblongata, pituitary gland, spinal cord, and peripheral tissues, D-Ser amounts were low compared to those in the frontal brain areas. Concerning the physiological fluids, D-Ser amounts in the plasma/serum of the rats and mice were around 2 nmol/mL, and those in the urine of the rats and mice were 3.2–8.8 nmol/mL and 54.1 nmol/mL, respectively.

1.2.2 Distribution of D-Asp

D-Asp was discovered in rats and humans in 1986 (Dunlop et al. 1986), and subsequently, the distribution of D-Asp in rodents has also been investigated (D'Aniello et al. 2000; Hamase et al. 1997; Hashimoto et al. 1995; Katane and Homma 2011; Morikawa et al. 2001). The amount of D-Asp in various tissues and physiological fluids of rats and mice obtained using the 2D-HPLC system are shown in Table 1.2 (Hamase et al. 2010; Han et al. 2011a, b, 2015). The distribution of D-Asp in mammalian tissues and physiological fluids was entirely different from that of D-Ser. D-Asp was mainly localized in the endocrine tissues such as the pineal gland, pituitary gland, adrenal gland, testis, and thymus of the rats and pineal gland, testis, and thymus of mice. Especially, a high amount of D-Asp was found in the rat pineal gland (1669.8–3910.4 nmol/g wet tissue). On the other hand, the levels of D-Asp in the mice pineal gland were different depending on the mouse strains. D-Asp amounts in the pineal gland of specific strains (C3H and CBA mice) were high, while those of other mouse strains such as ddY, C57BL, BALB/C, and ICR were low. Concerning other brain tissues such as the cerebrum, hippocampus, hypothalamus, cerebellum, and medulla oblongata, D-Asp amounts in both the rats and mice were lower than 60 nmol/g wet tissue. In the peripheral tissues of both species, D-Asp was widely present (3–200 nmol/g wet tissue). In the plasma and urine of both the rats and mice, low amounts of D-Asp were present; the amounts of plasma and urine were 0.4–0.9 nmol/mL and 0.6–4.1 nmol/mL, respectively.

1.2.3 Distribution of D-Ala

Similar to D-Ser, D-Ala is well known as a co-agonist of the NMDA receptor. The distribution of D-Ala in various brain tissues, peripheral organs, and physiological fluids has been investigated both in rats and mice (Miyoshi et al. 2009, 2011; Morikawa et al. 2003). Regarding its distribution in rats, a relatively high amount

Table 1.2 Determination of D-Asp in mammalian tissues and physiological fluids using the 2D-HPLC

	Cerebrum	Hippo-campus	Hypothalamus	Cerebellum	Medulla oblongata	Pineal gland	Pituitary gland	Spinal cord	Adrenal gland	Testis	Thymus	Pancreas	Spleen	Lung	Heart	Kidney	Liver	Muscle	Plasma	Urine
Rat	10.7–12.8	6.0	15.0	19.7–29.3	6.5	1669.8–3910.4	132.5	6.7	574.3	129.6–178.1	88.1	19.3	195.6	141.1	20.2	34.1	14.5	3.5	0.5–0.9	0.9–4.1
Mice	58.5	–	9.0	9.2–18.7	5.5	44.7–364.2[a]	27.6	5.1	13.5	193.2–318.6	144.3	12.2	160.8	98.0	5.7	11.6	40.7	9.4	0.4–0.6	0.6–3.9

Values represent nmol/g or mL
[a]Value represents pmol/whole pineal gland

of D-Ala was found in the pituitary gland and the pancreas. Especially, the highest amount was observed in the anterior pituitary gland (the amount was six times higher than that in the posterior pituitary gland). Concerning the distribution in mice, D-Ala was highly localized in the pancreas followed by the pituitary gland (the mouse pituitary gland is too small to separate the anterior and posterior lobes). In addition to its presence in mammals, D-Ala was reported to have clear circadian changes (Morikawa et al. 2003; Karakawa et al. 2013). The concentration of D-Ala was high during the daytime and closely related to the activity rhythms of the rats and mice. The physiological meanings of the distribution and circadian changes of D-Ala in mammals are highly expected to be clarified (Table 1.3).

1.2.4 Other D-Amino Acids Observed in Mammals

Other than D-Ser, D-Asp, and D-Ala, the presence of various D-amino acids in mammals has gradually been elucidated in spite of their low amounts. D-Pro and D-Leu were reported to exist in some brain tissues of mice, and the amounts of both amino acids were comparatively high in the pineal gland (Hamase et al. 2001, 2006). As for D-Leu, the content in rats was also investigated in some specific tissues (Hamase et al. 2007). The amounts of D-Leu in both rodents are quite similar. Regarding other branched aliphatic D-amino acids (D-Val, D-*allo*-Ile, and D-Ile), most of them were observed only in the urine. The agonist for the NMDA receptor, D-Glu, was widely observed in the rat brain, peripheral tissues, and physiological fluids (Han et al. 2011a). However, the amounts were much lower than those of D-Asp, which is the same neuroactive and acidic amino acid as D-Glu. Other D-amino acids, such as hydrophilic amino acids, D-Asn, D-Arg, and D-*allo*-Thr, were highly excreted in the urine (Hamase et al. 2010; Zhao et al. 2004).

1.3 Closing Remarks

It was long believed that only L-amino acids were present in the higher animals like mammals as the physiologically active substances. However, since the discovery of high amounts of D-Ser and D-Asp in the neuronal and endocrine tissues, and also along with the progress of highly sensitive and selective analytical technologies, D-amino acids were increasingly recognized as "intrinsic bioactive molecules." Their functions, diagnostic values, and regulation mechanisms have also been gradually clarified, and further contribution to the human health sciences is highly expected.

Table 1.3 Determination of D-Ala in mammalian tissues and physiological fluids using the 2D-HPLC

	Cerebrum	Corpus striatum	Olfactory bulb	Cerebellum	Medulla oblongata	Spinal cord	Hippo-campus	Hypo-thalamus	Pituitary gland	Anterior pituitary	Posterior pituitary	Pineal gland	Thyroid gland
Rat	3.3-7.0	4.8	3.6-4.2	0.5–0.7	0.9–1.5	1.5–1.8	4.8–5.4	3.3–5.0	–	55.3–86.4	14.7	15.9	7.4
Mice	4.7	–	3.6	**0.4**	1.0	–	3.6	5.7	15.1	–	–	–	–

	Adrenal gland	Pancreas	Testis	Thymus	Spleen	Bone marrow	Muscle	Heart	Lung	Kidney	Liver	Serum	Plasma	Urine
Rat	12.1	29.2–73.4	7.3	15.2	9.7	17.0	10.6	6.7	19.9	3.6–4.4	4.4–6.0	10.9	10.4–16.2	11.0–33.0
Mice	–	109.4	–	–	–	–	–	–	–	5.8	17.5	10.9	–	58.5

Values represent nmol/g or mL

References

Armstrong DW, Yang X, Han SM, Menges RA (1987) Direct liquid chromatographic separation of racemates with an α-cyclodextrin bonded phase. Anal Chem 59:2594–2596

Aswad DW (1984) Determination of D- and L-aspartate in amino acid mixtures by high-performance liquid chromatography after derivatization with a chiral adduct of *o*-phthaldialdehyde. Anal Biochem 137:405–409

Bhushan R, Brückner H (2011) Use of Marfey's reagent and analogs for chiral amino acid analysis: assessment and applications to natural products and biological systems. J Chromatogr B 879:3148–3161

Brückner H, Haasmann S, Langer M, Westhauser T, Wittner R, Godel H (1994) Liquid chromatographic determination of D- and L-amino acids by derivatization with *o*-phthaldialdehyde and chiral thiols; applications with reference to biosciences. J Chromatogr A 666:259–273

Buck RH, Krummen K (1984) Resolution of amino acid enantiomers by high-performance liquid chromatography using automated pre-column derivatisation with a chiral reagent. J Chromatogr 315:279–285

D'Aniello A, Di Fiore MM, Fisher GH, Milone A, Seleni A, D'Aniello S, Perna AF, Ingrosso D (2000) Occurrence of D-aspartic acid and *N*-methyl-D-aspartic acid in rat neuroendocrine tissues and their role in the modulation of luteinizing hormone and growth hormone release. FASEB J 14:699–714

Dunlop DS, Neidle A, McHale D, Dunlop DM, Lajtha A (1986) The presence of free D-aspartic acid in rodents and man. Biochem Biophys Res Commun 141:27–32

Einarsson S, Josefsson B, Möller P, Sanchez D (1987) Separation of amino acid enantiomers and chiral amines using precolumn derivatization with (+)-1-(9-fluorenyl)ethyl chloroformate and reversed-phase liquid chromatography. Anal Chem 59:1191–1195

Fujii N, Kaji Y, Fujii N (2011) D-Amino acids in aged proteins: analysis and biological relevance. J Chromatogr B 879:3141–3147

Gal J, Ames MM (1977) Gas-chromatographic resolution of the enantiomers of 3-(3,4-dihydroxyphenyl)alanine and its α-methyl analog. Anal Biochem 83:266–273

Hamase K, Homma H, Takigawa Y, Fukushima T, Santa T, Imai K (1997) Regional distribution and postnatal changes of D-amino acids in rat brain. Biochim Biophys Acta 1334:214–222

Hamase K, Inoue T, Morikawa A, Konno R, Zaitsu K (2001) Determination of free D-proline and D-leucine in the brains of mutant mice lacking D-amino acid oxidase activity. Anal Biochem 298:253–258

Hamase K, Morikawa A, Zaitsu K (2002) D-Amino acids in mammals and their diagnostic value. J Chromatogr B 781:73–91

Hamase K, Takagi S, Morikawa A, Konno R, Niwa A, Zaitsu K (2006) Presence and origin of large amounts of D-proline in the urine of mutant mice lacking D-amino acid oxidase activity. Anal Bioanal Chem 386:705–711

Hamase K, Morikawa A, Ohgusu T, Lindner W, Zaitsu K (2007) Comprehensive analysis of branched aliphatic D-amino acids in mammals using an integrated multi-loop two-dimensional column-switching high-performance liquid chromatographic system combining reversed-phase and enantioselective columns. J Chromatogr A 1143:105–111

Hamase K, Miyoshi Y, Ueno K, Han H, Hirano J, Morikawa A, Mita M, Kaneko T, Lindner W, Zaitsu K (2010) Simultaneous determination of hydrophilic amino acid enantiomers in mammalian tissues and physiological fluids applying a fully automated micro-two-dimensional high-performance liquid chromatographic concept. J Chromatogr A 1217:1056–1062

Han H, Miyoshi Y, Ueno K, Okamura C, Tojo Y, Mita M, Lindner W, Zaitsu K, Hamase K (2011a) Simultaneous determination of D-aspartic acid and D-glutamic acid in rat tissues and physiological fluids using a multi-loop two-dimensional HPLC procedure. J Chromatogr B 879: 3196–3202

Han H, Miyoshi Y, Oyama T, Konishi R, Mita M, Hamase K (2011b) Enantioselective micro-2D-HPLC determination of aspartic acid in the pineal glands of rodents with various melatonin contents. J Sep Sci 34:2847–2853

Han H, Miyoshi Y, Koga R, Mita M, Konno R, Hamase K (2015) Changes in D-aspartic acid and D-glutamic acid levels in the tissues and physiological fluids of mice with various D-aspartate oxidase activities. J Pharm Biomed Anal 116:47–52

Hashimoto A, Nishikawa T, Hayashi T, Fujii N, Harada K, Oka T, Takahashi K (1992a) The presence of free D-serine in rat brain. FEBS Lett 296:33–36

Hashimoto A, Nishikawa T, Oka T, Takahashi K, Hayashi T (1992b) Determination of free amino acid enantiomers in rat brain and serum by high-performance liquid chromatography after derivatization with *N-tert.*-butyloxycarbonyl-L-cysteine and *o*-phthaldialdehyde. J Chromatogr 582:41–48

Hashimoto A, Oka T, Nishikawa T (1995) Anatomical distribution and postnatal changes in endogenous free D-aspartate and D-serine in rat brain and periphery. Eur J Neurosci 7:1657–1663

Hoffmann CV, Pell R, Lämmerhofer M, Lindner W (2008) Synergistic effects on enantioselectivity of zwitterionic chiral stationary phases for separations of chiral acids, bases, and amino acids by HPLC. Anal Chem 80:8780–8789

Imai K, Watanabe Y (1981) Fluorimetric determination of secondary amino acids by 7-fluoro-4-nitrobenzo-2-oxa-l,3-diazole. Anal Chim Acta 130:377–383

Imai K, Fukushima T, Hagiwara K, Santa T (1995) Occurrence of D-aspartic acid in rat brain pineal gland. Biomed Chromatogr 9:106–109

Junge M, Huegel H, Marriott PJ (2007) Enantiomeric analysis of amino acids by using comprehensive two-dimensional gas chromatography. Chirality 19:228–234

Karakawa S, Miyoshi Y, Konno R, Koyanagi S, Mita M, Ohdo S, Hamase K (2013) Two-dimensional high-performance liquid chromatographic determination of day–night variation of D-alanine in mammals and factors controlling the circadian changes. Anal Bioanal Chem 405:8083–8091

Katane M, Homma H (2011) D-Aspartate-an important bioactive substance in mammals: a review from an analytical and biological point of view. J Chromatogr B 879:3108–3121

Kato S, Kito Y, Hemmi H, Yoshimura T (2011) Simultaneous determination of D-amino acids by the coupling method of D-amino acid oxidase with high-performance liquid chromatography. J Chromatogr B 879:3190–3195

Kitagawa F, Otsuka K (2011) Recent progress in capillary electrophoretic analysis of amino acid enantiomers. J Chromatogr B 879:3078–3095

Koga R, Miyoshi Y, Negishi E, Kaneko T, Mita M, Lindner W, Hamase K (2012) Enantioselective two-dimensional high-performance liquid chromatographic determination of *N*-methyl-D-aspartic acid and its analogues in mammals and bivalves. J Chromatogr A 1269:255–261

Marfey P (1984) Determination of D-amino acids. II. Use of a bifunctional reagent, 1,5-difluoro-2,4-dinitrobenzene. Carlsb Res Commun 49:591–596

Miao H, Rubakhin SS, Sweedler JV (2005) Subcellular analysis of D-aspartate. Anal Chem 77:7190–7194

Min JZ, Hatanaka S, Yu H, Higashi T, Inagaki S, Toyo'oka T (2011) Determination of DL-amino acids, derivatized with *R*(−)-4-(3-isothiocyanatopyrrolidin-1-yl)-7-(*N,N*-dimethylaminosulfonyl)-2,1,3-benzoxadiazole, in nail of diabetic patients by UPLC-ESI-TOF-MS. J Chromatogr B 879:3220–3228

Miyoshi Y, Hamase K, Tojo Y, Mita M, Konno R, Zaitsu K (2009) Determination of D-serine and D-alanine in the tissues and physiological fluids of mice with various D-amino-acid oxidase activities using two-dimensional high-performance liquid chromatography with fluorescence detection. J Chromatogr B 877:2506–2512

Miyoshi Y, Hamase K, Okamura T, Konno R, Kasai N, Tojo Y, Zaitsu K (2011) Simultaneous two-dimensional HPLC determination of free D-serine and D-alanine in the brain and periphery of mutant rats lacking D-amino-acid oxidase. J Chromatogr B 879:3184–3189

Miyoshi Y, Koga R, Oyama T, Han H, Ueno K, Masuyama K, Itoh Y, Hamase K (2012a) HPLC analysis of naturally occurring free D-amino acids in mammals. J Pharm Biomed Anal 69: 42–49

Miyoshi Y, Konno R, Sasabe J, Ueno K, Tojo Y, Mita M, Aiso S, Hamase K (2012b) Alteration of intrinsic amounts of D-serine in the mice lacking serine racemase and D-amino acid oxidase. Amino Acids 43:1919–1931

Miyoshi Y, Oyama T, Koga R, Hamase K (2013) Amino acid and bioamine separations. In: Fanali S, Haddad PR, Poole CF, Schoenmakers P, Lloyd D (eds) Liquid chromatography: applications. Elsevier, Amsterdam, pp 131–147

Miyoshi Y, Oyama T, Itoh Y, Hamase K (2014) Enantioselective two-dimensional high-performance liquid chromatographic determination of amino acids; analysis and physiological significance of D-amino acids in mammals. Chromatography 35:49–57

Morikawa A, Hamase K, Inoue T, Konno R, Niwa A, Zaitsu K (2001) Determination of free D-aspartic acid, D-serine and D-alanine in the brain of mutant mice lacking D-amino acid oxidase activity. J Chromatogr B 757:119–125

Morikawa A, Hamase K, Zaitsu K (2003) Determination of D-alanine in the rat central nervous system and periphery using column-switching high-performance liquid chromatography. Anal Biochem 312:66–72

Nagata Y, Akino T, Ohno K (1985) Microdetermination of serum D-amino acids. Anal Biochem 150:238–242

Nagata Y, Horiike K, Maeda T (1994) Distribution of free D-serine in vertebrate brains. Brain Res 634:291–295

Nimura N, Kinoshita T (1986) *o*-Phthalaldehyde-*N*-acetyl-L-cysteine as a chiral derivatization reagent for liquid chromatographic optical resolution of amino acid enantiomers and its application to conventional amino acid analysis. J Chromatogr 352:169–177

Nishikawa T (2011) Analysis of free D-serine in mammals and its biological relevance. J Chromatogr B 879:3169–3183

Ohgusu T, Hamase K, Tanaka H, Shoyama Y, Zaitsu K (2006) High-throughput determination of free D-aspartic acid in mammals by enzyme immunoassay using specific monoclonal antibody. Anal Biochem 357:15–20

Ôi N, Kitahara H, Matsumoto Y, Nakajima H, Horikawa Y (1989) (*R*)-*N*-(3,5-Dinitrobenzoyl)-1-naphthylglycine as a chiral stationary phase for the separation of enantiomers by high-performance liquid chromatography. J Chromatogr 462:382–386

Ota N, Rubakhin SS, Sweedler JV (2014) D-Alanine in the islets of Langerhans of rat pancreas. Biochem Biophys Res Commun 447:328–333

Pirkle WH, House DW, Finn JM (1980) Broad spectrum resolution of optical isomers using chiral high-performance liquid chromatographic bonded phases. J Chromatogr 192:143–158

Reischl RJ, Lindner W (2012) Methoxyquinoline labeling—a new strategy for the enantioseparation of all chiral proteinogenic amino acids in 1-dimensional liquid chromatography using fluorescence and tandem mass spectrometric detection. J Chromatogr A 1269: 262–269

Reischl RJ, Hartmanova L, Carrozzo M, Huszar M, Frühauf P, Lindner W (2011) Chemoselective and enantioselective analysis of proteinogenic amino acids utilizing *N*-derivatization and 1-D enantioselective anion-exchange chromatography in combination with tandem mass spectrometric detection. J Chromatogr A 1218:8379–8387

Schell MJ, Brady RO Jr, Molliver ME, Snyder SH (1997) D-Serine as a neuromodulator: regional and developmental localizations in rat brain glia resemble NMDA receptors. J Neurosci 17: 1604–1615

Schurig V (2011) Gas chromatographic enantioseparation of derivatized α-amino acids on chiral stationary phases – past and present. J Chromatogr B 879:3122–3140

Suzuki C, Murakami M, Yokobori H, Tanaka H, Ishida T, Horiike K, Nagata Y (2011) Rapid determination of free D-serine with chicken D-serine dehydratase. J Chromatogr B 879: 3326–3330

Tojo Y, Hamase K, Nakata M, Morikawa A, Mita M, Ashida Y, Lindner W, Zaitsu K (2008) Automated and simultaneous two-dimensional micro-high-performance liquid chromatographic determination of proline and hydroxyproline enantiomers in mammals. J Chromatogr B 875: 174–179

Visser WF, Verhoeven-Duif NM, Ophoff R, Bakker S, Klomp LW, Berger R, de Koning TJ (2011) A sensitive and simple ultra-high-performance-liquid chromatography-tandem mass spectrometry based method for the quantification of D-amino acids in body fluids. J Chromatogr A 1218:7130–7136

Waldhier MC, Almstetter MF, Nürnberger N, Gruber MA, Dettmer K, Oefner PJ (2011) Improved enantiomer resolution and quantification of free D-amino acids in serum and urine by comprehensive two-dimensional gas chromatography–time-of-flight mass spectrometry. J Chromatogr A 1218:4537–4544

Zhao H, Hamase K, Morikawa A, Qiu Z, Zaitsu K (2004) Determination of D- and L-enantiomers of threonine and *allo*-threonine in mammals using two-step high-performance liquid chromatography. J Chromatogr B 810:245–250

Part II
Physiological Functions and Pathophysiology of D-Serine

Chapter 2
Overview

Toru Nishikawa

Abstract The discovery of endogenous D-serine and its brain-preferring and NMDA type glutamate receptor (NMDAR)-like distribution by the Nishikawa research group has become a milestone opening a new avenue in the research fields of not only D-amino acids but also the glutamate neurotransmission. This chapter consists of an overview and seven review articles describing the following current extension of our knowledge on different aspects of physiological functions and pathophysiology of D-serine in the central nervous system (CNS). In accordance with the finding preceded to the intrinsic D-serine detection that D-serine, but not L-serine, acts as a coagonist for the NMDAR by stimulating its strychnine-insensitive glycine site, D-serine, but not another coagonist glycine, has been shown to be required for full activation of the NMDAR in vitro and in vivo forebrain preparations. More recently, D-serine is suggested to be an endogenous ligand for the δ-type glutamate receptor in the cerebellum. These interactions have been considered to play a pivotal role in the regulation of the higher-order brain functions because glutamate neurotransmission is essential for their expression and control. The first half of this chapter depicts major progresses in the research about the molecular and cellular mechanisms that regulate the extracellular and intracellular concentrations and the target molecules of D-serine. The later half discusses pathophysiological consequences of disturbed D-serine metabolism and signaling and their relevance to clarify the biological basis of and to develop a novel therapy for neuropsychiatric disorders such as schizophrenia, ischemia, neurodegenerative disorders, and epilepsy.

Keywords N-Methyl-D-aspartate-type glutamate receptor • δ2-type glutamate receptor • D-Serine metabolism, function, and pathophysiology • Neuropsychiatric disorders

T. Nishikawa (✉)
Department of Psychiatry and Behavioral Sciences, Graduate School of Medical and Dental Sciences, Tokyo Medical and Dental University, 1-5-45, Yushima, Bunkyo-ku, Tokyo 113-8519, Japan
e-mail: tnis.psyc@tmd.ac.jp

T. Yoshimura et al. (eds.), *D-Amino Acids*, DOI 10.1007/978-4-431-56077-7_2

The detection of the homeostatic presence of endogenous D-serine at high concentrations in mammalian brains by the Nishikawa research group (Hashimoto et al. 1992, 1993; Nishikawa 2011 for a review) was a groundbreaking discovery that required rethinking about the homochirality dogma that the free D-amino acid is uncommon in mammalian tissues. Until then, while D-amino acid residues of long-living proteins (Fujii et al. 2010 for a review) and the transient emergence of free D-aspartate during embryonic or early postnatal stages of development in brain and other organs in mammals (Dunlop et al. 1986; Neidle and Dunlop 1990) had been reported, these observations had been considered to be due to nonphysiological changes or an unusual phenomenon seen only in a very restricted developmental period. Endogenous D-serine has also attracted interests from the finding of its N-methyl-D-aspartate-type glutamate receptor (NMDAR)-like distribution in the brain that suggest the possibility of D-serine as a novel type of physiological modulator of the receptor (Hashimoto et al. 1992, 1993; Nishikawa 2011), because (1) D-serine had been known from earlier studies as a coagonist for the glycine modulatory site of the NMDAR that is needed for the activation of the NMDAR but does not produce the current of the receptor by itself (Danysz and Parsons 1998), (2) the NMDAR has been revealed to play essential roles in expression and regulation of the higher-order functions of the mammalian central nervous system (CNS) (Danysz and Parsons 1998), and (3) the pathophysiology of a variety of neuropsychiatric disorders such as brain ischemia, Alzheimer's disease, and schizophrenia has been documented to be related to malfunction of the NMDAR (Nishikawa 2011). Since the first paper of the detection of endogenous brain D-serine published in 1992, enormous efforts have focused and advanced our knowledge on the molecular and cellular mechanisms underlying the metabolism and functions of D-serine and their pathological changes (Nishikawa 2011). It is notable that recent studies have revealed extra-NMDAR target molecules for D-serine (Nishikawa 2011). This chapter contains the seven review articles on the important contributions to the CNS free D-serine research by selected experts in this field.

The first three chapters will center the topics on the physiological metabolism and functions of D-serine. Synaptic dynamics and interactions with other CNS physiologically active substances of D-serine will also be reviewed in Chap. 6 in the second half of this chapter.

In the beginning Chap. 3, the Mothet group of CNRS of the Aix-Marseille University, France, will illustrate differential roles between D-serine and glycine in the glutamate synapse of the brain during postnatal development and in adult period. Before the advent of endogenous D-serine, glycine was believed to be the only NMDA receptor coagonist acting at its glycine modulatory site. In vitro experiments indicated that the allosteric regulation site of the NMDA was saturated by intrinsic glycine. However, this conjecture is challenged by the additional augmentation of the NMDA receptor function by exogenous application of glycine or D-serine based on both in vivo and in vitro experiments. On the other hand, the respective types of neurons and glia that synthesize, store, and release D-serine in the brain are still a debatable matter. To answer these fundamental questions, they will provide the data depicting how the two coagonists share the region-specific, NMDAR subunit-related, synaptic, or extrasynaptic area-directed, and development-dependent actions at the NMDAR in the rodent brains. The neuronal and glial components of these actions, their regulation mechanisms, and cognitive functional contexts will also be also extensively discussed.

In the next Chap. 4, the Konno group of the International University of Health and Welfare, Japan, will describe the significance of D-amino acid oxidase (DAO) in the degradation of D-serine and its behavioral consequences. Although the existence of D-amino acid oxidase in mammals was identified in 1935 (Krebs), indwelling D-amino acids as substrates for the enzyme had long been uncertain until the detection of various D-amino acids including D-serine following the advances in separation measurement technology for chiral amino acids. The spontaneously DAO activity-deficient mice that the Konno group found have significantly contributed to analyses to clarify the involvement of DAO in the decomposition of D-serine in vivo. They will review the changes in the D-serine concentrations in various CNS portions and peripheral organs and in motions and drug-induced and cognitive behavior in the DAO mutant mice.

In Chap. 5, Yuzaki's group in Department of Physiology at Keio University, Japan, will present the fact that the δ-type glutamate receptors are second physiological binding targets for D-serine besides the NMDAR. The ligand-binding portion of the δ2-type glutamate receptor has been shown to bind D-serine and glycine with a low affinity (10^{-4} M order) (Naur et al. 2007). This observation fits with the presence of a specific non-NMDAR glycine site binding of [3H]D-serine in the brain (Matoba et al. 1997). Extremely higher contents of D-serine in the cerebellum during infancy compared to the adult period and cerebellum-preferring localization of the δ2 receptor have predicted a developmentally regulated functional link between the two molecules. They will summarize their results demonstrating that D-serine, but not glycine, plays a pivotal role through its action at the δ2 receptor during the induction of long-term depression that is associated with motor learning only in the neonatal, but not adult, period and that D-serine also interacts with another δ-type glutamate receptor, the δ1 receptor that is widely distributed in the brain. These findings extend a perspective to the endogenous D-serine functions.

The subsequent four chapters of this chapter will discuss the relationships between D-serine and neuropsychiatric disorders with special attention on schizophrenia, neurodegenerative diseases of Alzheimer's disease, amyotrophic lateral sclerosis (ALS), and epilepsy. The increased and decreased D-serine signalings have been postulated to underline the pathological NMDAR hypofunction and hyperfunction in schizophrenia and the latter three disorders, respectively.

In Chap. 6, Nishikawa's group of the Tokyo Medical and Dental University, Japan, will review the link of D-serine with schizophrenia to the detection of endogenous D-serine and their pioneering achievements about the neuroanatomical and biochemical nature of endogenous D-serine, its metabolism and related molecules, and mode of control of tissue and extracellular D-serine concentrations, chiefly in vivo. The potential modifications of D-serine signaling in schizophrenia will be pointed out in terms of the data of the above pieces of the D-serine system in the light of biological fluids, postmortem CNS tissues, or brain imaging of patients with schizophrenia. Moreover, the therapeutic efficacy of D-serine for schizophrenic symptoms will be outlined, and several target molecule candidates for the development of a new treatment will be mentioned.

In Chap. 7, Coyle's group of the Harvard Medical School, USA, will introduce their systematic approach to elucidate the pathological consequences of reduced D-serine signaling in schizophrenia by creating total serine racemase knockout mice. They will clearly delineate the abnormalities in the serine racemase-deficient mice that mimic those seen in schizophrenia. Furthermore, they will present the data obtained from their genetically engineered mice with the neuron- or astroglia-specific disruption of serine racemase, which provides a powerful clue to resolving the current disputes about cellular localization of D-serine and its synthesis, and analyze the disturbances of cell circuits in schizophrenia. A critical review about the clinical trials of D-serine and other NMDAR glycine site agonists that the Coyle group pioneered will also be included.

In Chap. 8, Mori's group of Toyama University, Japan, will purvey important lines of information concerning the modulatory roles of the basal level of the D-serine signal in the neurotoxicity and hyperexcitability mediated by the NMDAR in the brain based on the studies of their original serine racemase totally deficient mice. These studies provided a clue to clarifying the exact in vivo mechanisms of the excessive NMDAR activity that causes neuronal cell death or degeneration and convulsive seizures. Furthermore, their perspective will be extended to the molecular basis of control of serine racemase activity and the positioning of the serine racemase-null mutant mice.

Finally, in Chap. 9, Sasabe's group from the Department of Anatomy at Keio University, Japan, will summarize the findings obtained from animal models of and patients with ALS that suggest involvement of the aberrant D-serine metabolism in the loss of the motoneurons in ALS. By integrating the increased D-serine concentrations and/or DAO activity in ALS and its model, the association of the mutation of DAO with familiar ALS, and the degeneration of motoneurons in the mice with diminished DAO activity, they will discuss the plausible pathological pathways and a future therapy targeting the D-serine metabolism.

References

Danysz W, Parsons CG (1998) Glycine and N-methyl-D-aspartate receptors: physiological significance and possible therapeutic applications. Pharmacol Rev 50(4):597–664

Dunlop DS, Neidle A, McHale D et al (1986) The presence of free D-aspartic acid in rodents and man. Biochem Biophys Res Commun 141(1):27–32

Fujii N, Kaji Y, Fujii N et al (2010) Collapse of homochirality of amino acids in proteins from various tissues during aging. Chem Biodivers 7(6):1389–1397

Hashimoto A, Nishikawa T, Hayashi T et al (1992) The presence of free D-serine in rat brain. FEBS Lett 296:33–36

Hashimoto A, Nishikawa T, Oka T et al (1993) Endogenous D-serine in rat brain: N-Methyl-D-aspartate receptor-related distribution and aging. J Neurochem 60:783–786

Matoba M, Tomita U, Nishikawa T (1997) Characterization of 5, 7-dichlorokynurenate-insensitive [3H] D-serine binding to synaptosomal fraction isolated from rat brain tissues. J Neurochem 69:399–405

Naur P, Hansen KB, Kristensen AS et al (2007) Ionotropic glutamate-like receptor. delta2 binds D-serine and glycine. Proc Natl Acad Sci U S A 104(35):14116–14121

Neidle A, Dunlop DS (1990) Developmental changes in free D-aspartic acid in the chicken embryo and in the neonatal rat. Life Sci 46(21):1517–1522

Nishikawa T (2011) Analysis of free D-serine in mammals and its biological relevance. J Chromatogr B Anal Technol Biomed Life Sci 879:3169–3183

Chapter 3
Physiological Roles of D-Serine in the Central Nervous System

Jean-Pierre Mothet, Grégoire Mondielli, and Magalie Martineau

Abstract The N-methyl D-aspartic acid receptors (NMDARs) are key glutamate receptors that transduce glutamatergic signals throughout the developing and adult central nervous system (CNS). Despite diversity in their subunit composition, their subcellular localization, and their biophysical and pharmacological properties, activation of NMDARs always requires in addition to glutamate the binding of a co-agonist that has long been thought to be glycine. However, research over the last two decades has challenged this long-cherished model by showing that the D-isomer of serine is the preferential co-agonist for a large population of NMDARs in many areas of the adult brain. Nowadays, a totally new picture of glutamatergic synapses is emerging where both glycine and D-serine are involved in a complex interplay to regulate NMDAR functions in the CNS following time and space constraints. In this review, we focus on the particular contribution of D-serine relatively to glycine in orchestrating synapse formation, dynamics, and neuronal network activity in a time- and synapse-specific manner and its role in cognitive functions. We will discuss also how astroglia and neurons use different pathways to regulate levels of extracellular D-serine and how alterations in synaptic availability of this D-amino acid may contribute to cognitive deficits associated to healthy aging and therefore may open new avenues for therapies.

Keywords NMDA receptors • Synaptic transmission • Synaptic plasticity • Neurons • Astrocytes

The neurotransmitter glutamate once released in the synaptic cleft binds and acts on various transporters, ionotropic and metabotropic membrane receptors located onto neuronal elements (Traynelis et al. 2010), but also onto glial cells including

---

J.-P. Mothet (✉) • G. Mondielli
Team 'Gliotransmission and Synaptopathies', CRN2M UMR7286, Aix-Marseille Université, CNRS, 13344 cedex 15 Marseille, France
e-mail: jean-pierre.mothet@univ-amu.fr

M. Martineau
Interdisciplinary Institute for Neuroscience, UMR 5297, University of Bordeaux and Centre National de la Recherche Scientifique, F-33000 Bordeaux, France

T. Yoshimura et al. (eds.), *D-Amino Acids*, DOI 10.1007/978-4-431-56077-7_3

astrocytes and oligodendrocytes (Parpura and Verkhratsky 2013). The ionotropic glutamate-gated receptor (iGluR) family encodes 18 gene products or subunits subdivided into three main subtypes characterized by their preferential agonists: AMPA receptors (GluARs), kainate receptors (GluKRs), and NMDARs (GluNs) (Traynelis et al. 2010; Paoletti et al. 2013). AMPARs and KARs are cation-permeable ion channels which upon binding of glutamate mediate fast excitatory synaptic transmission, while NMDARs are calcium permeable and drive the transduction of specific temporal patterns of synaptic activity into long-term structural and functional changes. N-Methyl D-aspartic receptors (NMDARs) are ionotropic glutamate-gated receptors that are central to many physiological processes including learning and memory and are involved in neurotoxicity and psychiatric disorders (Traynelis et al. 2010; Paoletti et al. 2013).

3.1 The NMDA Receptors

3.1.1 Structure and Subunit Diversity of NMDARs in the CNS

NMDA receptors are heterotetramers associating two obligatory GluN1 subunits with two identical or different GluN2 or a mixture of GluN2 and GluN3 subunits (Traynelis et al. 2010; Paoletti et al. 2013) (Fig. 3.1). The GluN1 subunit exists as eight distinct splice variants of a single encoding gene (*GRIN1*). Four distinct GluN2 subunits (GluN2A-D) and two GluN3 subunits (GluN3A, B) have been identified which are encoded by four (*GRIN2A-D*) and two (*GRIN3A*, *B*) different genes, respectively (Traynelis et al. 2010; Paoletti et al. 2013). The large repertoire of diheteromeric and triheteromeric NMDARs display distinct trafficking, biophysical, and pharmacological properties and are differentially populating synapses depending on neuronal subpopulations and on development (Traynelis et al. 2010; Paoletti et al. 2013).

Similar to all iGluR subunits, every GluN subunit has four domains (Fig. 3.1). The extracellular region comprises the amino-terminal domain (NTD) involved in subunit association and the agonist-binding domain (ABD) formed by two segments S1 and S2 which bind the co-agonists glycine and D-serine on GluN1 and GluN3 subunits or glutamate on GluN2 subunits. The transmembrane domain (TMD) is formed by three helices (M1, M3, M4) and a hairpin (M2) forming the channel pore and conferring the ion selectivity. The cytoplasmic carboxy-terminal domain (CTD) is involved in protein-protein interactions which determine trafficking (membrane targeting, lateral diffusion, stabilization, recycling), posttranslational regulation, and coupling to specific intracellular signaling cascades. All domains except the CTD are the targets of multiple endogenous or synthetic allosteric modulators (Traynelis et al. 2010; Paoletti et al. 2013).

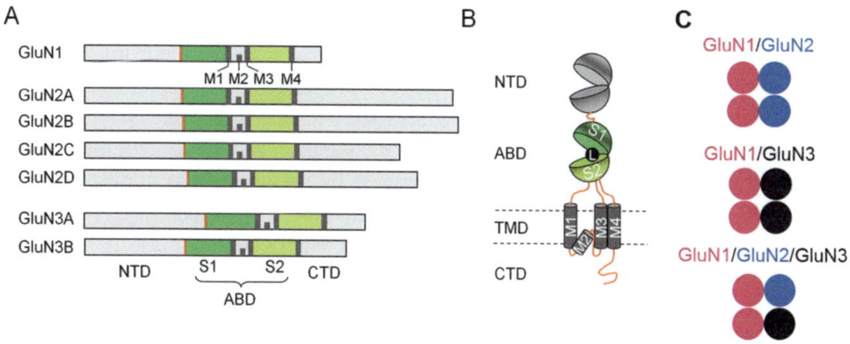

Fig. 3.1 The NMDA receptors. (**a**) Linear organization of monomeric subunits. Seven subunits (GluN) have been identified: GluN1, GluN2A–GluN2D, and GluN3A and GluN3B. M1–M4 refer to membrane segments. (**b**) Modular organization of a GluN subunit. The N-terminal domain (*NTD*); the agonist-binding domain (*ABD*) that binds to the ligand (L), i.e., glycine or D-serine in GluN1 and GluN3 and glutamate in GluN2; the transmembrane domain (*TMD*) containing the ion channel; and an intracellular C-terminal domain (*CTD*). With permission from Mothet et al. (2015) J Neurochem. (**c**) Possible subunits compositions for diheteromeric and triheteromeric receptors

3.1.2 NMDARs and the Glycine Modulatory Binding Site

NMDARs possess unique biophysical and pharmacological properties among the iGluR family (Traynelis et al. 2010; Paoletti et al. 2013). In contrast to GluKRs and GluARs, all NMDARs show much slower gating and deactivating kinetics, thus displaying a large single channel conductance (~40 pS) depending on the subunit composition. The NMDARs display high Ca^{2+} permeability (with subunit-specific variations) and behave as unique Hebbian-like coincidence detectors owing to their voltage-dependent Mg^{2+} block and the presence of many modulatory sites on the different domains of each subunit. But a major property of NMDAR activation of interest for us here is the need of a concomitant binding of glutamate on GluN2 subunits and of a co-agonist D-serine or glycine on GluN1 (Martineau et al. 2006; Traynelis et al. 2010; Paoletti et al. 2013).

The absolute necessity of a co-agonist together with glutamate for the activation of the NMDARs stemmed from the original observations that NMDA-evoked inward currents recorded in native NMDARs (Johnson and Ascher 1987) and from NMDARs expressed in oocytes (Kleckner and Dingledine 1988) require the addition of glycine at high nanomolar range. In their seminal work, Kleckner and Dingledine already showed that several glycine analogs such as D-serine or D-alanine could substitute in activating NMDARs (Kleckner and Dingledine 1988). This observation is consistent with early binding experiments and electrophysiological recordings showing that glycine and D-serine target NMDARs with relatively similar nanomolar affinity (Monaghan et al. 1988; Matsui et al. 1995). Cocrystal structures of the GluN1 S1-S2 ligand-binding core with glycine or D-serine reveal that agonist binding is critically dependent upon a series of hydrogen bonds to side-chain and main-chain atoms, as well as to water molecules in the binding pocket.

Because it makes three additional hydrogen bonds and displaces a water molecule, D-serine binds more tightly to the receptor in comparison to glycine (Furukawa and Gouaux 2003) and would therefore virtually make it more efficient to activate the NMDAR. Nevertheless, the affinity of GluN1 for D-serine or glycine is very similar ($EC_{50} \sim 0.1$–1 μM, (Priestley and Kemp 1994; Chen et al. 2008). The affinity of glycine or D-serine for GluN1 is in fact dictated by the identity of GluN2 subunit. Indeed, the binding sites of the co-agonist and glutamate are allosterically coupled. Accordingly, heterodimeric GluN2B-containing NMDARs display 10× higher affinity for glycine than those containing the GluN2A subunit ($EC_{50} \sim 0.1$ μM versus 1 μM). Conversely, the affinity of D-serine for GluN1 seems to be higher when the NMDARs are containing the GluN2A subunits (Priestley and Kemp 1994; Matsui et al. 1995). While the binding of D-serine at NMDARs was described more than 25 years ago, only glycine was thought to be the right co-agonist since D-serine was not expected to exist in higher living organisms. The discovery that D-amino acids including D-serine are present in organisms throughout evolution including in humans has forced neuroscientists to reconsider the mode of activation of NMDARs. In different tissues including the brain, D-amino acids have emerged as important signaling molecules (Ohide et al. 2011). The field has grown very fast since the pioneer observations made in early 1990s and results from intensive research and from many groups around the world now support that D-serine is the preferred endogenous ligand for the glycine site of NMDARs and thus that this unexpected D-amino acid is central to many brain functions.

3.2 Astrocytes and Neurons Use Different Pathways to Release D-Serine

3.2.1 D-Serine: Glial and Neuronal

In mammals, D-serine is exclusively synthesized from L-serine by serine racemase (SR) (Wolosker and Mori 2012), an enzyme widely distributed in the CNS. SR is a highly regulated enzyme reviewed in (Campanini et al. 2013; Martineau et al. 2014) that binds to several receptor-interacting proteins that modulate the production of D-serine. In addition, D-serine degradation is controlled by the flavoenzyme DAAO, a peroxisomal enzyme that eliminates the amino acid retrieved from the extracellular space (Martineau et al. 2006). The expression and activity of DAAO is widespread in the human and rodent CNS being expressed in both astrocytes and neurons (Verrall et al. 2007; Sasabe et al. 2014). Despite many progresses, our knowledge on the biochemical cascades modulating DAAO activity is still very limited. DAAO interacts with pLG72, a primate-specific protein present in mitochondria (Sacchi et al. 2008; Otte et al. 2014). The interaction with DAAO would involve interaction of the cytosolic fraction of DAAO with pLG72 potentially exposed at the surface of mitochondria as suggested by Sacchi and colleagues

(Sacchi et al. 2011). We do not know if an ortholog of pLG72 may exist in non-primate mammals. Proteomics analyses have revealed that DAAO interacts also with Bassoon and Piccolo, two proteins of the presynaptic active zone (Popiolek et al. 2011) supporting the neuronal localization for DAAO.

In which cell type D-serine is formed? Although a glial localization has been largely attributed to SR and D-serine (Schell et al. 1997; Wolosker et al. 1999; Panatier et al. 2006; Williams et al. 2006), allowing the amino acid to be considered as a gliotransmitter, several studies indicate that SR may be instead preferentially expressed in excitatory and inhibitory neurons in rodent and human brains rather than in glia (Miya et al. 2008; Ehmsen et al. 2013; Balu et al. 2014). Accordingly, neuronal-specific conditional deletion of SR induces a much larger decrease in cerebral and hippocampal D-serine contents compared to deletion limited to the astrocyte population (Benneyworth et al. 2012). However, variations in the cellular expression of SR and D-serine between species could not be totally ruled out. Accordingly, Suzuki et al. (2015), using SR-KO as negative controls, recently reported that astrocytes abundantly express SR in the human subiculum. The notion that D-serine could be formed in astroglia is further reinforced by the fact that SR interacts with DISC-1 in astrocytes (Ma et al. 2013). Furthermore, using multiplex single-cell RT-PCR, we observe that both neurons and astrocytes express SR at least at the levels of transcripts in the prefrontal cortex of adult rat (Dallerac, Turpin & Mothet, Personal communication). Thus, the cellular distribution of SR remains an open issue particularly since a recent transcriptome analysis of isolated purified brain cell groups shows that the enzyme is more expressed in glial cell lineage and notably astrocytes than in neurons in the mature cortex of mice (Zhang et al. 2014). As neurons are not capable of synthesizing much L-serine, neuronal SR activity would depend on the supply of the D-serine precursor which is primarily formed in astrocytes from glucose by 3-phosphoglycerate dehydrogenase (Yang et al. 2010; Ehmsen et al. 2013).

3.2.2 *Cell-Type-Specific Mechanisms of Release: Exocytosis Versus Hetero-exchange*

As shown so far, D-serine can be formed in both neurons and astrocytes (Martineau et al. 2014). Different elements of the literature support the notion that neurons and astroglia use different molecular machinery and signaling pathways to release D-serine into the extracellular space (Fig. 3.2). Early studies have shown that D-serine release is triggered by agonists of the ionotropic and metabotropic glutamate receptors (iGluR and mGluR, respectively; (Schell et al. 1995; Mothet et al. 2005; Martineau et al. 2008; Ishiwata et al. 2013). We have accumulated evidence that the release of D-serine by astrocytes is mainly dependent on an increase in cytosolic Ca^{2+}. Indeed, disrupting Ca^{2+} signaling inside astrocytes reduces D-serine release from astrocytes in culture and in hippocampal slices

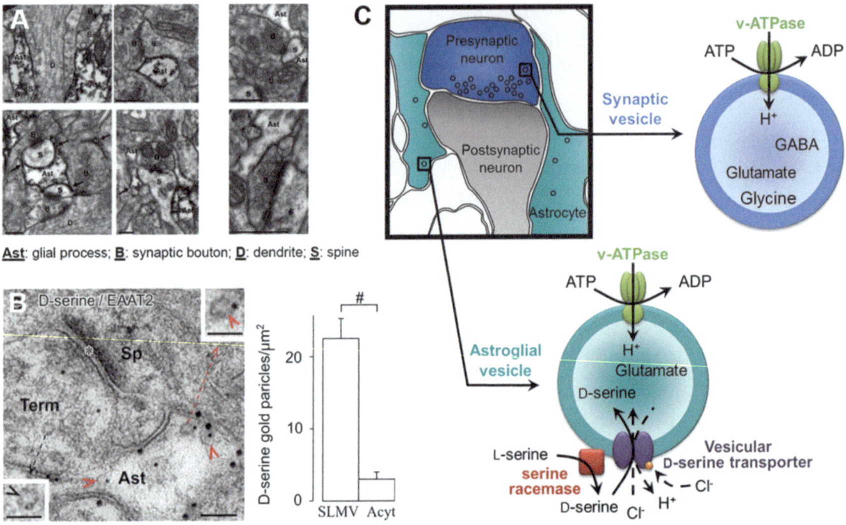

Fig. 3.2 Modes of release of D-serine. (**a**) Preembedding peroxidase immunohistochemistry showing that D-serine immunoreactivity is mainly distributed in astrocytic processes (*arrows*) at the level of asymmetric synapses in the adult rat cerebral cortex. Representative low-magnification (*top*) and high-magnification (*bottom*) electron micrographs showing that no peroxidase staining is detected with a preabsorbed anti-D-serine antibody. *Ast* astrocyte, *D* dendrite, *S* dendritic spine, *B* axonal bouton. *Scale bars*, 500 nm. Modified with permission from The Society of Neuroscience (Martineau et al. 2013). (**b**) Immunogold labeling (small gold particles) of D-serine in small synaptic-like vesicles (indicated by *red arrowheads*) in astrocytic processes (*Ast*) positive for EAAT2 (large gold particles) contacting asymmetric synapses (stars) between nerve terminals (Term) and dendritic spines (Sp). *Insets*: higher magnification showing the similarities between SLMVs (*red dotted arrows*) and SVs (*black dotted arrow*). Note that the astrocytic SLMVs are often localized in small clusters close to the plasma membrane. *Scale bars*: 100 and 50 nm in insets. Right immunogold quantification of D-serine gold particles in astrocytic processes. The *bar charts* show the mean number of D-serine gold particles/μm^2 ± SD in SLMVs and the cytoplasmic matrix of astrocytic processes (*Acyt*). Reprinted with permission from Oxford University Press (Bergersen et al 2012). (**c**) Graphical abstract showing that in astrocytes D-serine is mainly released by exocytosis of some vesicles containing also glutamate, while in neurons D-serine is released from the cytosol through asc-1. For explanation, see text

(Mothet et al. 2005; Henneberger et al. 2010; Shigetomi et al. 2013) but also from astrocytes in vivo (Takata et al. 2011). Calcium certainly represents the most critical signaling hub for D-serine release from astrocytes, even though we are still largely ignoring the spatial and temporal signature of these Ca^{2+} signals: activation of astrocytic transforming growth factor (TGF)-β (TGFR; (Diniz et al. 2012)), bradykinin-type2 (B2R; (Martineau et al. 2008)), adenosine-type 2 (A2R; (Scianni et al. 2013)), ephrinB3 (Zhuang et al. 2010), and muscarinic acetylcholine (Takata et al. 2011; Lopez-Hidalgo et al. 2012) receptors. The activation of these G protein-coupled receptors (GPCRs) is associated with the recruitment of Ca^{2+} from the intracellular stores mainly via inositol-1,4,5-trisphosphate receptors (IP3R) located on the endoplasmic reticulum (ER) (Zorec et al. 2012). But Ca^{2+} driving D-serine release could also originate from the extracellular space through channel-mediated transmembrane Ca^{2+} fluxes in astrocytes. For example, astrocytic transient receptor potential A1 (TRPA1) channels contribute to basal Ca^{2+} signals which are required for D-serine release (Shigetomi et al. 2013). These studies clearly established that astrocytes express a plethora of functional receptors which activation is coupled to the release of D-serine. Then, what could be the mechanisms downstream Ca^{2+} triggering D-serine release? Although not exclusive of another release mechanism, we have found that D-serine release from astrocytes is dependent on soluble N-ethylmaleimide-sensitive factor attachment protein receptor (SNARE) proteins, indicating that glial Ca^{2+}-regulated exocytosis is a main release mechanism used by astroglia to release D-serine. Indeed, astrocytes as most eukaryotic cells expressed synaptobrevins, also referred to as vesicle-associated proteins-VAMPs, or R (for arginine)-SNAREs) and the plasma membrane (syntaxins and synaptosome-associated protein of 23 kDa (SNAP23); Q (for glutamine)-SNAREs) (Jahn and Fasshauer 2012). The vesicle fusion is triggered by an increase in cytosolic Ca^{2+}, which presumably binds to vesicular synaptotagmins (Montana et al. 2006). Astroglial vesicles also possess proteins necessary for vesicular filling, such as the vacuolar type H^{+}-ATPase (V-ATPase) (Martineau et al. 2013) which provides the proton gradient necessary for intravesicular loading of gliotransmitters via appropriate transporter(s). Cleavage of Sb2 and cellubrevin by tetanus neurotoxin causes a strong inhibition of Ca^{2+}-dependent D-serine release (Mothet et al. 2005; Henneberger et al. 2010). Additionally, the blockade of D-serine vesicular uptake using a V-ATPase blocker inhibits D-serine release from astrocytes (Mothet et al. 2005). Indeed, based on electron microscopy (EM), D-serine accumulates in clear vesicles with a diameter of 36 nm in the perisynaptic processes of hippocampal and cortical astrocytes (Bergersen et al. 2012; Martineau et al. 2013). In the adult hippocampus, clear vesicles are organized in small groups of 2–15 vesicles preferentially located within 100 nm from the astrocytic plasma membrane (Bergersen et al. 2012) and observed at sites adjacent to neuronal elements bearing NMDARs (Bezzi et al. 2004). Thus, astrocytes possess small vesicles resembling those found at synaptic terminals, albeit at lower density. It should be noted, however, that the vesicular size in astrocytes appears to be more diverse than that described for neurons. In addition to Ca^{2+}- and SNARE-dependent exocytotic release of D-serine from astrocytes, this gliotransmitter can be released *via* alternative non-exocytotic conduits at the plasma membrane, including volume-regulated anion channels (VRAC; (Rosenberg et al. 2010)), alanine–serine–cysteine transporter (ASCT; (Ribeiro et al. 2002), but see (Maucler et al. 2013)) and likely through connexin 43 hemichannels (Stehberg et al. 2012). Although the vesicular transporter for D-serine has not been identified, the transport of D-serine inside astroglial vesicles was recently characterized (Martineau et al. 2013). While glutamate transport is observed in both synaptic and astroglial vesicles, the transport of D-serine is specific to astroglial vesicles (Martineau et al. 2013). Its apparent affinity is ~7 mM, consistent with the affinity of vesicular inhibitory amino acid transporter for γ-aminobutyric acid, another neutral amino acid (Chaudhry 2008; Martineau et al. 2013). Similar to glutamate, extravesicular chloride concentration modulates

D-serine transport into astroglial vesicles, reaching the maximum activity at 4 mM. Because D-serine transport induces vesicular acidification and critically relies on chloride, the vesicular D-serine transporter is proposed to be a D-serine/chloride co-transporter (Martineau et al. 2013) (Fig. 3.2). A spatial association of SR activity and D-serine vesicular transport was observed, resulting in a functional coupling between D-serine synthesis and uptake (Martineau et al. 2013). Finally, D-serine and glutamate vesicular loading exert a mutual stimulation which indicates a functional crosstalk between the two transporters (Martineau et al. 2013). This synergy at vesicular level can only be explained if both transporters reside on the same vesicle, indicating the co-storage and thus the co-release of both gliotransmitters. However, the immunogold colabeling of D-serine and glutamate in the adult hippocampus did not reveal a population of vesicles containing both gliotransmitters (Bergersen et al. 2012); this seemingly disparate findings could be a result of the limited sensitivity of the immuno-EM. Nonetheless, the mechanism underlying the vesicular synergy between D-serine and glutamate uptake requires additional investigation. The possible co-storage of glutamate and D-serine, at least in a subpopulation of astroglial vesicles, points to the possibility of interdependence of their dynamics and modulatory functions at synaptic and extrasynaptic sites.

Release of D-serine as a neurotransmitter operates through a different mechanism than those described before. Neuronal D-serine is mainly released in response to depolarization, induced by veratridine, a voltage-dependent Na^+ channel activator, or by increased extracellular potassium concentration, both in vitro and in vivo (Hashimoto et al. 2000; Rosenberg et al. 2010) (Fig. 3.2). Intriguingly, neuronal silencing by tetrodotoxin increased D-serine extracellular concentration in vivo (Maucler et al. 2013) although some studies show no effect (Hashimoto et al. 1995). Chelation of intracellular and extracellular Ca^{2+} does not affect depolarization-elicited D-serine release. In addition, inhibition of V-ATPase by bafilomycin A1 or cleavage of Sb2 by tetanus neurotoxin failed to inhibit D-serine release from neurons, excluding the vesicular release mechanism (Rosenberg et al. 2010) which is consistent with D-serine absence in the lumen of synaptic vesicles (Martineau et al. 2013), D-serine is released by neuronal presynaptic elements from a cytosolic pool through the Na^+-independent antiporter alanine–serine–cysteine transporter 1 (asc-1; (Rosenberg et al. 2010; Rosenberg et al. 2013); Fig. 3.2). Asc-1 is restricted to neurons and catalyzes neutral amino acid hetero-exchange (Fukasawa et al. 2000; Helboe et al. 2003; Matsuo et al. 2004), between D-serine and another neutral amino acid. Neuronal D-serine release could be triggered by neutral amino acids, such as L-serine, D-alanine, or D-isoleucine, through this hetero-exchange (Rosenberg et al. 2010, 2013). Two independent studies published in 2013 (Ishiwata et al. 2013; Maucler et al. 2013) have explored in vivo the dynamics of D-serine release in rat cortex. Using a second-by-second time scale microelectrode biosensors, Maucler et al. (2013) found that the application of the asc-1 transporter inhibitor, S-methyl-L-cysteine, depressed the release of D-serine evoked by L-serine, while L-asparagine, a substrate of ASCT2, did not induce D-serine release in vivo. Conversely, intra-medial frontal cortex infusion of S-methyl-L-cysteine, caused a concentration-dependent increase in the microdialysate contents of D-serine (Ishiwata et al. 2013). The reasons of these opposite results are not known but could reflect differences in timescale of both analytical methods used in the two studies but clearly show that asc-1 could work in both directions. Which direction (uptake or release) in vivo prevails is not known so far.

3.3 Functions of D-Serine: From Development to Cognition

What could be the function of D-serine in the CNS? Early studies in the 1980–1990s evidenced that exogenous D-serine can potentiate NMDARs activity (Danysz et al. 1990). Two pioneer observations in the 1990s put D-serine in the spotlight by revealing that endogenous D-serine may be linked to NMDARs activation. First, its distribution in the rodent brain mirrors the expression of NMDARs, as one might expect for an NMDAR co-agonist, during development and in the adult animal (Hashimoto et al. 1993; Schell et al. 1995, 1997).

3.3.1 Role of D-Serine in the Adult Brain

Although D-serine could serve as a co-agonist of NMDAR, the experimental evidence for it awaited for the development of specific pharmacological and genetic tools allowing manipulating the levels of the endogenous amino acid.

A first approach we introduced consisted in the use of purified DAAO on top cultured neurons or brain slices to deplete endogenous D-serine while recording synaptic events (Mothet et al. 2000). We then discovered that depleting D-serine with DAAO consistently reduces synaptic events and NMDAR-mediated currents just demonstrating that D-serine rather than glycine as initially proposed would be the endogenous ligand for synaptic NMDARS in the hippocampus and juvenile cerebellum (Mothet et al. 2000). Therefore, it was predicted that the functions of D-serine and glycine could be dependent on the location of specific synapses within different brain area. But, subsequently acute depletion of D-serine with recombinant *Rhodotorula gracilis* DAAO (*Rg*DAAO) or serine deaminase (Dsda), two enzymatic D-serine scavengers, consistently affects NMDAR activity on glutamatergic neurons in widespread area of the mature brain including the Schaeffer collateral (SC)-CA1 synapse in the hippocampus (Papouin et al. 2012; Le Bail et al. 2015), medial prefrontal cortex (Fossat et al. 2012), hypothalamic supraoptic nucleus (Panatier et al. 2006), nucleus accumbens (Curcio et al. 2013), the medial nucleus of the trapezoid body (MNTB) (Reyes-Haro et al. 2010), or even the retina (Stevens et al. 2003). These observations lead us to jump on the conclusion that D-serine and not glycine is the right endogenous co-agonist for synaptic NMDARs at most central synapses. This conclusion was reinforced by the observations that a concomitant removal of endogenous glycine with *Bacillus subtilis* GO (*Bs*GO) has only weak or no effects at the SC-CA1 glutamatergic synapse in the hippocampus (Papouin et al. 2012; Le Bail et al. 2015) or in the hypothalamus (Panatier et al. 2006). Conversely, in some case like the nucleus tractus solitarii (NTS), a structure located in the brainstem, glycine acts as the primary co-agonist for synaptic receptors (Panatier et al. 2006). In the structures where D-serine action predominates, it is thought that extrasynaptic glia-derived glycine should be unable to reach synaptic sites due to efficient capture by GlyT1 (Fossat et al. 2012; Papouin et al. 2012; Le Bail et al. 2015). Nevertheless, this dichotomic view is challenged by the fact that a substantial fraction of the synaptic NMDAR responses remained systematically resistant to the enzymatic degradation of D-serine. In addition, neuronal-derived glycine released through Asc-1 physiologically modulates NMDAR activation (Rosenberg et al. 2013) at the SC-CA1 hippocampal synapses. Introduction of animal models for D-serine represented another important opening to define the function of D-serine. A first model was developed early by Konno and collaborators which is a mutant mouse strain (ddY/DAO$^-$) lacking DAO activity (Konno and Yasumura 1983). The ddY/DAO$^-$ mice have a missense mutation (G181R) that causes a complete loss of enzyme activity (Sasaki et al. 1992). As expected, these mutant mice exhibit larger NMDA/AMPA ratio due to a higher occupancy of the glycine site of NMDARs in diverse areas like the hippocampus (Maekawa et al. 2005) and the retina (Romero et al. 2014) and also show enhanced hippocampal LTP (Maekawa et al. 2005). More recently, specific DAAO inhibitors have been developed by different companies. Duplantier and colleagues (2011) show that inhibition of DAAO increases NMDA receptor-mediated synaptic currents in primary neuronal cultures from rat hippocampus and resulted in a significant increase in evoked hippocampal theta rhythm, an in vivo electrophysiological model of hippocampal activity (Strick et al. 2011). Likewise, inhibition of DAAO was reported to increase NMDAR activity and LTP formation in the mature hippocampus (Hopkins et al. 2013a, b; Le Bail et al. 2015) and to normalize the activity of impaired NMDARs in the supraoptic nucleus of lactating rats (Hopkins et al. 2013a) by increasing D-serine levels. These results predict that DAAO inhibition would have a positive effect on cognition, most probably through its ability to augment NMDAR-mediated currents. In 2009, Coyle and colleagues introduced the first genetically invalidation of SR in mice (Basu et al. 2009). Despite some subtle behavioral and structural deficits (see Sects. 3.3.2 and 3.3.3), electrophysiological phenotyping of NMDARs activity and synaptic plasticity intriguingly revealed that a 90 % decrease in forebrain D-serine content (Basu et al. 2009), had only a moderate impact on NMDARs at SC-CA1 synapses. Indeed, the amplitude of NMDA-mediated synaptic currents was normal; only the decay kinetics of evoked NMDAR-mediated currents were slower in hippocampal slices in SR-KO mice (Basu et al. 2009) due certainly to a compensatory mechanism by glycine. The most striking effect in SR-KO is the absence of LTP induced by a pairing protocol (Basu et al. 2009; Rosenberg et al. 2013; Le Bail et al. 2015). It is worth noticing that the selective depletion of both D-serine and glycine reduce NMDAR synaptic responses in the lateral nucleus of the amygdala (LNA) (Li et al. 2013). Therefore, the contribution of the two co-agonists in activating

NMDARs shows general overlap in the brain, but regional differences in the preference for one co-agonist over the other certainly occur.

Could the preference for a co-agonist be related to the molecular composition of synaptic NMDARs? We have recently discovered that in the mature hippocampus, D-serine is the main co-agonist at SC-CA1 hippocampal synapses whereas glycine is preferentially concerned at connections within the dentate gyrus (Le Bail et al. 2015). The role of D-serine prevails in brain regions where the GluN2A-containing subtype of NMDARs predominates such as hippocampal SC-CA1 and prefrontal cortex synapses (Fossat et al. 2012; Papouin et al. 2012; Le Bail et al. 2015), while glycine acts as the preferred co-agonist in regions where the contribution of the GluN2B subunit is favored, as seen in the dentate gyrus (Le Bail et al. 2015). This segregation would explain why SR-deficient mice display moderate synaptic dysfunctions at the medial perforant path-dentate gyrus synapses (Balu et al. 2013).

But in addition to be synapse specific, the function of D-serine is developmentally regulated (Le Bail et al. 2015) (Fig. 3.3). Indeed, we have found that glycine is the preferred co-agonist at immature SC-CA1 synapses and demonstrated that a switch from glycine to D-serine occurs at these synapses during postnatal development (Le Bail et al. 2015) that parallels a progressive replacement of GluN2B by GluN2A subunits in NMDAR assemblies (Rodenas-Ruano et al. 2012). Nevertheless, a strict association between the identity of the co-agonist and the NMDAR molecular composition is not a general rule since D-serine acts as the main co-agonist in the supraoptic nucleus (Panatier et al. 2006) where GluN2B-containing receptors predominate but also in the nucleus accumbens and the prefrontal cortex (Fossat et al. 2012; Curcio et al. 2013), where both GluN2 subunits are equally expressed. Additionally, one has to keep in mind that a substantial fraction of synaptic NMDARs are triheteromers containing both GluN2A and GluN2B subunits (reviewed in (Traynelis et al. 2010; Paoletti et al. 2013)) and are thus expected to be regulated by both D-serine and glycine.

Besides the impact of the co-agonist regional specificity, a segregated role for D-serine and glycine at the synaptic cleft has also been reported. Indeed, NMDARs are not only present synaptic sites but are also located at extra- and peri-synaptic sites (Petralia 2012) where they may have different roles in cell death or survival programs (Hardingham and Bading 2010) and may be involved in different forms of synaptic plasticity (Papouin et al 2012; Mothet et al 2015). Enzymatic depletion of D-serine with *Rg*DAAO consistently reduces NMDAR synaptic events at mature SC-CA1 connections (Papouin et al. 2012; Le Bail et al. 2015), while the pharmacologically isolated extrasynaptic NMDAR activation or the NMDAR-driven tonic glutamate current remain mostly unaffected by changes in D-serine levels (Papouin et al. 2012). Opposite effects were observed when endogenous glycine was depleted using *Bacillus subtilis* GO (*Bs*GO), indicating that at those hippocampal synapses, D-serine would gate preferentially the synaptic GluN2A-NMDARs while glycine would target extrasynaptic receptors (Papouin et al. 2012). The segregated role of the two co-agonists at gating synaptic and extrasynaptic NMDARs at SC-CA1 synapses delineates their respective role in synaptic plasticity, with D-serine

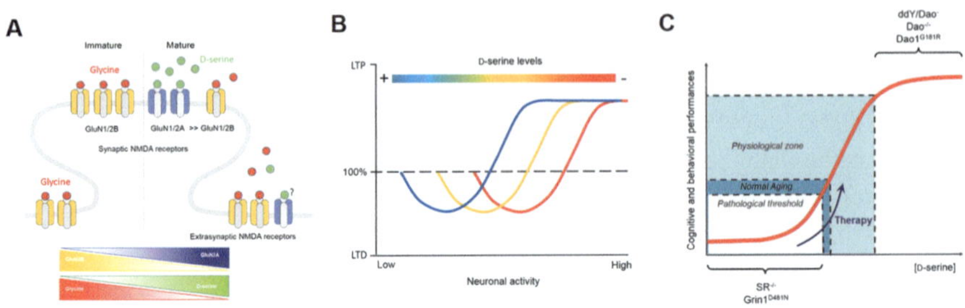

Fig. 3.3 Functions of D-serine. (**a**) Immature versus mature synapses. At immature synapse where only GluN1/2B receptors are present, glycine serves as the co-agonist. During postnatal development, the replacement of GluN2B- by GluN2A-containing NMDARs at synapses parallels a change in the identity of the co-agonist from glycine to D-serine. Conversely, glycine is the main co-agonist for extrasynaptic NMDA receptors. (**b**) D-serine and metaplasticity. D-serine levels govern the activity of NMDA receptor-dependent synaptic plasticity. Decreasing the levels of D-serine shifts the relationship to higher activity values, whereas increasing them has the opposite effect. Reprinted with permission from Cell Press (Panatier et al. 2006). (**c**) D-serine and cognition. D-serine levels drive cognitive and behavioral performances. Decreased levels as observed in normal aging and below a pathological threshold as observed in SR–KO or GrinD481N mice induce deficits. Conversely, increased levels of D-serine as observed under DAAO inactivation increases performances. Current therapy interventions aimed to normalize D-serine levels

specifically required for the expression of LTP and both D-serine and glycine acting for the induction of long-term depression (LTD) (Papouin et al. 2012). Our observation for a role of D-serine in the formation of hippocampal LTD in the mature brain expands earlier observations made by Xu and colleagues (2008) that the D-amino acid regulates LTD in the hippocampal CA1 region in a "bell-shaped" concentration-dependent manner (Zhang et al. 2008). The retina is offering another model to appreciate the relative functions of D-serine and glycine at synaptic and extrasynaptic sites. The group of Miller have nourished evidence that AMPA receptor-dependent and light-evoked NMDA receptor-mediated currents recorded at retinal ganglion cells are shaped by the release of D-serine from Muller cells (Stevens et al. 2003, 2010; Gustafson et al. 2007), while glycine would rather activate extrasynaptic NMDARs (Sullivan and Miller 2012). However, this appealing dichotomic synaptic versus extrasynaptic segregation of the functions of D-serine and glycine has limitations. Several studies demonstrate that GluN2B-NMDARs preferring glycine are also required to induce LTP (reviewed in (Paoletti et al. 2013)). Furthermore, the repetitive observation that LTP at SC-CA1 synapses is only partly affected in SR-KO mice displaying trace levels of D-serine (Basu et al. 2009; Rosenberg et al. 2013) supports the idea that glycine could compensate for any D-serine synaptic dysfunction. Indeed, a couple of very recent studies have shown that the two endogenous co-agonists are necessary for LTP since both treatments with *Rg*DAAO and *Bs*GO result in significantly reduced LTP expression in CA1 hippocampal area and dentate gyrus, as well as in the LNA (Li et al. 2013; Le Bail et al. 2015). Interestingly, a robust LTP may be induced in the LNA of

SR-KO mice by increasing the strength of presynaptic activation. This LTP is then consistently reduced under glycine depletion by *Bs*GO (Li et al. 2013). These results indicate that independent of their synaptic location in the brain, both D-serine and glycine may be recruited for activating NMDARs during enhanced synaptic activity. Accordingly, treatments with *Rg*DAAO but not *Bs*GO reduces the NMDAR component of spontaneous excitatory postsynaptic currents in the LNA while synaptic responses evoked by enhanced afferent stimulation are impacted by *Bs*GO (Li et al. 2013). Ambient D-serine could therefore drive tonic activation of NMDARs under low-synaptic activity in this structure, while the involvement of glycine should rely on increased afferent activity. In the hippocampus, although D-serine and glycine preferentially drive low-frequency-evoked NMDAR synaptic responses at mature SC-CA1 and dentate gyrus synapses, respectively, they both contribute to LTP expression driven by high-frequency stimulation at both hippocampal synapses (Le Bail et al. 2015) suggesting that both co-agonists may be necessary for memory formation. Therefore, although the identity of the NMDAR co-agonist may be synapse specific under low basal conditions, it appears that the activation by the two co-agonists is concerned when a sustained recruitment of synaptic activity is promoted.

NMDARs are also present at presynaptic sites where they regulate the release of glutamate by nerve terminals (Sjostrom et al. 2003; Corlew et al. 2008; Larsen et al. 2014). The mode of activation by either glycine or D-serine remained to be defined. Jones and colleagues, by using *Rg*DAAO and *Bs*GO to deplete D-serine and glycine, respectively, have recently reported that in the mature entorhinal cortex, these presynaptic NMDA receptors are tonically activated by D-serine but not by glycine, suggesting that D-serine is the endogenous co-agonist of presynaptic NMDARs (Lench et al. 2014).

Finally, D-serine and glycine synaptic functions closely rely on a fine coupling between astrocytes and neurons. Indeed, the profile of the tripartite synapse could also impact the identity of the predominant co-agonist acting on NMDAR. As mentioned earlier, a progressive developmental switch from glycine to D-serine occurs at SC-CA1 synapses that could reflect to some extent postnatal changes in the intimate neuron–glial coupling (Le Bail et al. 2015). At immature synapses, pharmacological blockade of astrocytic GlyT1 does not increase NMDAR activation as it does at mature connections, confirming that ambient glycine is able to saturate NMDAR binding sites early after birth. These elevated levels of glycine at birth presumably rely on a weak expression of GlyT1 (Zafra et al. 1995) related to the lack of a prominent astrocyte network that only progressively develops during the first postnatal weeks (Yang et al. 2013). The fact that inactivation of astrocyte activity with the gliotoxin fluoroacetate (FAC) does not significantly alter NMDAR activation at immature synapses further supports the view that this functional aspect of neuron–glial coupling is weakly effective early after birth (Le Bail et al. 2015). On the other hand, the postnatal formation of the tripartite synapse does not only result in a higher clearance of glycine by astrocytic GlyT1 but also in a better supply of D-serine either from astrocyte or through activation of the serine shuttle (Wolosker 2011). Further lines of experimental evidence show that D-serine derived from glia regulates NMDARs. Rusakov and colleagues reported that in the mature hippocampus, astrocyte processes contacting SC-CA1 synapses retain the ability to control Hebbian LTP within or near their individual territories involving Ca^{2+}-and SNARE protein-dependent D-serine release (Henneberger et al. 2010) confirming earlier evidence for a vesicular release of D-serine from astrocytes (Mothet et al. 2005; Martineau et al. 2008). Because astrocytes express a plethora of receptors, they could link different neurotransmitter systems through the release of D-serine. Accordingly, potentiating effect of nicotine on hippocampal synaptic transmission and long-term synaptic plasticity recorded at SC-CA1 synapses depends on D-serine furnished by astrocytes (Lopez-Hidalgo et al. 2012). Noteworthy, α7 nAChR gene deletion leads to impaired synaptic NMDAR functions through specific loss of D-serine in the neocortex (Lin et al. 2014). Likewise, astrocyte TRPA1 channels contribute to basal Ca^{2+} levels and are required for constitutive D-serine release into the extracellular space, which in turn contributes to NMDA receptor-dependent LTP (Shigetomi et al. 2013). In layer 2/3 of the neocortex, activation of either astroglial cannabinoid or α1-adrenoreceptors receptors induces Ca^{2+}-signaling and synaptic plasticity through notably the release of D-serine (Rasooli-Nejad et al. 2014; Pankratov and Lalo 2015). Such a relationship between the degree of the astrocyte–neuron coupling and the prevalence of D-serine as a co-agonist agrees with the differential impact of the amino acid in activating NMDARs at hypothalamus synapses during lactation where D-serine efficiency is linearly related to the extent of astrocytic coverage of synapses (Panatier et al. 2006). Indeed in virgin rats, astrocytic coverage of synapses is maximal guaranteeing an optimal activity of NMDARs and enabling formation of LTP by D-serine. Conversely, lactation induces loss of synapses coverage by glia. This anatomical remodeling leads to a decrease of occupancy by D-serine of NMDARs (Panatier et al. 2006). As a consequence, the activity dependence of LTP and LTD whose induction depends on NMDAR activation is modified by this neuron–glial remodeling. The levels of D-serine shift the activity dependence of long-term changes toward higher activity (Fig. 3.3).

But, astrocyte-derived D-serine could play also a critical role in neurovascular coupling. Indeed astrocytes modulate hemodynamism of microcirculation by providing a physical linkage from synapses to arterioles and through notably the release of vasoactive gliotransmitters (Haydon and Carmignoto 2006). In two elegant studies, Anderson and colleagues have discovered that endogenous D-serine released by astrocytes contributes to the vasodilatory response of penetrating cortical arterioles produced by astrocyte activation (LeMaistre et al. 2012; Stobart et al. 2013). Still, we could also predict that the progressive setup of a close astrocyte–neuron interaction will favor the astrocyte-derived precursor L-serine to be provided to neurons to boost D-serine synthesis. According to this possibility, the FAC-induced decrease in NMDAR activation recorded at mature SC-CA1 synapses is rescued by increasing synaptic availability of L-serine (Le Bail et al. 2015).

Adult neurogenesis offers another example where the functions of D-serine have been recently explored. In the adult mammalian brain, neurogenesis occurs mainly in the subventricular zone (SVZ) and the subgranular zone (SGZ) of DG in the hippocampus (Aimone et al. 2014). NSCs are the progenitor cells in the brain, which are capable of continuous self-renewal and differentiation into neurons, astrocytes, and oligodendrocytes (Aimone et al. 2014). The local environment may dictate the fate choice of NSCs. Nevertheless, we are still largely ignoring the identity of the extrinsic mediators that regulate proliferation, survival, migration, and differentiation of NSCs. Hu and colleagues have shown that forebrain neural stem cells (NSCs) in vitro could synthesize and release D-serine and that D-serine supports the proliferation and neuronal differentiation of NSCs (Huang et al. 2012). Furthermore, they show that D-serine promotes neurogenesis through acting on NMDARs and subsequently regulating the Ca^{2+}-related signaling pathways, including the phosphorylation of ERK1/2-CREB and GSK-3β. These in vitro observations were later expanded by Toni and colleagues (2013) who reported that D-serine increased adult hippocampal neurogenesis in vivo and in vitro and increased the density of neural stem cells and transit amplifying progenitors. Furthermore, D-serine increased the survival of newborn neurons (Sultan et al. 2013).

3.3.2 Role of D-Serine in the Developing Brain

Even that D-serine function may preferentially dominate at mature synapses, this co-agonist could already play an important role in the developing brain by controlling neuronal migration and synaptogenesis. Indeed, series of evidence indicate that D-serine and its enzymes are already present early during embryonic and postnatal development in different like the retina (Romero et al. 2014) and the vestibular nuclei (Puyal et al. 2006) where it could critically influence postnatal development. Therefore defects in neonatal D-serine could lead to severe brain defaults. Mice with a 3-phosphoglycerate dehydrogenase deficiency lack both L-serine and D-serine throughout their lives (Yang et al. 2010), but maternal D-serine supplementation can completely reverse the abnormal neurological phenotype (Fuchs et al. 2006).

During neocorticogenesis, migrating neurons can adopt different types of trajectories and a large proportion of neurons migrate radially, along radial glial cells, from the germinative zone to their final place (Yacubova and Komuro 2003). Glutamate acting on NMDARs exerts a crucial role by acting as a motility-promoting signal for immature neurons. Blocking NMDARs impairs neuronal migration during postnatal development (Yacubova and Komuro 2003). Radial migration of immature granule cells in the developing cerebellum, along the Bergmann glia, is one of the best studied models. How NMDARs of migrating immature neurons are activated remains unknown. Because D-serine levels peak at P14 in the cerebellum at the time of intense granule cell migration (Schell et al. 1997) and due to the absolute requirement of NMDARs activation for normal migration, one can imagine that D-serine is involved in this process. Snyder and colleagues (2005) have demonstrated that D-serine released by Bergmann glial cells promotes the migration of granule cells from the external to the internal granular layer through activation of NMDARs (Kim et al. 2005). The physiological influence of D-serine on neuronal migration involves the activation of SR by GRIP.

The motility-promoting role of D-serine is probably not restricted to the postnatal development of the brain and may be involved earlier during fetal development. Indeed SR is present in the perireticular nucleus, a transient structure of the human fetal brain supposed to be engaged in the lamination of the forebrain (Martineau et al. 2006). Furthermore, D-serine and NMDARs are already present at embryonic day 14 (Paoletti et al. 2013; Martineau et al. 2006). Thus D-serine is well positioned both spatially and temporally to control NMDAR-mediated neuronal migration and synaptogenesis. Although SR knockout mice develop normally and show no obvious deficit in lamination or ectopic neurons, the mice have a significantly reduced cortical volume (Balu et al. 2012, 2013). Detailed structural analysis of the CNS of SR knockout mice reveals that SR deletion when achieved before weaning results in altered cortical dendritic morphology of pyramidal neurons in the prefrontal cortex and in the primary somatosensory cortex (DeVito et al. 2011) supporting the hypothesis that D-serine through the modulation of NMDARs could contribute to neuronal differentiation and therefore to synaptogenesis. Noteworthy, SR knockout mice have a reduction in total brain-derived neurotrophic factor (BDNF) protein levels (Balu et al. 2012), an activity-regulated neurotrophin that plays a key role in promoting synapse formation and maturation and in regulating the functional development of neuronal circuits (Park and Poo 2013). Accordingly, Gomes and colleagues (2012) have discovered that astrocytes induce the formation of glutamatergic synapses through *transforming growth factor* (TGF)-β1 pathway and the TGF-β1 synaptogenic property is dependent on D-serine signaling (Diniz et al. 2012). Indeed, the growth factor increases D-serine release from astrocytes, and blocking the synthesis of D-serine or its function prevents TGF-β1 to promote the development of functional glutamatergic synapses in the cerebral cortex. Another study demonstrates that differentiation of P19 cells induces SR expression and D-serine formation and that D-serine shapes synaptogenesis, potentially by preventing widespread untargeted synaptogenesis (Fuchs et al. 2012). These results contribute to expand evidence for a role of D-serine in synaptic shaping and wiring of neuronal circuitry during brain development.

3.3.3 Role of D-Serine in Cognition and Healthy Aging

In addition to their role in shaping neuronal circuitries and in synapse dynamics, NMDARs are essential for memory formation and for higher cognitive functions and social interactions (Paoletti et al. 2013). Genetically or pharmacologically driven hypofunction of NMDARs leads to profound deficits in cognition and social activity (Paoletti et al. 2013). For example, mice with a point mutation in Grin1 (D481N) showed abnormally persistent latent inhibition (a measure of information-processing deficit), reduced social approach behaviors, and enhanced startle reactivity (measure of sensory gatting) and impairments in nonassociative spatial object recognition task (Labrie et al. 2008). These deficits are reversed by administration of agents that enhance NMDAR glycine site function like D-serine (Labrie et al. 2008). Various regimens of systemic administration of low-dose D-serine (50 mg/kg/day) increase performances on BALB/c mice performing object recognition, T-maze alternation, and open-field exploration tasks (Bado et al. 2011). Analysis of SR-KO mice revealed that the animals exhibited a significantly disrupted representation of the order of events in distinct experimental paradigms as shown by object recognition and odor sequence tests (DeVito et al. 2011); However, SR mutant mice showed normal detection of novel objects and in spatial displacement and showed intact relational memory in a test of transitive inference. SR mice exhibited normal sociability and preference for social novelty (DeVito et al. 2011). Analysis of mice with inactive DAAO further offers complementary evidence that D-serine is crucial for normal cognition and social behavior. First, Roder and colleagues have generated mice carrying point mutations in DAO ($Dao1^{G181R}$). These mice have no active DAAO and display elevated brain levels of D-serine. Introducing the $Dao1^{G181R}$ mutations in Grin1(D481N) mice corrects all cognitive deficits and normalizes social behavior deficit normally observed in Grin1(D481N) mice (Labrie et al. 2010). Second, $Dao^{-/-}$ mice generated more recently by Peirson and colleagues (2015) demonstrate enhanced spatial recognition memory performance, improved odor recognition memory performance, and enhanced spontaneous alternation in the T-maze (Pritchett et al. 2015). Overall, these observations are fully consistent with, and extend, findings in the natural mutant ddY/Dao^{-} line introduced and largely popularized by Konno and colleagues (Almond et al. 2006; Zhang et al. 2011). Thus, an increased level of D-serine resulting from decreased catalysis (i.e., DAAO inactivation) increases the cognitive performance and behavior of mice in striking contrast to SR deficient mice (Fig. 3.3). Building on this ground, inhibitors for DAAO show remarkable benefit in improving cognitive performance in animal models of NMDAR hypofunction (Smith et al. 2009; Strick et al. 2011; Hopkins et al. 2013a, b). Nicotine has long been known to improve long-term hippocampal-dependent memory in both laboratory animals and humans (Changeux 2010). García-Colunga and colleagues (2012) have found that the facilitating effect of nicotine on long-term memory evaluated with the one-trial step-through inhibitory-avoidance task depends on glial D-serine (Lopez-Hidalgo et al. 2012). Accordingly, nicotine activates nAChRs present on glial cells stimulating the release of D-serine and then the activation of NMDARs in the hippocampus. These data further expand the earlier observation that glial cells through the release of D-serine contribute to spatial memory retrieval (Zhang et al. 2008).

Non-pathological brain aging is certainly one of the best documented situation where accumulated experimental evidence have shown that deficits in synaptic plasticity and cognitive functions notably memory and learning result from reduced activation of NMDARs. In collaboration with Jean-Marie Billard, we have discovered that those deficits are caused by a reduction in the levels of D-serine produced by SR during aging (Mothet et al. 2006; Turpin et al. 2011) thus resulting in a hypofunction of NMDARs (Fig. 3.3). More detailed analysis revealed that oxidative stress as occurring during aging affects SR and then the production of D-serine and that treatment with the reducing agent N-acetyl cysteine prevents hippocampal synaptic plasticity deficits by protecting D-serine-dependent NMDA receptor activation (Haxaire et al. 2012). Recently the notion that D-serine contributes to age-related cognitive decline has been expanded by Saitoe and colleagues who demonstrated that age-related memory impairments in *Drosophila melanogaster* are a result of reduced D-serine production by glia and that D-serine feeding suppresses these deficits (Yamazaki et al. 2014) thus showing that D-serine metabolism and functions may be conserved throughout evolution (Schell 2004).

3.4 Conclusion

D-serine has overturned fundamental axioms of biology. Intensive research during the last two decades has shown that this D-amino acid has emerged as a very important brain messenger with implication for Human disease. Despite many progresses since the breakthrough of Nishikawa (1992) and Snyder (1995) groups, there are still some opened questions regarding the functions of D-serine in the CNS and in the peripheral nervous system. The discovery that NMDAR co-activation takes more than glycine illustrates perfectly there's no simple rule governing the regulation of NMDA receptors. There is still a great need for efforts to discriminate the relative contribution of these two NMDAR neuromodulators in the healthy and diseased nervous system.

Acknowledgments We thank our collaborators who help to expand the field. We apologize to our colleagues whose work could not be cited due to space limitations. Work in Dr Mothet's lab is supported by operating grants from the Centre National de la Recherche Scientifique, France Alzheimer, the Fondation pour la Recherche Médicale, Direction Générale de l'Armement and the Université Aix-Marseille.

References

Aimone JB, Li Y, Lee SW et al (2014) Regulation and function of adult neurogenesis: from genes to cognition. Physiol Rev 94:991–1026

Almond SL, Fradley RL, Armstrong EJ et al (2006) Behavioral and biochemical characterization of a mutant mouse strain lacking D-amino acid oxidase activity and its implications for schizophrenia. Mol Cell Neurosci 32:324–334

Bado P, Madeira C, Vargas-Lopes C et al (2011) Effects of low-dose D-serine on recognition and working memory in mice. Psychopharmacology 218:461–470

Balu DT, Basu AC, Corradi JP et al (2012) The NMDA receptor co-agonists, D-serine and glycine, regulate neuronal dendritic architecture in the somatosensory cortex. Neurobiol Dis 45:671–682

Balu DT, Li Y, Puhl MD et al (2013) Multiple risk pathways for schizophrenia converge in serine racemase knockout mice, a mouse model of NMDA receptor hypofunction. Proc Natl Acad Sci U S A 110:31

Balu DT, Takagi S, Puhl MD et al (2014) D-serine and serine racemase are localized to neurons in the adult mouse and human forebrain. Cell Mol Neurobiol 34:419–435

Basu AC, Tsai GE, Ma CL et al (2009) Targeted disruption of serine racemase affects glutamatergic neurotransmission and behavior. Mol Psychiatry 14:719–727

Benneyworth MA, Li Y, Basu AC et al (2012) Cell selective conditional null mutations of serine racemase demonstrate a predominate localization in cortical glutamatergic neurons. Cell Mol Neurobiol 32:613–624

Bergersen LH, Morland C, Ormel L et al (2012) Immunogold detection of L-glutamate and D-serine in small synaptic-like microvesicles in adult hippocampal astrocytes. Cereb Cortex 22:1690–1697

Bezzi P, Gundersen V, Galbete JL et al (2004) Astrocytes contain a vesicular compartment that is competent for regulated exocytosis of glutamate. Nat Neurosci 7:613–620

Campanini B, Spyrakis F, Peracchi A et al (2013) Serine racemase: a key player in neuron activity and in neuropathologies. Front Biosci 18:1112–1128

Changeux JP (2010) Allosteric receptors: from electric organ to cognition. Annu Rev Pharmacol Toxicol 50:1–38

Chaudhry FA, Boulland JL, Jenstad M et al (2008) Pharmacology of neurotransmitter transport into secretory vesicles. Handb Exp Pharmacol 184:77–106

Chen PE, Geballe MT, Katz E et al (2008) Modulation of glycine potency in rat recombinant NMDA receptors containing chimeric NR2A/2D subunits expressed in Xenopus laevis oocytes. J Physiol 586:227–245

Corlew R, Brasier DJ, Feldman DE et al (2008) Presynaptic NMDA receptors: newly appreciated roles in cortical synaptic function and plasticity. Neuroscientist 14:609–625

Curcio L, Podda MV, Leone L et al (2013) Reduced D-serine levels in the nucleus accumbens of cocaine-treated rats hinder the induction of NMDA receptor-dependent synaptic plasticity. Brain 136:1216–1230

Danysz W, Fadda E, Wroblewski JT et al (1990) [3H]D-serine labels strychnine-insensitive glycine recognition sites of rat central nervous system. Life Sci 46:155–164

DeVito LM, Balu DT, Kanter BR et al (2011) Serine racemase deletion disrupts memory for order and alters cortical dendritic morphology. Genes Brain Behav 10:210–222

Diniz LP, Almeida JC, Tortelli V et al (2012) Astrocyte-induced synaptogenesis is mediated by transforming growth factor beta signaling through modulation of D-serine levels in cerebral cortex neurons. J Biol Chem 287:41432–41445

Ehmsen JT, Ma TM, Sason H et al (2013) D-serine in glia and neurons derives from 3-phosphoglycerate dehydrogenase. J Neurosci 33:12464–12469

Fossat P, Turpin FR, Sacchi S et al (2012) Glial D-serine gates NMDA receptors at excitatory synapses in prefrontal cortex. Cereb Cortex 22:595–606

Fuchs SA, Dorland L, de Sain-van der Velden MG et al (2006) D-serine in the developing human central nervous system. Ann Neurol 60:476–480

Fuchs SA, Roeleveld MW, Klomp LW et al (2012) D-serine influences synaptogenesis in a p19 cell model. JIMD Rep 6:47–53

Fukasawa Y, Segawa H, Kim JY et al (2000) Identification and characterization of a Na(+)-independent neutral amino acid transporter that associates with the 4F2 heavy chain and exhibits substrate selectivity for small neutral D- and L-amino acids. J Biol Chem 275:9690–9698

Furukawa H, Gouaux E (2003) Mechanisms of activation, inhibition and specificity: crystal structures of the NMDA receptor NR1 ligand-binding core. EMBO J 22:2873–2885

Gustafson EC, Stevens ER, Wolosker H et al (2007) Endogenous D-serine contributes to NMDA-receptor-mediated light-evoked responses in the vertebrate retina. J Neurophysiol 98:122–130

Hardingham GE, Bading H (2010) Synaptic versus extrasynaptic NMDA receptor signalling: implications for neurodegenerative disorders. Nat Rev Neurosci 11:682–696

Hashimoto A, Nishikawa T, Oka T et al (1993) Endogenous D-serine in rat brain: N-methyl-D-aspartate receptor-related distribution and aging. J Neurochem 60:783–786

Hashimoto A, Oka T, Nishikawa T (1995) Extracellular concentration of endogenous free D-serine in the rat brain as revealed by in vivo microdialysis. Neuroscience 66:635–643

Hashimoto A, Kanda J, Oka T (2000) Effects of N-methyl-D-aspartate, kainate or veratridine on extracellular concentrations of free D-serine and L-glutamate in rat striatum: an in vivo microdialysis study. Brain Res Bull 53:347–351

Haxaire C, Turpin FR, Potier B et al (2012) Reversal of age-related oxidative stress prevents hippocampal synaptic plasticity deficits by protecting D-serine-dependent NMDA receptor activation. Aging Cell 11:336–344

Haydon PG, Carmignoto G (2006) Astrocyte control of synaptic transmission and neurovascular coupling. Physiol Rev 86:1009–1031

Helboe L, Egebjerg J, Moller M et al (2003) Distribution and pharmacology of alanine-serine-cysteine transporter 1 (asc-1) in rodent brain. Eur J Neurosci 18:2227–2238

Henneberger C, Papouin T, Oliet SH et al (2010) Long-term potentiation depends on release of D-serine from astrocytes. Nature 463:232–236

Hopkins SC, Heffernan ML, Saraswat LD et al (2013a) Structural, kinetic, and pharmacodynamic mechanisms of D-amino acid oxidase inhibition by small molecules. J Med Chem 56:3710–3724

Hopkins SC, Campbell UC, Heffernan ML et al (2013b) Effects of D-amino acid oxidase inhibition on memory performance and long-term potentiation in vivo. Pharmacol Res Perspect 1:1

Huang X, Kong H, Tang M et al (2012) D-Serine regulates proliferation and neuronal differentiation of neural stem cells from postnatal mouse forebrain. CNS Neurosci Ther 18:4–13

Ishiwata S, Umino A, Umino M et al (2013) Modulation of extracellular D-serine content by calcium permeable AMPA receptors in rat medial prefrontal cortex as revealed by in vivo microdialysis. Int J Neuropsychopharmacol 16:1395–1406

Jahn R, Fasshauer D (2012) Molecular machines governing exocytosis of synaptic vesicles. Nature 490:201–207

Johnson JW, Ascher P (1987) Glycine potentiates the NMDA response in cultured mouse brain neurons. Nature 325:529–531

Kim PM, Aizawa H, Kim PS et al (2005) Serine racemase: activation by glutamate neurotransmission via glutamate receptor interacting protein and mediation of neuronal migration. Proc Natl Acad Sci U S A 102:2105–2110

Kleckner NW, Dingledine R (1988) Requirement for glycine in activation of NMDA-receptors expressed in Xenopus oocytes. Science 241:835–837

Konno R, Yasumura Y (1983) Mouse mutant deficient in D-amino acid oxidase activity. Genetics 103:277–285

Labrie V, Lipina T, Roder JC (2008) Mice with reduced NMDA receptor glycine affinity model some of the negative and cognitive symptoms of schizophrenia. Psychopharmacology (Berlin) 200:217–230

Labrie V, Wang W, Barger SW et al (2010) Genetic loss of D-amino acid oxidase activity reverses schizophrenia-like phenotypes in mice. Genes Brain Behav 9:11–25

Larsen RS, Smith IT, Miriyala J et al (2014) Synapse-specific control of experience-dependent plasticity by presynaptic NMDA receptors. Neuron 83:879–893

Le Bail M, Martineau M, Sacchi S et al (2015) Identity of the NMDA receptor coagonist is synapse specific and developmentally regulated in the hippocampus. Proc Natl Acad Sci U S A 112:30

LeMaistre JL, Sanders SA, Stobart MJ et al (2012) Coactivation of NMDA receptors by glutamate and D-serine induces dilation of isolated middle cerebral arteries. J Cereb Blood Flow Metab 32:537–547

Lench AM, Massey PV, Pollegioni L et al (2014) Astroglial D-serine is the endogenous co-agonist at the presynaptic NMDA receptor in rat entorhinal cortex. Neuropharmacology 83:118–127

Li Y, Sacchi S, Pollegioni L et al (2013) Identity of endogenous NMDAR glycine site agonist in amygdala is determined by synaptic activity level. Nat Commun 4:1760

Lin H, Hsu FC, Baumann BH et al (2014) Cortical synaptic NMDA receptor deficits in alpha7 nicotinic acetylcholine receptor gene deletion models: implications for neuropsychiatric diseases. Neurobiol Dis 63:129–140

Lopez-Hidalgo M, Salgado-Puga K, Alvarado-Martinez R et al (2012) Nicotine uses neuron-glia communication to enhance hippocampal synaptic transmission and long-term memory. PLoS ONE 7:21

Ma TM, Abazyan S, Abazyan B et al (2013) Pathogenic disruption of DISC1-serine racemase binding elicits schizophrenia-like behavior via D-serine depletion. Mol Psychiatry 18:557–567

Maekawa M, Watanabe M, Yamaguchi S et al (2005) Spatial learning and long-term potentiation of mutant mice lacking D-amino-acid oxidase. Neurosci Res 53:34–38

Martineau M, Baux G, Mothet JP (2006) D-serine signalling in the brain: friend and foe. Trends Neurosci 29:481–491

Martineau M, Galli T, Baux G et al (2008) Confocal imaging and tracking of the exocytotic routes for D-serine-mediated gliotransmission. Glia 56:1271–1284

Martineau M, Shi T, Puyal J et al (2013) Storage and uptake of D-serine into astrocytic synaptic-like vesicles specify gliotransmission. J Neurosci 33:3413–3423

Martineau M, Parpura V, Mothet JP (2014) Cell-type specific mechanisms of D-serine uptake and release in the brain. Front 6:12. doi:10.3389/fnsyn.2014.00012

Matsui T, Sekiguchi M, Hashimoto A et al (1995) Functional comparison of D-serine and glycine in rodents: the effect on cloned NMDA receptors and the extracellular concentration. J Neurochem 65:454–458

Matsuo H, Kanai Y, Tokunaga M et al (2004) High affinity D- and L-serine transporter Asc-1: cloning and dendritic localization in the rat cerebral and cerebellar cortices. Neurosci Lett 358:123–126

Maucler C, Pernot P, Vasylieva N et al (2013) In vivo D-serine hetero-exchange through alanine-serine-cysteine (ASC) transporters detected by microelectrode biosensors. ACS Chem Neurosci 4:772–781

Miya K, Inoue R, Takata Y et al (2008) Serine racemase is predominantly localized in neurons in mouse brain. J Comp Neurol 510:641–654

Monaghan DT, Olverman HJ, Nguyen L et al (1988) Two classes of N-methyl-D-aspartate recognition sites: differential distribution and differential regulation by glycine. Proc Natl Acad Sci U S A 85:9836–9840

Montana V, Malarkey EB, Verderio C et al (2006) Vesicular transmitter release from astrocytes. Glia 54:700–715

Mothet JP, Parent AT, Wolosker H et al (2000) D-serine is an endogenous ligand for the glycine site of the N-methyl-D-aspartate receptor. Proc Natl Acad Sci U S A 97:4926–4931

Mothet JP, Pollegioni L, Ouanounou G et al (2005) Glutamate receptor activation triggers a calcium-dependent and SNARE protein-dependent release of the gliotransmitter D-serine. Proc Natl Acad Sci U S A 102:5606–5611

Mothet JP, Rouaud E, Sinet PM et al (2006) A critical role for the glial-derived neuromodulator D-serine in the age-related deficits of cellular mechanisms of learning and memory. Aging Cell 5:267–274

Mothet JP, Le Bail M, Billard JM (2015) Time and space profilling of NMDA receptor co-agonist functions. J Neurochem. doi:10.1111/jnc.13204

Ohide H, Miyoshi Y, Maruyama R et al (2011) D-Amino acid metabolism in mammals: biosynthesis, degradation and analytical aspects of the metabolic study. J Chromatogr B Anal Technol Biomed Life Sci 879:3162–3168

Otte DM, Rasko T, Wang M et al (2014) Identification of the mitochondrial MSRB2 as a binding partner of LG72. Cell Mol Neurobiol 34:1123–1130

Panatier A, Theodosis DT, Mothet JP et al (2006) Glia-derived D-serine controls NMDA receptor activity and synaptic memory. Cell 125:775–784

Pankratov Y, Lalo U (2015) Role for astroglial alpha1-adrenoreceptors in gliotransmission and control of synaptic plasticity in the neocortex. Front 9:230

Paoletti P, Bellone C, Zhou Q (2013) NMDA receptor subunit diversity: impact on receptor properties, synaptic plasticity and disease. Nat Rev Neurosci 14:383–400

Papouin T, Ladepeche L, Ruel J et al (2012) Synaptic and extrasynaptic NMDA receptors are gated by different endogenous coagonists. Cell 150:633–646

Park H, Poo MM (2013) Neurotrophin regulation of neural circuit development and function. Nat Rev Neurosci 14:7–23

Parpura V, Verkhratsky A (2013) Astroglial amino acid-based transmitter receptors. Amino Acids 44:1151–1158

Petralia RS (2012) Distribution of extrasynaptic NMDA receptors on neurons. ScientificWorldJournal 2012:30

Popiolek M, Ross JF, Charych E et al (2011) D-amino acid oxidase activity is inhibited by an interaction with bassoon protein at the presynaptic active zone. J Biol Chem 286:28867–28875

Priestley T, Kemp JA (1994) Kinetic study of the interactions between the glutamate and glycine recognition sites on the N-methyl-D-aspartic acid receptor complex. Mol Pharmacol 46:1191–1196

Pritchett D, Hasan S, Tam SK et al (2015) D-amino acid oxidase knockout (Dao(−/−)) mice show enhanced short-term memory performance and heightened anxiety, but no sleep or circadian rhythm disruption. Eur J Neurosci 41:1167–1179

Puyal J, Martineau M, Mothet JP et al (2006) Changes in D-serine levels and localization during postnatal development of the rat vestibular nuclei. J Comp Neurol 497:610–621

Rasooli-Nejad S, Palygin O, Lalo U et al (2014) Cannabinoid receptors contribute to astroglial Ca (2)(+)-signalling and control of synaptic plasticity in the neocortex. Philos Trans R Soc Lond B Biol Sci 369:0077

Reyes-Haro D, Muller J, Boresch M et al (2010) Neuron-astrocyte interactions in the medial nucleus of the trapezoid body. J Gen Physiol 135:583–594

Ribeiro CS, Reis M, Panizzutti R et al (2002) Glial transport of the neuromodulator D-serine. Brain Res 929:202–209

Rodenas-Ruano A, Chavez AE, Cossio MJ et al (2012) REST-dependent epigenetic remodeling promotes the developmental switch in synaptic NMDA receptors. Nat Neurosci 15:1382–1390

Romero GE, Lockridge AD, Morgans CW et al (2014) The postnatal development of D-serine in the retinas of two mouse strains, including a mutant mouse with a deficiency in D-amino acid oxidase and a serine racemase knockout mouse. ACS Chem Neurosci 5:848–854

Rosenberg D, Kartvelishvily E, Shleper M et al (2010) Neuronal release of D-serine: a physiological pathway controlling extracellular D-serine concentration. FASEB J 24:2951–2961

Rosenberg D, Artoul S, Segal AC et al (2013) Neuronal D-serine and glycine release via the Asc-1 transporter regulates NMDA receptor-dependent synaptic activity. J Neurosci 33:3533–3544

Sacchi S, Cappelletti P, Giovannardi S et al (2008) Evidence for the interaction of D-amino acid oxidase with pLG72 in a glial cell line. Mol Cell Neurosci 48:20–28

Sacchi S, Cappelletti P, Giovannardi S et al (2011) Evidence for the interaction of D-amino acid oxidase with pLG72 in a glial cell line. Mol Cell Neurosci 48(1):20–28

Sasabe J, Suzuki M, Imanishi N et al (2014) Activity of D-amino acid oxidase is widespread in the human central nervous system. Front Synaptic Neurosci 6:14

Sasaki M, Konno R, Nishio M et al (1992) A single-base-pair substitution abolishes D-amino-acid oxidase activity in the mouse. Biochim Biophys Acta 1139:315–318

Schell MJ (2004) The N-methyl D-aspartate receptor glycine site and D-serine metabolism: an evolutionary perspective. Philos Trans R Soc Lond B Biol Sci 359:943–964

Schell MJ, Molliver ME, Snyder SH (1995) D-serine, an endogenous synaptic modulator: localization to astrocytes and glutamate-stimulated release. Proc Natl Acad Sci U S A 92:3948–3952

Schell MJ, Brady RO Jr, Molliver ME et al (1997) D-serine as a neuromodulator: regional and developmental localizations in rat brain glia resemble NMDA receptors. J Neurosci 17:1604–1615

Scianni M, Antonilli L, Chece G et al (2013) Fractalkine (CX3CL1) enhances hippocampal N-methyl-D-aspartate receptor (NMDAR) function via D-serine and adenosine receptor type A2 (A2AR) activity. J Neuroinflammation 10:1742–2094

Shigetomi E, Jackson-Weaver O, Huckstepp RT et al (2013) TRPA1 channels are regulators of astrocyte basal calcium levels and long-term potentiation via constitutive D-serine release. J Neurosci 33:10143–10153

Sjostrom PJ, Turrigiano GG, Nelson SB (2003) Neocortical LTD via coincident activation of presynaptic NMDA and cannabinoid receptors. Neuron 39:641–654

Smith SM, Uslaner JM, Yao L et al (2009) The behavioral and neurochemical effects of a novel D-amino acid oxidase inhibitor compound 8 [4H-thieno [3,2-b]pyrrole-5-carboxylic acid] and D-serine. J Pharmacol Exp Ther 328:921–930

Stehberg J, Moraga-Amaro R, Salazar C et al (2012) Release of gliotransmitters through astroglial connexin 43 hemichannels is necessary for fear memory consolidation in the basolateral amygdala. FASEB J 26:3649–3657

Stevens ER, Esguerra M, Kim PM et al (2003) D-serine and serine racemase are present in the vertebrate retina and contribute to the physiological activation of NMDA receptors. Proc Natl Acad Sci U S A 100:6789–6794

Stevens ER, Gustafson EC, Sullivan SJ et al (2010) Light-evoked NMDA receptor-mediated currents are reduced by blocking D-serine synthesis in the salamander retina. Neuroreport 21:239–244

Stobart JL, Lu L, Anderson HD et al (2013) Astrocyte-induced cortical vasodilation is mediated by D-serine and endothelial nitric oxide synthase. Proc Natl Acad Sci U S A 110:3149–3154

Strick CA, Li C, Scott L et al (2011) Modulation of NMDA receptor function by inhibition of D-amino acid oxidase in rodent brain. Neuropharmacology 61:1001–1015

Sullivan SJ, Miller RF (2012) AMPA receptor-dependent, light-evoked D-serine release acts on retinal ganglion cell NMDA receptors. J Neurophysiol 108:1044–1051

Sultan S, Gebara EG, Moullec K et al (2013) D-serine increases adult hippocampal neurogenesis. Front 7:155

Suzuki M, Sasabe J, Miyoshi Y et al (2015) Glycolytic flux controls D-serine synthesis through glyceraldehyde-3-phosphate dehydrogenase in astrocytes. Proc Natl Acad Sci U S A 112:13

Takata N, Mishima T, Hisatsune C et al (2011) Astrocyte calcium signaling transforms cholinergic modulation to cortical plasticity in vivo. J Neurosci 31:18155–18165

Traynelis SF, Wollmuth LP, McBain CJ et al (2010) Glutamate receptor ion channels: structure, regulation, and function. Pharmacol Rev 62:405–496

Turpin FR, Potier B, Dulong JR et al (2011) Reduced serine racemase expression contributes to age-related deficits in hippocampal cognitive function. Neurobiol Aging 32:1495–1504

Verrall L, Walker M, Rawlings N et al (2007) D-Amino acid oxidase and serine racemase in human brain: normal distribution and altered expression in schizophrenia. Eur J Neurosci 26:1657–1669

Williams SM, Diaz CM, Macnab LT et al (2006) Immunocytochemical analysis of D-serine distribution in the mammalian brain reveals novel anatomical compartmentalizations in glia and neurons. Glia 53:401–411

Wolosker H (2011) Serine racemase and the serine shuttle between neurons and astrocytes. Biochim Biophys Acta 2011:1558–1566

Wolosker H, Mori H (2012) Serine racemase: an unconventional enzyme for an unconventional transmitter. Amino Acids 43:1895–1904

Wolosker H, Blackshaw S, Snyder SH (1999) Serine racemase: a glial enzyme synthesizing D-serine to regulate glutamate-N-methyl-D-aspartate neurotransmission. Proc Natl Acad Sci U S A 96:13409–13414

Yacubova E, Komuro H (2003) Cellular and molecular mechanisms of cerebellar granule cell migration. Cell Biochem Biophys 37:213–234

Yamazaki D, Horiuchi J, Ueno K et al (2014) Glial dysfunction causes age-related memory impairment in Drosophila. Neuron 84:753–763

Yang JH, Wada A, Yoshida K et al (2010) Brain-specific Phgdh deletion reveals a pivotal role for L-serine biosynthesis in controlling the level of D-serine, an N-methyl-D-aspartate receptor co-agonist, in adult brain. J Biol Chem 285:41380–41390

Yang Y, Higashimori H, Morel L (2013) Developmental maturation of astrocytes and pathogenesis of neurodevelopmental disorders. J Neurodev Disord 5:1866–1955

Zafra F, Gomeza J, Olivares L et al (1995) Regional distribution and developmental variation of the glycine transporters GLYT1 and GLYT2 in the rat CNS. Eur J Neurosci 7:1342–1352

Zhang Z, Gong N, Wang W et al (2008) Bell-shaped D-serine actions on hippocampal long-term depression and spatial memory retrieval. Cereb Cortex 18:2391–2401

Zhang M, Ballard ME, Basso AM et al (2011) Behavioral characterization of a mutant mouse strain lacking D-amino acid oxidase activity. Behav Brain Res 217:81–87

Zhang Y, Chen K, Sloan SA et al (2014) An RNA-sequencing transcriptome and splicing database of glia, neurons, and vascular cells of the cerebral cortex. J Neurosci 34:11929–11947

Zhuang Z, Yang B, Theus MH et al (2010) EphrinBs regulate D-serine synthesis and release in astrocytes. J Neurosci 30:16015–16024

Zorec R, Araque A, Carmignoto G et al (2012) Astroglial excitability and gliotransmission: an appraisal of Ca2+ as a signalling route. ASN Neuro 4:2

Chapter 4
Behaviors of Mutant Mice Lacking D-Amino-Acid Oxidase Activity

Hiroaki Sakaue, Hiroko Ohide, Masahiro Yamanaka, and Ryuichi Konno

Abstract D-Amino-acid oxidase (DAO) catalyzes oxidative deamination of D-amino acids, including D-serine. DAO is present in the brain, liver, and kidney of higher animals. A large amount of D-serine is known to be present in the mammalian brains. D-Serine is a co-agonist of *N*-methyl-D-aspartate-subtype glutamate receptors (NMDARs) and potentiates NMDAR activity. NMDARs are involved in many physiological and pathological processes, and an abnormal NMDAR activity can thus have a severe effect on normal functioning. For instance, NMDAR hypofunction is considered one of the causes of schizophrenia, and over-excitation of NMDARs causes neuronal cell death, which can lead to amyotrophic lateral sclerosis. Therefore, D-serine is considered to be deeply involved in both diseases. Since DAO degrades D-serine, DAO is also associated with these illnesses. Accordingly, mutant mice lacking DAO activity have been attracting much attention of the researchers. In this article, we review currently available studies on behaviors of DAO mutant mice, especially those associated with schizophrenia. As of yet, accumulated behavioral data have not been necessarily consistent to draw a clear conclusion, the reasons for which are considered and discussed.

Keywords D-Amino-acid oxidase • Mutant mice • Behavior • D-Serine • NMDA receptor

4.1 Introduction

In 2002, Chumakov et al. (2002) published an astonishing article that D-amino-acid oxidase (DAO, EC 1.4.3.3) is involved in the etiology of schizophrenia. Since then, many articles that approve or disapprove of this finding have been presented (see Verrall et al. 2010). More recently, Mitchell et al. (2010) published another interesting article that DAO is involved in the etiology of amyotrophic lateral sclerosis (ALS). DAO is considered to be responsible to these neuropsychiatric

H. Sakaue (✉) • H. Ohide • M. Yamanaka • R. Konno
Department of Pharmaceutical Sciences, International University of Health and Welfare, 2600-1 Kitakanemaru, Ohtawara 324-8501, Tochigi, Japan
e-mail: sakaue@iuhw.ac.jp

T. Yoshimura et al. (eds.), *D-Amino Acids*, DOI 10.1007/978-4-431-56077-7_4

diseases through its metabolism of D-serine which potentiates *N*-methyl-D-aspartate receptors (NMDARs). These researches have spotlighted mutant mice lacking DAO activity. Here, we review the properties and behaviors of these mutant mice.

4.2 D-Amino-Acid Oxidase (DAO)

DAO catalyzes oxidative deamination of D-amino acids, stereoisomers of naturally occurring L-amino acids, which yield the corresponding 2-oxo acids, hydrogen peroxide and ammonia (Krebs 1935). DAO is present in a wide variety of organisms, from yeast to humans. In higher animals, it is primarily present in the kidney, the liver (with the exception of the mouse liver), and the brain (Konno et al. 1997). DAO has a wide range of substrate specificity and oxidizes many neutral and basic D-amino acids. Because D-amino acids are considered rare in higher animals, physiological role of DAO has remained unclear for some time. However, the recent developments of analytical instruments and improvements in analytical techniques have revealed the presence of D-amino acids in higher animals, including humans. The discovery of D-serine in the brains was a landmark event in the research of the physiological function of DAO.

4.3 D-Serine and NMDA Receptors

High levels of D-serine have been found in the brains of mammals (Hashimoto et al. 1992, 1993). D-Serine is produced from L-serine by serine racemase (SR) (Wolosker et al. 1999). D-Serine functions as a co-agonist of NMDARs, a subclass of ionotropic glutamate receptors (Mothet et al. 2000). For full activation, NMDARs require the binding of a co-agonist (D-serine or glycine) in addition to glutamate. NMDARs are involved in many related physiological and pathological processes, including synaptic plasticity, learning, memory, and neural diseases (Wolosker et al. 2008; Santangelo et al. 2012).

4.4 Mutant Mice Lacking DAO

In order to uncover the physiological function of DAO, we established a mutant strain (ddY/DAO$^-$) that lacks DAO activity (Konno and Yasumura 1983). These ddY/DAO$^-$ mice carry a missense mutation (Gly181Arg) that causes a complete loss of DAO activity (Sasaki et al. 1992).

Labrie et al. (2009a) and Sasabe et al. (2012) independently crossed mutant ddY/DAO$^-$ mice with C57BL/6 J mice and continuously backcrossed their offspring to C57BL/6 J mice in order to transfer the Gly181Arg mutation to a C57BL/

6 J genetic background. Then, two strains have been established: the *Dao1*^{G181R} strain (Labric et al. 2009a) and the ^{B6}DAO$^{-/-}$ strain (Sasabe et al. 2012).

More recently, a knockout strain (DAO-KO) has been established (Rais et al. 2012; Schweimer et al. 2014). DAO-KO mice have a deletion of exons 6–9 of the DAO gene in a 129SvEV genetic background.

4.5 Properties of Mutant Mice Lacking DAO Activity

Mutant ddY/DAO^{-} mice contain large quantities of D-amino acids, including D-serine in their organs and body fluids (see Hashimoto et al. 2005; Yamanaka et al. 2012). Despite this, ddY/DAO^{-} mice do not show overt abnormality and exhibited normal development, longevity, and reproduction (Konno and Yasumura 1983; Konno et al. 2010). In addition to this, ddY/DAO^{-} mice have not manifested obvious compensatory changes in D-serine-associated proteins (Almond et al. 2006).

Detailed analyses have shown that the ddY/DAO^{-} mice display an enhancement of NMDAR-mediated synaptic transmission in the hippocampus and spinal cord (Wake et al. 2001; Maekawa et al. 2005). Function and expression of NMDARs in the retina are also different in the ddY/DAO^{-} mice (Gustafson et al. 2013).

Dao1^{G181R} mice do not show any gross abnormalities, either (Labrie et al. 2009b), and are physically indistinguishable from wild-type mice in terms of their reflexes, vision, and fur and whisker condition (Labrie et al. 2010).

^{B6}DAO$^{-/-}$ mice, on the other hand, have been reported to develop an abnormal limb reflex: they retracted the hind limbs toward the trunk when lifted up by their tails, which was thought to be due to the degeneration of motoneurons of the lumber ventral horn (Sasabe et al. 2012). The authors also found that aged mice exhibited muscle atrophy.

4.6 Behaviors of Mutant Mice Lacking DAO Activity

Hypofunction of NMDARs is one of the most plausible causes of schizophrenia (see Verall et al. 2010; Labrie et al. 2012). DAO metabolizes D-serine, which potentiates NMDAR activity. Since some behaviors are known to be reliant on NMDAR activity, the behaviors of mutant mice lacking DAO have been examined, especially those relating to schizophrenia.

While preliminary tests have failed to show significant differences between the ddY/DAO^{-} and ddY/DAO^{+} mice in the swim test, grip strength, or time taken to cross a beam, ddY/DAO^{-} mice have been shown to commit fewer overall footslips than ddY/DAO^{+} mice when walking across a slender, round, wooden beam (Almond et al. 2006). The ddY/DAO^{-} mice committed fewer overall footslips compared to the ddY/DAO^{+} mice. This may be due to high content of D-serine in the cerebellum; D-serine is involved in learning, memory, and motor coordination through δ2 glutamate receptors (Kakegawa et al. 2011).

Findings that *Dao1*^{G181R} mice show similar latencies to find a food pellet buried beneath corncob bedding as wild-type mice indicate that they have normal olfactory activity (Labrie et al. 2010). The authors also found no differences in motor coordination, balance, and motor learning between *Dao1*^{G181R} mice and wild-type mice, as measured by the performance of the accelerating rotarod task. Latencies to fall from the rotating axle were not different between the two groups.

4.6.1 Locomotor Activity

Tests of locomotor activity in mutant mice lacking DAO activity have yielded variable results. Hashimoto et al. (2008) observed that, compared to ddY/DAO^{+} mice, ddY/DAO^{-} mice were hyperlocomotive. On the contrary, Almond et al. (2006) reported that ddY/DAO^{-} mice were hypolocomotive in a novel environment. However, Labrie et al. (2009a) did not see the difference between ddY/DAO^{-} and control mice in locomotion on the elevated plus maze test. Horizontal and vertical locomotion did not differ between ddY/DAO^{-} and wild-type mice. Zhang et al. (2011) observed that the ddY/DAO^{-} mice were hypoactive and traveled less than the ddY/DAO^{+} mice in enclosed boxes but that they were as active as the ddY/DAO^{+} mice though they spent more time on the periphery in the Barnes maze.

While *Dao1*^{G181R} mice and wild-type mice have previously been found to show no differences in their horizontal and vertical locomotor activity in the open-field test (Labrie et al. 2009a), subsequent research has reported a greater vertical locomotion in *Dao1*^{G181R} mice than in wild-type mice (Labrie et al. 2010). It has also been found that these mice do not present locomotion differences in the elevated plus maze test or in the novel object test (Labrie et al. 2009a).

4.6.2 Novel Object Test

For this test, a novel object was placed in the center of the home cage, and time spent for a test mouse to explore it, the number of contacts, and the duration of locomotor activity were recorded. Labrie et al. (2009a) have shown that *Dao1*^{G181R} mice and wild-type mice do not show differences in their response to a novel object.

4.6.3 *Spatial and Nonspatial Object Recognition Test*

A test mouse was habituated to four different objects placed near each corner of the square field, after which two of four objects were switched while the mouse was set aside. When the mouse was returned to the field, its reaction to the displaced and non-displaced objects was recorded. Next, one of four objects was replaced by a novel object while the mouse was set aside. When the mouse was again returned to the field, its reaction to the novel object and the three familiar objects was recorded.

In the first test, Labrie et al. (2010) found that *Dao1*^{G181R} mice and wild-type mice spent more time exploring the objects with a novel spatial configuration than the objects that remained stationary. In the second test, *Dao1*^{G181R} mice and wild-type mice spent more time investigating the novel object than the familiar objects. Because there were no differences between the two strains, the authors concluded that *Dao1*^{G181R} mice had normal recognition abilities.

4.6.4 *Social Interaction Test*

Social behaviors in *Dao1*^{G181R} mice were also examined by Labrie et al. (2010) using a social interaction test with three chambers. Similar to wild-type mice, *Dao1* ^{G181R} mice preferred the chamber containing an unfamiliar conspecific mouse to the empty cage chamber. In addition to this, both groups were examined for their preference to social novelty. Both groups of mice spent more time in the chamber containing a newly introduced mouse (stranger 2) than in the chamber containing a now familiar mouse (stranger 1).

4.6.5 *Morris Water Maze*

Maekawa et al. (2005) have reported that ddY/DAO^{-} mice performed better in the Morris water maze test than did ddY/DAO^{+} mice, which suggests that they have a better spatial working memory. Indeed, long-term potentiation (LTP) in the CA1 area of the hippocampus, the basis of learning and memory, was augmented in ddY/DAO^{-} mice.

Conversely, Zhang et al. (2011) found that ddY/DAO^{-} mice could not perform the Morris water maze test due to their hypoactivity; they did not readily swim and instead just floated.

Labrie et al. (2009b) did not find a difference between *Dao1*^{G181R} mice and wild-type mice in task performance. The two groups showed a similar path length and latency to reach the platform, and swim speed, floating time, and thigmotaxis duration were not significantly different.

In an extension of the Morris water maze test, Labrie et al. (2009b) changed the platform position and found that the extinction of the learned experience occurred earlier in the *Dao1*^{G181R} mice. The memory for the new platform location was substantially enhanced in *Dao1*^{G181R} mice in the reversal probe test. Therefore, they concluded that the loss of DAO activity enhances extinction and reversal learning.

4.6.6 Barnes Maze

The ddY/DAO^{-} mice showed a shortened latency and a higher velocity to enter the tunnel (Zhang et al. 2011). However, when the tunnel was removed, there was no significant difference in total distance traveled on the Barnes maze area between the ddY/DAO^{-} and ddY/DAO^{+} mice. The authors suggest that the better performance of the ddY/DAO^{-} mice was due to high edge activity as these mice spent much more time on the periphery, which enabled them to get into the tunnel easily.

4.6.7 24-h Inhibitory Avoidance Test

Memory consolidation was examined using a two-compartment chamber (Zhang et al. 2011). A mouse was placed into the light chamber. When the mouse entered the adjoining dark chamber, a foot shock was presented. Twenty-four hours later, the mouse was placed into the light chamber again, and the latency to enter the dark chamber was measured. During training, there was longer latency in the ddY/DAO^{-} mice due to their hypoactivity. However, no difference was observed between the ddY/DAO^{-} and ddY/DAO^{+} mice when they were tested for recall after 24 h, which suggests that ddY/DAO^{-} mice have an unimpaired memory consolidation.

4.6.8 Novelty-Suppressed Feeding Test

In order to investigate novelty-suppressed feeding in ddY/DAO^{-} mice, Zhang et al. (2011) gave food to a starved mouse in a novel environment and measured the feeding onset latency. The authors found that the feeding onset latency and the time spent interacting with the food in a novel environment did not differ between ddY/DAO^{-} mice and ddY/DAO^{+} mice, which suggests that ddY/DAO^{-} mice have a normal level of anxiety.

4.6.9 Contextual Fear-Conditioning Test

Using a contextual fear-conditioning test, Labrie et al. (2009b) found no differences in freezing time between *Dao1*^{G181R} mice and wild-type mice; both groups showed a similar learning of the context-shock association. However, *Dao1*^{G181R} mice displayed an enhanced rate of extinction of contextual fear memory. The extinction in the cued/tone fear-conditioning test, on the other hand, was not different between *Dao1*^{G181R} mice and wild-type mice.

4.6.10 Open-Field Test

Female ddY/DAO^{-} mice display greater anxiety in the open-field test than their male counterparts, spending less time in the center area of the open field (Labrie et al. 2009a). However, the authors found that horizontal and vertical locomotion of both male and female ddY/DAO^{-} mice was not different from that of wild-type mice. The time for freezing and grooming was also similar between ddY/DAO^{-} mice and wild-type mice.

Female *Dao1*^{G181R} mice also display anxiety-like behaviors in the open field like the female ddY/DAO^{-} mice above (Labrie et al. 2009a).

Anxiety has also been reported in DAO-KO mice, which were found by Rais et al. (2012) to show a decreased center path in the open-field test. However, they did not show anxiety behavior in the elevated plus maze test.

4.6.11 Elevated Plus Maze

In the elevated plus maze, both male and female ddY/DAO^{-} mice have been found to enter the open arms less frequently and to spend less time in the open arms compared to wild-type mice, indicating an increased anxiety in the mutant mice (Labrie et al. 2009a).

Labrie et al. (2009a) also found that female *Dao1*^{G181R} mice, but not male mice, exhibited more anxiety than wild-type females and spent less time and made fewer entries into the open arms of the elevated plus maze.

However, Zhang et al. (2011) did not find a difference in the time spent in the open arms between the ddY/DAO^{-} mice and ddY/DAO^{+} mice. Nonetheless, the authors reported a significant reduction of closed arm entries and total arm entries in the ddY/DAO^{-} mice, which they considered the reflection of their hypoactivity of these mice.

4.6.12 Acoustic Startle

There was a significant difference in the acoustic startle response of ddY/DAO$^-$ compared to ddY/DAO$^+$ mice (Almond et al. 2006). The ddY/DAO$^-$ mice showed a greater startle response (startle amplitude) for several sound stimuli.

While male mice of both *Dao1*^{G181R} and wild-type strains showed an increased startle response in comparison to female counterparts, no difference between these two strains was found (Labrie et al. 2010).

4.6.13 Prepulse Inhibition (PPI)

PPI to an acoustic startle response has been reported to be markedly reduced in patients with schizophrenia (see Swerdlow et al. 2008). While Almond et al. (2006) did not find a difference in PPI between ddY/DAO$^-$ and ddY/DAO$^+$ mice, Zhang et al. (2011) found that ddY/DAO$^-$ mice did indeed exhibit an enhanced PPI response compared to ddY/DAO$^+$ mice.

Labrie et al. (2010) did not find a difference in PPI between *Dao1*^{G181R} mice and wild-type mice.

4.6.14 Tail-Flick Test

The latency to flick the tail following immersion of the tail in hot water was significantly longer in ddY/DAO$^-$ mice compared to ddY/DAO$^+$ mice (Zhao et al. 2008).

However, the tail-flick test in *Dao1*^{G181R} mice revealed no significant differences in tail-flick latency from wild-type mice, suggesting that these mice have equivalent pain sensitivity (Labrie et al. 2009b).

4.6.15 Hot-Plate Test

In the hot-plate test (at 55 °C), Zhao et al. (2008) found that the latencies of licking, biting, or jumping were significantly prolonged in ddY/DAO$^-$ mice compared to ddY/DAO$^+$ mice.

4.6.16 Formalin Test

Subcutaneous injection of formalin (20 μl of 8 % formaldehyde) into the hind paw produced a biphasic pain response with an early phase of acute pain and a late phase of chronic pain. No significant difference in licking behaviors during the early phase of acute pain was observed between ddY/DAO$^-$ and ddY/DAO$^+$ mice, but the licking response was significantly augmented during the late phase of chronic pain in ddY/DAO$^-$ mice compared to ddY/DAO$^+$ mice (Wake et al. 2001).

Similarly, Zhao et al. (2008) did not find a difference in the early phase response after formalin injection (10 μl of 5 % formaldehyde) between ddY/DAO$^-$ mice and ddY/DAO$^+$ mice but did report that ddY/DAO$^-$ mice showed a significant reduction of the licking/biting duration during the late phase compared to ddY/DAO$^+$ mice

4.6.17 Acetic Acid-Induced Writhing Test

Zhao et al. (2008) have also reported the effect of intraperitoneal injection of 1.5 % acetic acid (6 ml/kg). The resulting writhing response was significantly lesser in ddY/DAO$^-$ mice than in ddY/DAO$^+$ mice.

4.7 Responses to Inhibitors and Psychotropics

4.7.1 Phencyclidine, a Noncompetitive Inhibitor of NMDAR

Administration of phencyclidine (PCP) induces schizophrenia-like symptoms in healthy subjects and exacerbates symptoms of patients with schizophrenia (Javitt and Zukin 1991). It has been found that administration of PCP (3 mg/kg) caused ddY/DAO$^+$ mice to become hyperlocomotive, an effect that is not seen in ddY/DAO$^-$ mice (Almond et al. 2006). The authors also found that while PCP induced hyperactivity in both strains, its effects were attenuated in the ddY/DAO$^-$ mice.

4.7.2 MK-801, a Noncompetitive Inhibitor of NMDAR

Administration of MK-801 is known to induce schizophrenic behaviors in mice and rats (Contreras 1990). For instance, administration of MK-801 (0.4 mg/kg) has been found to induce stereotypy (sniffing, head weaving, and tumbling) and ataxia (awkward-jerky movements and falling) in both ddY/DAO$^-$ mice and ddY/DAO$^+$

mice but with attenuated effects in ddY/DAO$^-$ mice (Hashimoto et al. 2005). However, the authors found no difference in locomotor activity between MK-801 injected ddY/DAO$^-$ and ddY/DAO$^+$ mice.

MK-801 (0.3 mg/kg) significantly increased the startle response in ddY/DAO$^-$ mice compared to ddY/DAO$^+$ mice and disrupted PPI in both ddY/DAO$^-$ and ddY/DAO$^+$ mice; this disruptive effect was stronger in ddY/DAO$^-$ mice (Zhang et al. 2011).

4.7.3 SDZ 220–581, a Competitive NMDAR Antagonist

SDZ 220–581 is a competitive NMDAR antagonist that binds to the glutamate-recognition site on the outside of the channel. It has been found that this antagonist (5 mg/kg) disrupted the PPI response in both ddY/DAO$^-$ and ddY/DAO$^+$ mice, although the disruptive effect was stronger in the ddY/DAO$^-$ mice (Zhang et al. 2011).

4.7.4 L-701,324, an Antagonist to NMDAR Co-agonist-Binding Site

Almond et al. (2006) found that L-701,324 administration (0.5 mg/kg) increased both the number of footslips and falls during beam walking in the ddY/DAO$^+$ mice but did not significantly impair the performance of the ddY/DAO$^-$ mice. The ddY/DAO$^-$ mice were considered to be protected from ataxic effects of L-701,324 due to the occupancy of the co-agonist-binding site by increased D-serine levels.

4.7.5 Harmaline, a Psychoactive Alkaloid

After administration of harmaline (80 mg/kg), the cGMP level in the cerebellum was increased to a greater extent in ddY/DAO$^-$ mice than in ddY/DAO$^+$ mice (Almond et al. 2006), which indicates that NMDAR activity is enhanced in ddY/DAO$^-$ mice.

4.7.6 *Methamphetamine, a Psychostimulant*

Administration of methamphetamine (5 mg/kg) induced stereotypy and increased locomotion in both ddY/DAO$^-$ mice and ddY/DAO$^+$ mice (Hashimoto et al. 2008). However, compared to ddY/DAO$^+$ mice, ddY/DAO$^-$ mice exhibited a marked reduction of stereotypy and a drastic augmentation of locomotor activity.

4.8 Conclusions

At early stages of research into the effects of DAO in mouse behavior, behaviors of ddY/DAO$^-$ mice seemed to result from an enhanced NMDAR activity. These mice also showed resistance to NMDAR antagonists, which was considered to be due to the increased D-serine content resulting from the lack of DAO activity. However, subsequent research has yielded inconsistent results. Possible reasons for this are considered in more detail.

One reason for the inconsistent behavioral findings of mutant mice lacking DAO activity may be their different genetic backgrounds, which likely influence behavior. The ddY/DAO$^-$ strain was established from ddY mice (Konno and Yasumura 1983). The ddY mice have been maintained as a closed colony and are not inbred mice. These mice may have tolerance to changing environments. On the other hand, *DAO1*^{G181R} mice and ^{B6}DAO$^{-/-}$ mice have been produced by transferring the *G181R* mutation carried by ddY/DAO$^-$ mice to a C57BL/6 J background (Labrie et al. 2009a; Sasabe et al. 2012). These mice may be more sensitive to environmental changes. Another reason for inconsistent behavioral findings may be the differences in experimental environments; behavioral experiments described in this review have been carried out in many different places, which may have affected mouse behaviors.

Inconsistent behaviors have also been observed in wild-type mice and rats that were treated with DAO inhibitors (see Verrall et al. 2010; Ferraris and Tsukamoto 2011). Mice of several SR-KO strains also displayed inconsistent behaviors (see Labrie et al. 2012). These mice lack nearly 90 % of normal level of D-serine in the forebrain. They do not necessarily manifest the behaviors opposite to the mutant mice lacking DAO activity which contain very high level of D-serine.

It is also worth bearing in mind that changes in DAO activity and D-serine content may not cause clear changes in behaviors. This seems consistent with the finding that a lack of DAO does not cause compensatory changes in D-serine-related proteins (Almond et al. 2006). DAO and D-serine may produce subtle changes that lead to schizophrenia, which is consistent with the idea that schizophrenia is likely to be caused by the gradual accumulation of small genetic and environmental cues.

Compared with ddY/DAO$^+$ mice, D-serine content in ddY/DAO$^-$ mice was significantly elevated in the cerebellum, but not in the cerebrum (Hashimoto et al. 1993; Hashimoto et al. 2005). Similar results have been observed in *Dao1* ^{G181R} mice (Labrie et al. 2009a) and DAO-KO mice (Rais et al. 2012). Together, these results are consistent with the distribution of DAO in the brain; DAO is predominantly present in the cerebellum and brain stem and is essentially absent in the forebrain, though different results have been obtained (see Verrall et al. 2010). Because schizophrenia arises from an abnormality in the cerebrum, the underlying mechanism that connects DAO and schizophrenia is not known. Recently, Schweimer et al. (2014) have found that DAO modulates dopaminergic neuron activity. Since it is well known that aberrant dopamine system is implicated in schizophrenia, this finding may be a steppingstone. This possibility needs to be extensively investigated.

In addition to mutant mice, mutant rats lacking DAO (Konno et al. 2009; Miyoshi et al. 2011) would also be useful for further behavioral studies.

References

Almond SL, Fradley RL, Armstrong EJ, Heavens RB, Rutter AR, Newman RJ, Chiu CS, Konno R, Hutson PH, Brandon NJ (2006) Behavioral and biochemical characterization of a mutant mouse strain lacking D-amino acid oxidase activity and its implications for schizophrenia. Mol Cell Neurosci 32:324–334

Chumakov I, Blumenfeld M, Guerassimenko O, Cavarec L, Palicio M et al (2002) Genetic and physiological data implicating the new human gene G72 and the gene for D-amino acid oxidase in schizophrenia. Proc Natl Acad Sci USA 99:13675–13680

Contreras PC (1990) D-Serine antagonized phencyclidine- and MK-801-induced stereotyped behavior and ataxia. Neuropharmacology 29:291–293

Ferraris DV, Tsukamoto T (2011) Recent advances in the discovery of D-amino acid oxidase inhibitors and their therapeutic utility in schizophrenia. Curr Pharm Des 17:103–111

Gustafson EC, Morgans CW, Tekmen M, Sullivan SJ, Esguerra M, Konno R, Miller RF (2013) Retinal NMDA receptor function and expression are altered in a mouse lacking D-amino acid oxidase. J Neurophysiol 110:2718–2726

Hashimoto A, Nishikawa T, Hayashi T, Fujii N, Harada K, Oka T, Takahashi K (1992) The presence of free D-serine in rat brain. FEBS Lett 296:33–36

Hashimoto A, Nishikawa T, Konno R, Niwa A, Yasumura Y, Oka T, Takahashi K (1993) Free D-serine, D-aspartate and D-alanine in central nervous system and serum in mutant mice lacking D-amino acid oxidase. Neurosci Lett 152:33–36

Hashimoto A, Yoshikawa M, Niwa A, Konno R (2005) Mice lacking D-amino acid oxidase activity display marked attenuation of stereotypy and ataxia induced by MK-801. Brain Res 1033:210–215

Hashimoto A, Konno R, Yano H, Yoshikawa M, Tamaki R, Matsumoto H, Kobayashi H (2008) Mice lacking D-amino acid oxidase activity exhibit marked reduction of methamphetamine-induced stereotypy. Eur J Pharmacol 586:221–225

Javitt DC, Zukin SR (1991) Recent advances in the phencyclidine model of schizophrenia. Am J Psychiatry 148:1301–1308

Kakegawa W, Miyoshi Y, Hamase K, Matsuda S, Matsuda K, Kohda K, Emi K, Motohashi J, Konno R, Zaitsu K, Yuzaki M (2011) D-Serine regulates cerebellar LTD and motor coordination through the δ2 glutamate receptor. Nat Neurosci 14:603–611

Konno R, Yasumura Y (1983) Mouse mutant deficient in D-amino acid oxidase activity. Genetics 103:277–285

Konno R, Sasaki M, Asakura S, Fukui K, Enami J, Niwa A (1997) D-Amino-acid oxidase is not present in the mouse liver. Biochim Biophys Acta 1335:173–181

Konno R, Okamura T, Kasai N, Summer KH, Niwa A (2009) Mutant rat strain lacking D-amino-acid oxidase. Amino Acids 37:367–375

Konno R, Hamase K, Maruyama R, Zaitsu K (2010) Mutant mice and rats lacking D-amino acid oxidase. Chem Biodivers 7:1450–1458

Krebs HA (1935) Metabolism of amino-acids: deamination of amino-acids. Biochem J 29:1620–1644

Labrie V, Wong AH, Roder JC (2012) Contributions of the D-serine pathway to schizophrenia. Neuropharmacology 62:1484–1503

Labrie V, Clapcote SJ, Roder JC (2009a) Mutant mice with reduced NMDA-NR1 glycine affinity or lack of D-amino acid oxidase function exhibit altered anxiety-like behaviors. Pharmacol Biochem Behav 91:610–620

Labrie V, Duffy S, Wang W, Barger SW, Baker GB, Roder JC (2009b) Genetic inactivation of D-amino acid oxidase enhances extinction and reversal learning in mice. Learn Mem 16:28–37

Labrie V, Wang W, Barger SW, Baker GB, Roder JC (2010) Genetic loss of D-amino acid oxidase activity reverses schizophrenia-like phenotypes in mice. Genes Brain Behav 9:11–25

Maekawa M, Watanabe M, Yamaguchi S, Konno R, Hori Y (2005) Spatial learning and long-term potentiation of mutant mice lacking D-amino-acid oxidase. Neurosci Res 53:34–38

Mitchell J, Paul P, Chen HJ, Morris A, Payling M, Falchi M, Habgood J, Panoutsou S, Winkler S, Tisato V, Hajitou A, Smith B, Vance C, Shaw C, Mazarakis ND, de Belleroche J (2010) Familial amyotrophic lateral sclerosis is associated with a mutation in D-amino acid oxidase. Proc Natl Acad Sci U S A 107:7556–7561

Miyoshi Y, Hamase K, Okamura T, Konno R, Kasai N, Tojo Y, Zaitsu K (2011) Simultaneous two-dimensional HPLC determination of free D-serine and D-alanine in the brain and periphery of mutant rats lacking D-amino-acid oxidase. J Chromatogr B 879:3184–3189

Mothet JP, Parent AT, Wolosker H, Brady RO Jr, Linden DJ, Ferris CD, Rogawski MA, Snyder SH (2000) D-Serine is an endogenous ligand for the glycine site of the *N*-methyl-D-aspartate receptor. Proc Natl Acad Sci USA 97:4926–4931

Rais R, Thomas AG, Wozniak K, Wu Y, Jaaro-Peled H, Sawa A, Strick CA, Engle SJ, Brandon NJ, Rojas C, Slusher BS, Tsukamoto T (2012) Pharmacokinetics of oral D-serine in D-amino acid oxidase knockout mice. Drug Metab Dispos 40:2067–2073

Santangelo RM, Acker TM, Zimmerman SS, Katzman BM, Strong KL, Traynelis SF, Liotta DC (2012) Novel NMDA receptor modulators: an update. Expert Opin Ther Pat 22:1337–1352

Sasabe J, Miyoshi Y, Suzuki M, Mita M, Konno R, Matsuoka M, Hamase K, Aiso S (2012) D-Amino acid oxidase controls motoneuron degeneration through D-serine. Proc Natl Acad Sci USA 109:627–632

Sasaki M, Konno R, Nishio M, Niwa A, Yasumura Y, Enami J (1992) A single-base-pair substitution abolishes D-amino-acid oxidase activity in the mouse. Biochim Biophys Acta 1139:315–318

Schweimer JV, Coullon GS, Betts JF, Burnet PW, Engle SJ, Brandon NJ, Harrison PJ, Sharp T (2014) Increased burst-firing of ventral tegmental area dopaminergic neurons in D-amino acid oxidase knockout mice in vivo. Eur J Neurosci 40:2999–3009

Swerdlow NR, Weber M, Qu Y, Light GA, Braff DL (2008) Realistic expectations of prepulse inhibition in translational models for schizophrenia research. Psychopharmacology (Berl) 199:331–388

Verrall L, Burnet PW, Betts JF, Harrison PJ (2010) The neurobiology of D-amino acid oxidase and its involvement in schizophrenia. Mol Psychiatry 15:122–137

Wake K, Yamazaki H, Hanzawa S, Konno R, Sakio H, Niwa A, Hori Y (2001) Exaggerated responses to chronic nociceptive stimuli and enhancement of *N*-methyl-D-aspartate receptor-mediated synaptic transmission in mutant mice lacking D-amino-acid oxidase. Neurosci Lett 297:25–28

Wolosker H, Sheth KN, Takahashi M, Mothet JP, Brady RO Jr, Ferris CD, Snyder SH (1999) Purification of serine racemase: biosynthesis of the neuromodulator D-serine. Proc Natl Acad Sci USA 96:721–725

Wolosker H, Dumin E, Balan L, Foltyn VN (2008) D-Amino acids in the brain: D-serine in neurotransmission and neurodegeneration. FEBS J 275:3514–3526

Yamanaka M, Miyoshi Y, Ohide H, Hamase K, Konno R (2012) D-Amino acids in the brain and mutant rodents lacking D-amino-acid oxidase activity. Amino Acids 43:1811–1821

Zhang M, Ballard ME, Basso AM, Bratcher N, Browman KE, Curzon P, Konno R, Meyer AH, Rueter LE (2011) Behavioral characterization of a mutant mouse strain lacking D-amino acid oxidase activity. Behav Brain Res 217:81–87

Zhao W, Konno R, Zhou XJ, Yin M, Wang YX (2008) Inhibition of D-amino-acid oxidase activity induces pain relief in mice. Cell Mol Neurobiol 28:581–591

Chapter 5
Physiological Functions of D-Serine Mediated Through δ2 Glutamate Receptors in the Cerebellum

Wataru Kakegawa and Michisuke Yuzaki

Abstract D-Serine (D-Ser), a major gliotransmitter, acts as an endogenous co-agonist for the *N*-methyl-D-aspartate-type ionotropic glutamate receptor (NMDA receptor) which regulates synaptic plasticity underlying certain types of learning and memory in the central nervous system. While D-Ser is abundant in the forebrain throughout life, it exists only transiently in the immature cerebellum because D-amino acid oxidase (DAAO), a D-Ser-degrading enzyme, begins to be highly expressed in the mature cerebellum. However, it remains unclear why and how D-Ser impacts physiological functions in the developing cerebellum. Recently, D-Ser has been reported to bind to δ2-type ionotropic glutamate receptors (δ2 receptors), which are exclusively expressed on the excitatory synapses between cerebellar granule cell axons (parallel fibers) and Purkinje cells, and control the induction of long-term depression (LTD), a form of synaptic plasticity underlying cerebellar motor learning. In the developing cerebellum, we found that D-Ser, released from Bergmann glia in response to neuronal activity, binds to δ2 receptors, thereby conveying a critical signal for LTD and motor learning. Interestingly, while D-Ser activates NMDA receptor-coupled channels, it evoked non-channel functions of δ2 receptors. In addition to δ2 receptors, their homolog δ1 receptors are widely expressed in various brain regions throughout life and also bind to D-Ser; therefore, D-Ser-δ receptor interactions may be a general signaling mechanism by which glia communicate with neurons, an important facet of the tripartite synapse.

Keywords D-serine • Ionotropic glutamate receptor • Synapse • Long-term depression (LTD) • Learning and memory • Cerebellum

W. Kakegawa (✉) • M. Yuzaki
Department of Physiology, Keio University School of Medicine, 35 Shinanomachi, Shinjyuku, Tokyo 160-8582, Japan
e-mail: wkake@z7.keio.jp; myuzaki@a5.kcio.jp

T. Yoshimura et al. (eds.), *D-Amino Acids*, DOI 10.1007/978-4-431-56077-7_5

5.1 D-Serine in the Cerebellum

With the development and improvement of measurement techniques, including microdialysis and high-performance liquid chromatography (HPLC), it has been established that D-amino acids exist in various brain regions in mammals (see Chap. 1). Among these, D-serine (D-Ser) is abundant in the forebrain, such as in the cerebral cortex, striatum, and hippocampus, with high tissue concentrations (of the order of 10^{-3} M) throughout development and adulthood (Hashimoto et al. 1992, 1995; Hashimoto and Oka 1997; Matsui et al. 1995). D-Ser is reportedly the endogenous co-agonist for the *N*-methyl-D-aspartate-type ionotropic glutamate receptor (NMDA receptor; Fig. 5.1), a member of the ionotropic glutamate receptor (iGluR) family. Importantly, D-Ser has been reported to regulate synaptic transmission and synaptic plasticity, including long-term potentiation (LTP) and long-term depression (LTD), commonly accepted as molecular bases of learning and memory (Henneberger et al. 2010; Oliet and Mothet 2009; Panatier et al. 2006; Mothet et al. 2000) (see Chap. 3). In addition, many lines of evidence have indicated that D-Ser is crucially involved—NMDA receptor dependently—in various neurodegenerative and psychiatric disorders, such as Alzheimer's disease, amyotrophic lateral sclerosis, and schizophrenia (Billard 2008, 2013; Hashimoto and Oka 1997) (see Chaps. 6, 7, 8, 9).

In the cerebellum, a brain area essential for motor coordination and motor learning, high concentrations of D-Ser are also found during the early postnatal period. However, D-Ser becomes undetectable after the first postnatal month in rodents because D-amino acid oxidase (DAAO), an enzyme that degrades D-Ser, is highly expressed in later developmental stages (Hashimoto and Oka 1997; Schell et al. 1997; Wang and Zhu 2003; Weimar and Neims 1977). Although D-Ser activation of NMDA receptors reportedly contributes to the migration of cerebellar granule cells during development (Kim et al. 2005), the cytoarchitecture of cerebella lacking serine racemase, a major synthesizing enzyme for D-Ser (Wolosker et al. 1999a, b), is grossly normal (Miya et al. 2008). Therefore, whether and how D-Ser plays a role in the developing cerebellum have hitherto remained unclear. In this review, we discuss the functions of D-Ser in the cerebellum. Specifically, we focus on physiological roles of D-Ser binding to the orphan iGluRs known as δ-type glutamate receptors.

5.2 δ2 Glutamate Receptor: A New D-Ser Binding Receptor

Recently, X-ray crystallographic analysis has revealed that D-Ser binds to δ2-type glutamate receptors (δ2 receptors), members of the iGluR family (Naur et al. 2007). δ2 receptors, which were cloned in 1993 independently by Seeburg's and Mishina's groups (Araki et al. 1993; Lomeli et al. 1993), are supposedly homomeric tetramers consisting of GluD2 subunits from the δ-type subfamily (δ1 and δ2 receptors) of

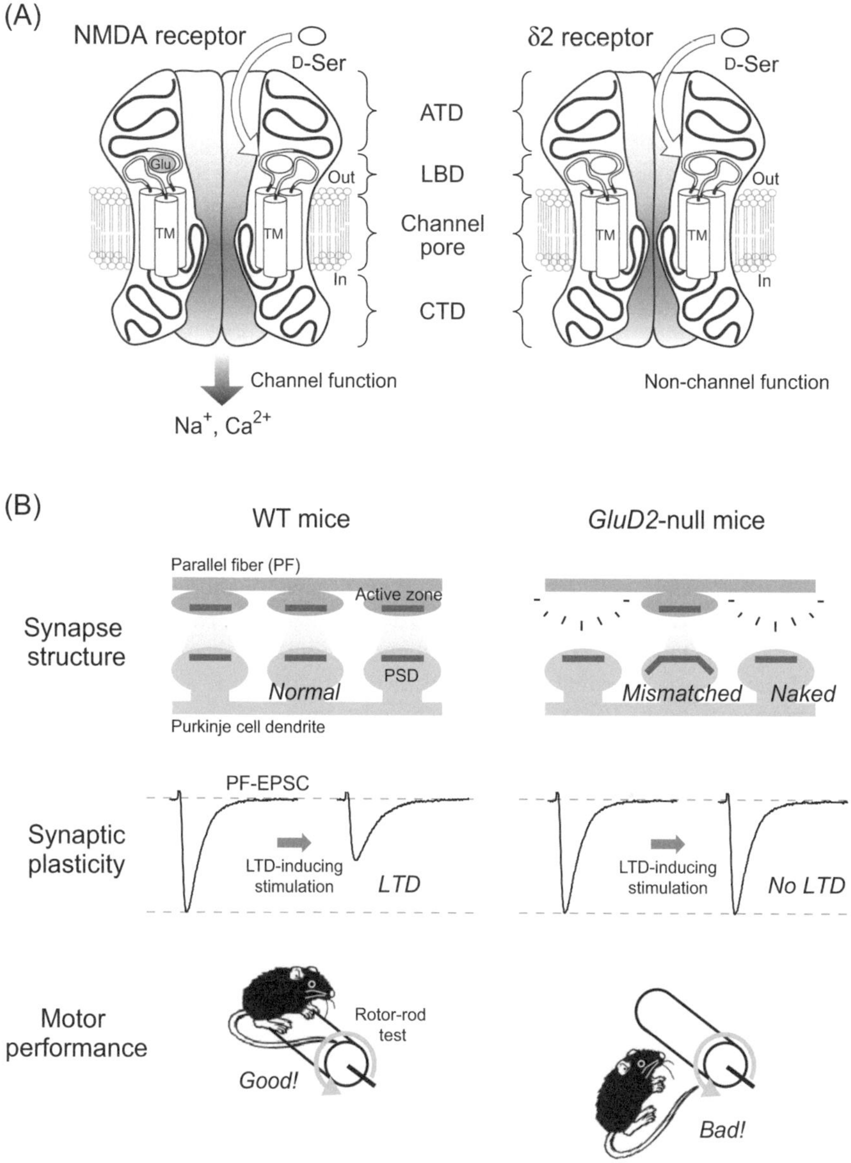

Fig. 5.1 δ2 receptor structure and the phenotypes of *GluD2*-null mice. (**A**) Membrane topology of the δ2 receptor. Like other iGluRs, including NMDA receptors (*left*), δ2 receptors (*right*) are tetrameric complexes of GluD2 subunits consisting of the ATD and LBD on the extracellular region, three transmembrane domains (TM1, TM3, and TM4), an ion channel-forming reentrant loop segment (TM2), and the CTD in the intracellular region. Although D-Ser binds to the LBD of GluD2 subunit in a similar way to other iGluR agonists, δ2 receptors do not show any ion channel activity in vitro. (**B**) Abnormal phenotypes of *GluD2*-null mice. These null mice show impaired PF synapse structures, which are called a "naked" synapse, which lacks a presynaptic-like contact or a "mismatched" synapse, in which the PSD is abnormally elongated compared to the active zone. In addition, *GluD2*-null mice exhibit abrogated LTD and impaired motor learning

iGluRs (Dingledine et al. 1999; Traynelis et al. 2010; Yuzaki 2003). δ2 receptors are exclusively expressed at the postsynaptic sites of cerebellar parallel fiber (PF, an axon of a granule cell)-Purkinje cell synapses throughout life (Kurihara et al. 1997; Landsend et al. 1997). On the basis of amino acid sequence similarity and computer-based analysis of transmembrane regions, the topology of GluD2 subunit in the cell membrane is predicted to be similar to that of other iGluR subunits [i.e., α-amino-3-hydroxy-5-methyl-4-isoxazolepropionic acid (AMPA), NMDA, and kainate receptor subunits], which are composed of an amino-terminal domain (ATD), a bipartite ligand-binding domain (LBD) on the extracellular side of the plasma membrane, three transmembrane (TM) domains (TM1, TM3, and TM4), an ion channel-forming reentrant loop segment (TM2 or P-loop), and a cytoplasmic carboxyl-terminal domain (CTD; Fig. 5.1). Although δ2 receptors are structurally similar to other iGluRs, heterologous cells expressing δ2 receptors display no ion channel activity in response to known iGluR agonists, such as glutamate, glycine, and D-Ser. Therefore, the δ2 receptor has long been classified as "an orphan receptor" (Yuzaki 2003).

However, mutant mice in which *GluD2* genes were disrupted (*GluD2*-null mice) exhibited cerebellar ataxia, motor discoordination, and severe synaptic abnormalities at PF-Purkinje cell synapses (Fig. 5.1): synapse malformation, abnormal synaptic structures, and abrogated LTD, a form of synaptic plasticity underlying motor learning and memory (Kashiwabuchi et al. 1995). Furthermore, cerebellum-dependent motor learning was markedly impaired in *GluD2*-null mice (Kashiwabuchi et al. 1995). These results indicate that δ2 receptors are crucial for synaptic integrity and cerebellar functions. Together with the findings that D-Ser is the endogenous ligand for δ2 receptors, it raises the intriguing possibility that D-Ser regulates these physiological functions in vivo.

5.3 D-Ser Modulates PF Synaptic Transmission Through δ2 Receptors

To unveil the mystery of δ2 receptors, we previously exploited a "transgenic rescue" approach (Yuzaki 2005). *GluD2*-null mice that expressed the wild-type (WT) *GluD2* transgene (designated Tg-WT rescue mice) or a mutant transgene, in which D-Ser binding was disrupted by replacing arginine (R) with lysine (K) in the LBD (Hansen et al. 2009; Hirai et al. 2005a) (designated Tg-R/K rescue mice), were generated and we analyzed their phenotypes. Because a polyclonal antibody against the LBD of GluD2 subunit was previously shown to induce the endocytosis of AMPA receptors at PF-Purkinje cell synapses in mature cerebellar slices (Hirai et al. 2003), we first electrophysiologically determined whether exogenous D-Ser can replicate this phenomenon. Whole-cell patch-clamp recordings were performed on Purkinje cells in acute cerebellar slices prepared from both Tg rescue mice (Fig. 5.2). In the presence of NMDA receptor blockers (D-AP5 and MK801), when

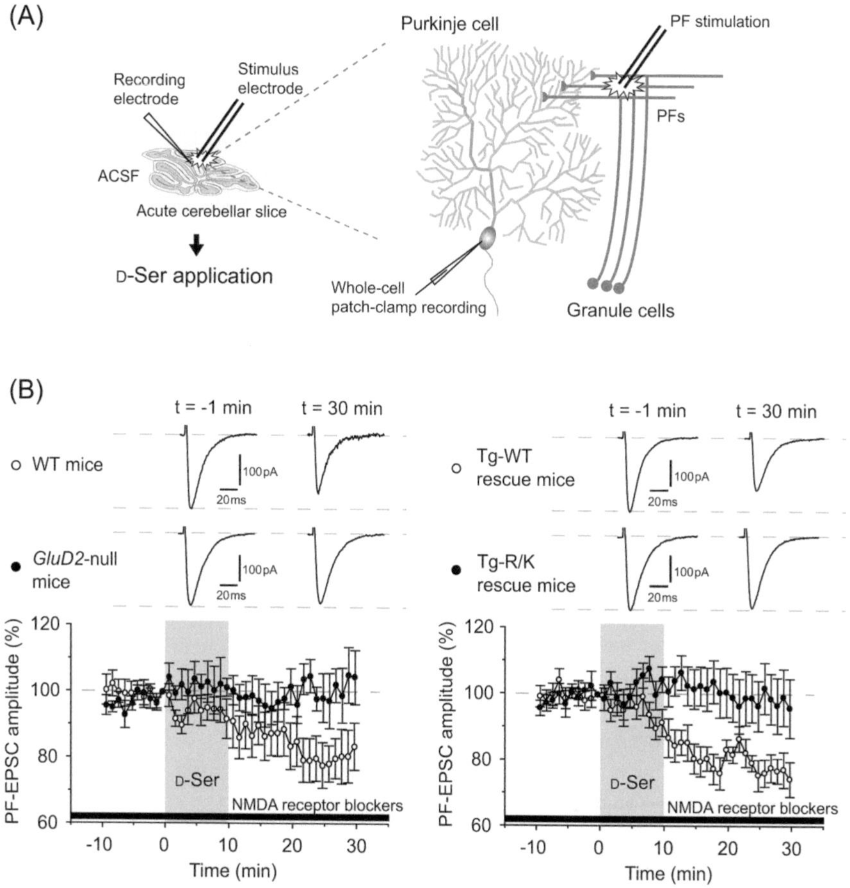

Fig. 5.2 Exogenous D-Ser causes PF-EPSC rundown through δ2 receptors. (**A**) Schematic of the recording and stimulus electrodes positioned in an acute cerebellar slice. D-Ser was applied to the artificial cerebrospinal fluid (ACSF) after the PF-EPSC became stable for 10 min. *Right*, enlarged cellular view of the stimulation paradigm. (**B**) D-Ser-mediated PF-EPSC rundown experiments from WT and *GluD2*-null mice (*left*) or from Tg-WT and Tg-R/K rescue mice (*right*)

D-Ser was applied to the cerebellar slices from WT or Tg-WT rescue mice, PF-evoked excitatory postsynaptic currents (PF-EPSCs), which consist of only an AMPA receptor component (Kano and Kato 1987), gradually decreased, and this effect persisted after D-Ser washout (Fig. 5.2). This inhibitory effect of D-Ser on PF-EPSCs increased in a dose-dependent manner with an apparent EC_{50} of 147 μM, becoming saturated at approximately 1 mM. These values were largely consistent with the affinity of the purified LBD of GluD2 subunit to D-Ser (Naur et al. 2007), indicating that, compared to NMDA receptors, δ2 receptors have extremely low affinities for D-Ser ($K_d = 7.02$ μM for GluN1, 643 nM for GluN3A, and 163 nM for GluN3B) (Yao et al. 2008). Similar to our results using an antibody against GluD2 subunit (Hirai et al. 2003), this D-Ser-mediated EPSC rundown was likely to be caused by the endocytosis of postsynaptic AMPA receptors since (i) decoy peptides disrupting GluA2 subunit-containing AMPA receptor endocytosis (Lee et al. 2002) completely blocked this rundown; (ii) this effect was mutually occlusive with LTD, which also occurs by AMPA receptor endocytosis (Man et al. 2000; Matsuda et al. 2000); and (iii) D-Ser application to cultured Purkinje cells dramatically decreased the surface expression level of AMPA receptors. Surprisingly, this D-Ser-mediated EPSC rundown was not observed in cerebellar slices from *GluD2*-null or Tg-R/K rescue mice (Fig. 5.2). Thus, by binding to δ2 receptors in Purkinje cells, exogenous D-Ser regulates postsynaptic AMPA receptor endocytosis (Kakegawa et al. 2011).

5.4 Activity-Dependent D-Ser Release from Glia in the Developing Cerebellum

In vivo microdialysis studies showed that D-Ser in the cerebrospinal fluid was present in rodents at concentrations of several micromolar (Hashimoto et al. 1995; Kakegawa et al. 2011; Matsui et al. 1995), which is too low to activate δ2 receptors. Therefore, δ2 receptors may be activated by D-Ser released in response to increased neuronal activity, such as during LTD-inducing stimulation. To examine whether endogenous D-Ser is released in an activity-dependent manner, a micro two-dimensional HPLC (2D-HPLC) system was used to measure the concentration of D-Ser released from cerebellar slices into extracellular recording solutions ($[\text{D-Ser}]_{out}$) (Miyoshi et al. 2009) (see Chap. 1). When burst stimulation was repetitively applied to the PFs in immature WT cerebellar slices, $[\text{D-Ser}]_{out}$ dramatically increased as well as $[\text{L-Ser}]_{out}$ and $[\text{Gly}]_{out}$. In contrast, although similar stimulations increased $[\text{L-Ser}]_{out}$ and $[\text{Gly}]_{out}$ in mature slices, it failed to increase $[\text{D-Ser}]_{out}$ (Fig. 5.3), consistent with earlier studies reporting that endogenous D-Ser becomes undetectable after the first postnatal month (Hashimoto and Oka 1997; Schell et al. 1997; Wang and Zhu 2003; Weimar and Neims 1977). These results indicate that D-Ser is preferentially released by an increase in neuronal activity in immature, but not mature, cerebellum (Kakegawa et al. 2011).

The cellular origin of D-Ser release remains contentious despite longstanding research efforts. D-Ser was originally proposed to be synthesized and released from glial cells as a "gliotransmitter" (Billard 2008; Martineau et al. 2006; Newman 2003; Oliet and Mothet 2009; Wolosker 2007). On the other hand, recent studies have indicated that serine racemase is expressed in both neurons and glia or rather selectively expressed in neurons (Balu et al. 2014; Ehmsen et al. 2013; Kartvelishvily et al. 2006; Miya et al. 2008; Yoshikawa et al. 2007). In cerebellar slice preparations, we found that PF-evoked D-Ser release was significantly reduced by sodium fluoroacetate (NaFAC), which preferentially inhibits glial metabolism and the release of D-Ser in hippocampal slices (Zhang et al. 2008). Similarly, D-Ser

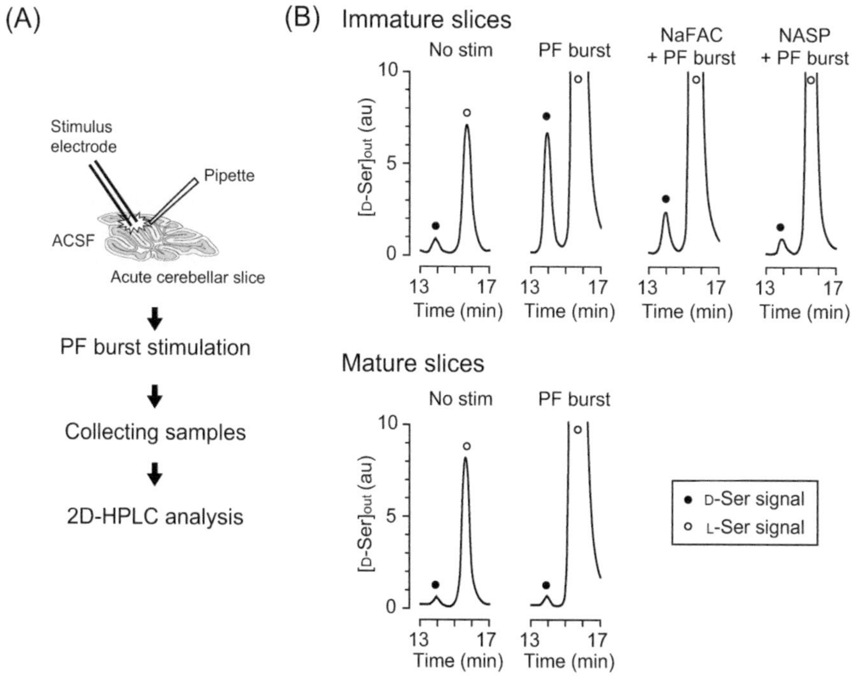

Fig. 5.3 PF burst stimulation induces endogenous D-Ser release from glia in immature cerebellar slices. (**A**) Procedure of 2D-HPLC experiments. (**B**) HPLC signal peaks obtained from immature (*upper*) and mature (*lower*) WT cerebellar slices in each stimulus condition. NaFAC is a potent inhibitor of glial-specific metabolism. NASP is a Ca^{2+}-permeable AMPA receptor blocker. *au* arbitrary unit

release was inhibited by 1-naphtyl acetyl spermine (NASP), a blocker of Ca^{2+}-permeable AMPA receptors, which are specifically expressed on Bergmann glia (Iino et al. 2001; Koike et al. 1997) (Fig. 5.3). Furthermore, although the mechanisms of how D-Ser is released are not completely clear, the adenovirally mediated expression of tetanus neurotoxin (TeNT, light-chain form), which blocks soluble *N*-ethylmaleimide-sensitive factor attachment protein receptor (SNARE)-dependent vesicular release of neurotransmitters, in Bergmann glia effectively suppressed D-Ser release. Similarly, D-Ser release from astrocytes was blocked by TeNT treatment in the hippocampus (Henneberger et al. 2010). Furthermore, D-Ser immunoreactivity is enriched in Bergmann glia (Schell et al. 1997; Williams et al. 2006; Wolosker et al. 1999a). Therefore, even if D-Ser is synthesized in neurons, glia likely play important roles in the activity-dependent release of D-Ser in immature cerebellar slices (Kakegawa et al. 2011; Kim et al. 2005; Martineau et al. 2013; Mothet et al. 2005; Schell et al. 1995). One possibility is that D-Ser might accumulate in glia via D-Ser transporters, such as neutral amino acid transporter (ASCT)-type transporters (Yamamoto et al. 2004).

5.5 Endogenous D-Ser Activates δ2 Receptors to Promote LTD and Motor Learning

D-Ser is absolutely required for the LTP and LTD at excitatory synapses expressing NMDA receptors in various brain regions (Henneberger et al. 2010; Oliet and Mothet 2009; Panatier et al. 2006; Mothet et al. 2000). Although synaptic plasticity at PF-Purkinje cell synapses is crucial for cerebellar motor learning and memory, this synapse lacks functional NMDA receptors (Kakegawa et al. 2003; Kano and Kato 1987). Therefore, as a substitute for NMDA receptors, δ2 receptors may be activated by endogenous D-Ser to regulate synaptic plasticity. In particular, LTD at these synapses has been intensively studied and is induced by the excitatory inputs from both PFs and climbing fiber (CF), another excitatory input from the inferior olive in the medulla that delivers an "error signal" to Purkinje cells (Ito 1989; Yuzaki 2013). Experimentally, LTD is detectable in cerebellar slices by simultaneously stimulating PFs and depolarizing Purkinje cell soma, which mimics a strong CF input. As described above, accumulating evidence shows that LTD is caused by the endocytosis of postsynaptic AMPA receptors, which is initiated by the phosphorylation of Ser880 located at the C-terminus of GluA2 subunit by protein kinase C (PKC) to dissociate it from the anchoring protein GRIP (Man et al. 2000; Matsuda et al. 2000).

This simultaneous stimulation consisting of a PF burst, which triggered D-Ser release in immature cerebellar slices, together with Purkinje cell depolarization (CJ-stim) led to a robust LTD in immature WT or Tg-WT rescue mice that depended on postsynaptic AMPA receptor trafficking (Fig. 5.4). This LTD induction was completely blocked by pretreating cerebellar slices with purified DAAO (Fig. 5.4), NaFAC, or NASP. In addition, slices expressing TeNT in Bergmann glia showed abrogated LTD, indicating that this LTD is regulated by glial-derived D-Ser. Furthermore, in the presence of NMDA receptor blockers, CJ-stim failed to induce LTD in immature *GluD2*-null or Tg-R/K rescue mice (Fig. 5.4). Altogether, glial-derived D-Ser, released in response to neuronal activity, promoted LTD in immature cerebellar slices by binding to δ2 receptors (Kakegawa et al. 2011).

We further examined whether D-Ser influenced cerebellar motor functions by performing a rotor-rod test on Tg-WT and Tg-R/K rescue mice during development. As expected, Tg-R/K rescue mice consistently exhibited a poorer performance than the Tg-WT rescue mice (Fig. 5.4). These results indicate that the specific activation of δ2 receptors by endogenous D-Ser is likely caused by the transiently high concentrations in the immature cerebellum, which may help animals to acquire motor coordination and learn more efficiently during early postnatal periods (Kakegawa et al. 2011).

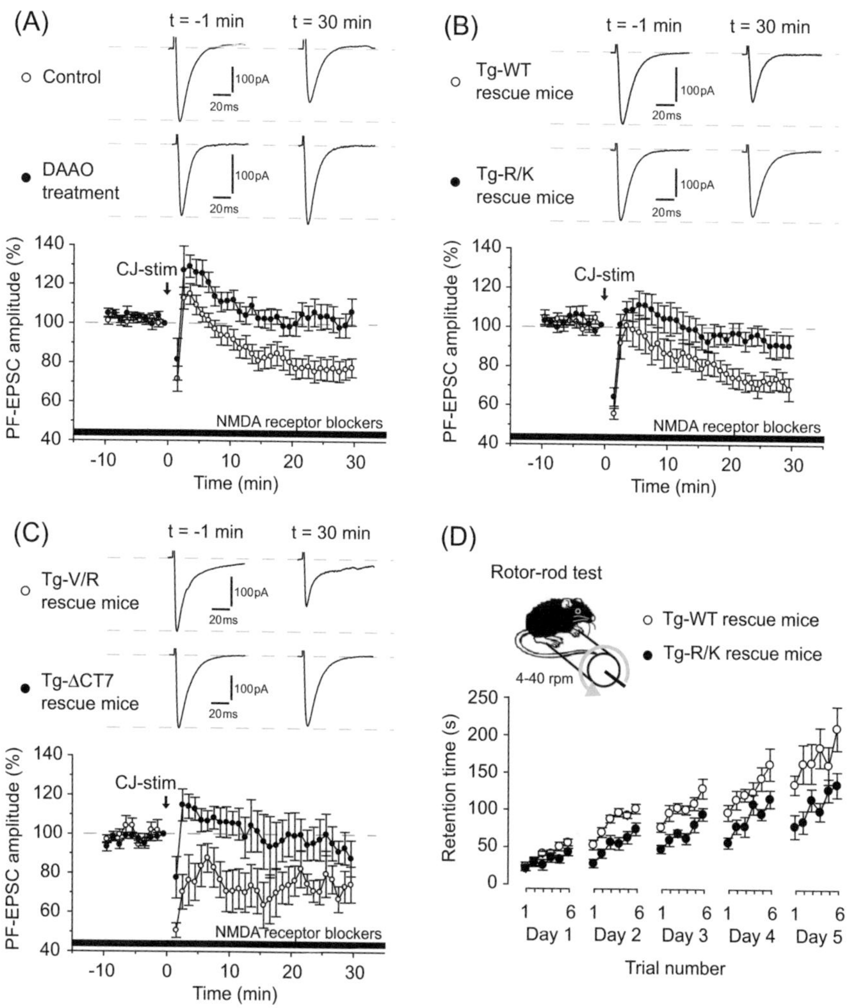

Fig. 5.4 Endogenous D-Ser regulates LTD and cerebellar motor learning by activating δ2 receptors. (**A–C**) CJ-stim-evoked LTD data from immature WT cerebellar slices treated with or without purified DAAO (**A**), from slices in Tg-WT and Tg-R/K rescue mice (**B**) and from slices in Tg-V/R and Tg-ΔCT7 rescue mice (**C**). (**D**) Rotor-rod test performance in immature Tg-WT and Tg-R/K rescue mice

5.6 δ2 Receptors Work as a Non-ion Channel In Vivo

Although it was initially discovered that D-Ser does not evoke channel activity in heterologous cells expressing δ2 receptors in vitro, it remained unclear how D-Ser regulated LTD and motor functions in the developing cerebellum in vivo. To determine how D-Ser binding to δ2 receptors regulates these physiological functions, we again utilized the transgenic rescue approach. We introduced two additional mutant *GluD2* transgenes into *GluD2*-null Purkinje cells: Tg-V/R, in which the putative channel pore was disrupted (Kakegawa et al. 2007), and Tg-ΔCT7, in which the most C-terminal seven amino acids necessary for binding to intracellular PDZ proteins (such as PSD-93, PTPMEG, S-SCAM, and delphilin) were deleted (Kakegawa et al. 2008).

Virally mediated expression of Tg-V/R into immature *GluD2*-null Purkinje cells restored abrogated LTD in the presence of NMDA receptor blockers, as well as Tg-WT (Fig. 5.4). Therefore, D-Ser-dependent LTD is unlikely to require the channel functions of δ2 receptors in the developing cerebellum. In contrast, immature *GluD2*-null mice genetically expressing Tg-ΔCT7 (designated as Tg-ΔCT7 rescue mice) (Kakegawa et al. 2008) showed impaired D-Ser-dependent cerebellar LTD in the presence of NMDA receptor blockers (Fig. 5.4). Furthermore, this LTD was effectively suppressed by the intracellular loading of inhibitory peptides for PKC, which phosphorylates GluA2-Ser880 to trigger the endocytosis of GluA2 subunit-containing AMPA receptors (Man et al. 2000; Matsuda et al. 2000). These results imply that D-Ser binding to δ2 receptors conveys signals through the cytoplasmic CTD, which converge onto PKC signals, thereby regulating AMPA receptor endocytosis during LTD in immature Purkinje cells (Fig. 5.5). Interestingly, such non-channel functions of iGluRs have also been reported for AMPA (Hayashi et al. 1999), NMDA (Nabavi et al. 2013), and kainate receptors (Rodriguez-Moreno and Lerma 1998).

Recently, we have shown that PTPMEG, a tyrosine phosphatase expressed in Purkinje cells that binds to the C-terminus of δ2 receptors (Hironaka et al. 2000;

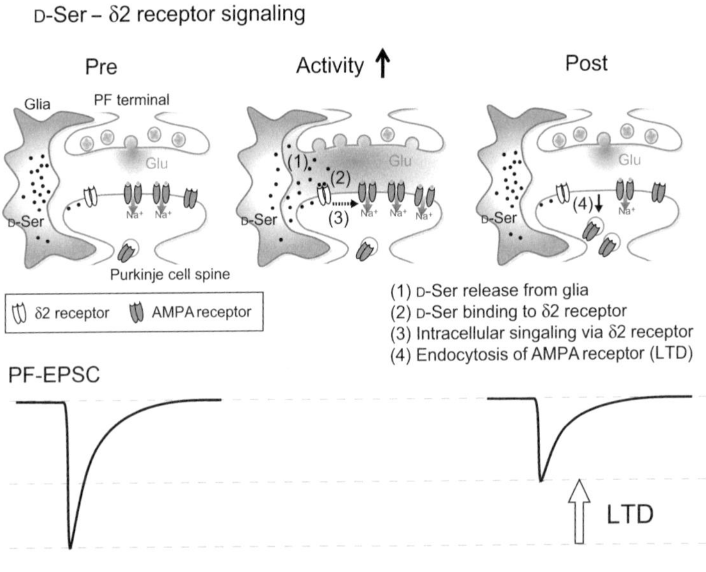

Fig. 5.5 D-Ser-δ2 receptor interaction linking neurons and glia. Details are discussed in the text

Kina et al. 2007), directly dephosphorylates Tyr876 located at the GluA2 C-terminus. Interestingly, dephosphorylation of Tyr876 of GluA2 subunit was sufficient to restore Ser880 phosphorylation and LTD induction in mature *GluD2*-null Purkinje cells. These results indicate that δ2 receptors serve as a "gatekeeper" for cerebellar LTD by coordinating interactions between the two phosphorylation sites of GluA2 subunit, Ser880 and Tyr876 (Kohda et al. 2013). Future studies are warranted to clarify whether and how D-Ser binding to δ2 receptors activates PTPMEG-GluA2 dephosphorylation signaling in Purkinje cells.

5.7 D-Ser-δ Receptor Interactions Outside the Cerebellum: Focusing on δ1 Receptors

Like δ2 receptors, δ1 receptors are reported to bind D-Ser in vitro (Yadav et al. 2011). Using specific antibodies against GluD1 subunit, we recently showed that δ1 receptors are widely expressed in the adult mouse brain, with high levels in forebrain regions, including the cerebral cortex, striatum, and hippocampus (Konno et al. 2014), where D-Ser is abundant throughout life. More recently, we analyzed *GluD1*-null mice and found severe abnormalities in excitatory synapse morphology and synaptic functions in the hippocampus (unpublished data). In addition, *GluD1*-null mice show impaired social behavior and certain forms of learning (Yadav et al. 2012; Yadav et al. 2013). Indeed, recent human genetic studies indicate a strong association of the *GluD1* gene with several neuropsychiatric disorders, such as schizophrenia and autism spectrum disorders (Cooper et al. 2011; Glessner et al. 2009; Guo et al. 2007; Smith et al. 2009; Treutlein et al. 2009). Thus, D-Ser binding to δ1 receptors may contribute to synaptic integrity involved in physiological functions at relevant brain regions.

5.8 Interactive Effects of Two Endogenous Ligands for δ Receptors: D-Ser and Cbln

We previously identified a Cbln1 as another endogenous ligand for δ2 receptors (Matsuda et al. 2010). Cbln1, which is a C1q family protein produced in cerebellar granule cells throughout life and released from PF terminals, is essential for PF-Purkinje cell synapse integrity by forming Cbln1-δ2 receptor-neurexin complexes: *Cbln1*-null mice exhibited severe cerebellar ataxia, synapse malformation, and impaired LTD, very similar to the phenotypes of *GluD2*-null mice (Hirai et al. 2005b). In the forebrain, not only Cbln1 but also its family members Cbln2-4 are highly expressed (Miura et al. 2006), and these Cbln proteins can also bind to δ1 receptors to establish functional synapses in vitro (Kuroyanagi et al. 2009;

Matsuda et al. 2010; Yasumura et al. 2012). Since both D-Ser and Cbln proteins are likely to be released in an activity-dependent manner, binding of D-Ser and Cbln to the LBD and ATD of GluD1/2 subunits, respectively, may display an allosteric interaction to regulate synaptic plasticity and other physiological functions of δ1/2 receptors. Similar allosteric interactions between the ATD and LBD have been observed in NMDA receptors (Traynelis et al. 2010).

5.9 Summary

The physiological significance of the transient expression of D-Ser in the developing cerebellum had long been unclear. We have recently demonstrated that D-Ser is released from glia in a neuronal activity-dependent manner and binds to δ2 receptors to promote cerebellar LTD and motor learning in developing mice. Unlike the binding of D-Ser to NMDA receptors, D-Ser binding to δ2 receptors does not evoke any channel activity but rather intracellular signaling leading to AMPA receptor endocytosis through the C-terminus. An understanding of such non-channel functions of δ2 receptors will provide a unified view of iGluR functions regulating physiological and pathological events at synapses. In addition, D-Ser binds not only to δ2 receptors but also to δ1 receptors, which are expressed in various brain regions throughout life. Therefore, D-Ser-δ receptor interactions may be a general signaling mechanism by which glia communicate with neurons, one facet of the tripartite synapse.

Acknowledgments This work was supported by a Grant-in-Aid for Scientific Research from the Ministry of Education, Culture, Sports, Science and Technology of Japan (MEXT) and the Japan Society for the Promotion of Science (JSPS); the Core Research for Evolutional Science and Technology (CREST) from the Japan Science and Technology Agency (JST); the Keio Gijuku Fukuzawa Memorial Fund for the Advancement of Education and Research; the Naito Foundation; the Kanae Foundation; the Nakajima Foundation; the Inamori Foundation; the Yamada Science Foundation; and the Takeda Science Foundation.

References

Araki K, Meguro H, Kushiya E et al (1993) Selective expression of the glutamate receptor channel δ2 subunit in cerebellar Purkinje cells. Biochem Biophys Res Commun 197:1267–1276

Balu DT, Takagi S, Puhl MD et al (2014) D-serine and serine racemase are localized to neurons in the adult mouse and human forebrain. Cell Mol Neurobiol 34:419–435

Billard JM (2008) D-serine signalling as a prominent determinant of neuronal-glial dialogue in the healthy and diseased brain. J Cell Mol Med 12:1872–1884

Billard JM (2013) Serine racemase as a prime target for age-related memory deficits. Eur J Neurosci 37:1931–1938

Cooper GM, Coe BP, Girirajan S et al (2011) A copy number variation morbidity map of developmental delay. Nat Genet 43:838–846

Dingledine R, Borges K, Bowie D et al (1999) The glutamate receptor ion channels. Pharmacol Rev 51:7–61

Ehmsen JT, Ma TM, Sason H et al (2013) D-serine in glia and neurons derives from 3-phosphoglycerate dehydrogenase. J Neurosci 33:12464–12469

Glessner JT, Wang K, Cai G et al (2009) Autism genome-wide copy number variation reveals ubiquitin and neuronal genes. Nature 459:569–573

Guo SZ, Huang K, Shi YY et al (2007) A case–control association study between the GRID1 gene and schizophrenia in the Chinese Northern Han population. Schizophr Res 93:385–390

Hansen KB, Naur P, Kurtkaya NL et al (2009) Modulation of the dimer interface at ionotropic glutamate-like receptor δ2 by D-serine and extracellular calcium. J Neurosci 29:907–917

Hashimoto A, Oka T (1997) Free D-aspartate and D-serine in the mammalian brain and periphery. Prog Neurobiol 52:325–353

Hashimoto A, Nishikawa T, Hayashi T et al (1992) The presence of free D-serine in rat brain. FEBS Lett 296:33–36

Hashimoto A, Oka T, Nishikawa T (1995) Extracellular concentration of endogenous free D-serine in the rat brain as revealed by in vivo microdialysis. Neuroscience 66:635–643

Hayashi T, Umemori H, Mishina M et al (1999) The AMPA receptor interacts with and signals through the protein tyrosine kinase Lyn. Nature 397:72–76

Henneberger C, Papouin T, Oliet SH et al (2010) Long-term potentiation depends on release of D-serine from astrocytes. Nature 463:232–236

Hirai H, Launey T, Mikawa S et al (2003) New role of δ2-glutamate receptors in AMPA receptor trafficking and cerebellar function. Nat Neurosci 6:869–876

Hirai H, Miyazaki T, Kakegawa W et al (2005a) Rescue of abnormal phenotypes of the δ2 glutamate receptor-null mice by mutant δ2 transgenes. EMBO Rep 6:90–95

Hirai H, Pang Z, Bao D et al (2005b) Cbln1 is essential for synaptic integrity and plasticity in the cerebellum. Nat Neurosci 8:1534–1541

Hironaka K, Umemori H, Tezuka T et al (2000) The protein-tyrosine phosphatase PTPMEG interacts with glutamate receptor δ2 and epsilon subunits. J Biol Chem 275:16167–16173

Iino M, Goto K, Kakegawa W et al (2001) Glia-synapse interaction through Ca^{2+}-permeable AMPA receptors in Bergmann glia. Science 292:926–929

Ito M (1989) Long-term depression. Annu Rev Neurosci 12:85–102

Kakegawa W, Tsuzuki K, Iino M et al (2003) Functional NMDA receptor channels generated by NMDAR2B gene transfer in rat cerebellar Purkinje cells. Eur J Neurosci 17:887–891

Kakegawa W, Kohda K, Yuzaki M (2007) The δ2 'ionotropic' glutamate receptor functions as a non-ionotropic receptor to control cerebellar synaptic plasticity. J Physiol 584:89–96

Kakegawa W, Miyazaki T, Emi K et al (2008) Differential regulation of synaptic plasticity and cerebellar motor learning by the C-terminal PDZ-binding motif of GluRδ2. J Neurosci 28:1460–1468

Kakegawa W, Miyoshi Y, Hamase K et al (2011) D-serine regulates cerebellar LTD and motor coordination through the δ2 glutamate receptor. Nat Neurosci 14:603–611

Kano M, Kato M (1987) Quisqualate receptors are specifically involved in cerebellar synaptic plasticity. Nature 325:276–279

Kartvelishvily E, Shleper M, Balan L et al (2006) Neuron-derived D-serine release provides a novel means to activate N-methyl-D-aspartate receptors. J Biol Chem 281:14151–14162

Kashiwabuchi N, Ikeda K, Araki K et al (1995) Impairment of motor coordination, Purkinje cell synapse formation, and cerebellar long-term depression in GluRδ2 mutant mice. Cell 81:245–252

Kim PM, Aizawa H, Kim PS et al (2005) Serine racemase: activation by glutamate neurotransmission via glutamate receptor interacting protein and mediation of neuronal migration. Proc Natl Acad Sci U S A 102:2105–2110

Kina S, Tezuka T, Kusakawa S et al (2007) Involvement of protein-tyrosine phosphatase PTPMEG in motor learning and cerebellar long-term depression. Eur J Neurosci 26:2269–2278

Kohda K, Kakegawa W, Matsuda S et al (2013) The δ2 glutamate receptor gates long-term depression by coordinating interactions between two AMPA receptor phosphorylation sites. Proc Natl Acad Sci U S A 110:E948–E957

Koike M, Iino M, Ozawa S (1997) Blocking effect of 1-naphthyl acetyl spermine on Ca^{2+}-permeable AMPA receptors in cultured rat hippocampal neurons. Neurosci Res 29:27–36

Konno K, Matsuda K, Nakamoto C et al (2014) Enriched expression of GluD1 in higher brain regions and its involvement in parallel fiber-interneuron synapse formation in the cerebellum. J Neurosci 34:7412–7424

Kurihara H, Hashimoto K, Kano M et al (1997) Impaired parallel fiber-->Purkinje cell synapse stabilization during cerebellar development of mutant mice lacking the glutamate receptor delta2 subunit. J Neurosci 17:9613–9623

Kuroyanagi T, Yokoyama M, Hirano T (2009) Postsynaptic glutamate receptor delta family contributes to presynaptic terminal differentiation and establishment of synaptic transmission. Proc Natl Acad Sci U S A 106:4912–4916

Landsend AS, Amiry-Moghaddam M, Matsubara A et al (1997) Differential localization of δ glutamate receptors in the rat cerebellum: coexpression with AMPA receptors in parallel fiber-spine synapses and absence from climbing fiber-spine synapses. J Neurosci 17:834–842

Lee SH, Liu L, Wang YT et al (2002) Clathrin adaptor AP2 and NSF interact with overlapping sites of GluR2 and play distinct roles in AMPA receptor trafficking and hippocampal LTD. Neuron 36:661–674

Lomeli H, Sprengel R, Laurie DJ et al (1993) The rat delta-1 and delta-2 subunits extend the excitatory amino acid receptor family. FEBS Lett 315:318–322

Man HY, Lin JW, Ju WH et al (2000) Regulation of AMPA receptor-mediated synaptic transmission by clathrin-dependent receptor internalization. Neuron 25:649–662

Martineau M, Baux G, Mothet JP (2006) D-serine signalling in the brain: friend and foe. Trends Neurosci 29:481–491

Martineau M, Shi T, Puyal J et al (2013) Storage and uptake of D-serine into astrocytic synaptic-like vesicles specify gliotransmission. J Neurosci 33:3413–3423

Matsuda S, Launey T, Mikawa S et al (2000) Disruption of AMPA receptor GluR2 clusters following long-term depression induction in cerebellar Purkinje neurons. EMBO J 19:2765–2774

Matsuda K, Miura E, Miyazaki T et al (2010) Cbln1 is a ligand for an orphan glutamate receptor δ2, a bidirectional synapse organizer. Science 328:363–368

Matsui T, Sekiguchi M, Hashimoto A et al (1995) Functional comparison of D-serine and glycine in rodents: the effect on cloned NMDA receptors and the extracellular concentration. J Neurochem 65:454–458

Miura E, Iijima T, Yuzaki M et al (2006) Distinct expression of Cbln family mRNAs in developing and adult mouse brains. Eur J Neurosci 24:750–760

Miya K, Inoue R, Takata Y et al (2008) Serine racemase is predominantly localized in neurons in mouse brain. J Comp Neurol 510:641–654

Miyoshi Y, Hamase K, Tojo Y et al (2009) Determination of D-serine and D-alanine in the tissues and physiological fluids of mice with various D-amino-acid oxidase activities using two-dimensional high-performance liquid chromatography with fluorescence detection. J Chromatogr B Anal Technol Biomed Life Sci 877:2506–2512

Mothet JP, Parent AT, Wolosker H et al (2000) D-serine is an endogenous ligand for the glycine site of the N-methyl-D-aspartate receptor. Proc Natl Acad Sci U S A 97:4926–4931

Mothet JP, Pollegioni L, Ouanounou G et al (2005) Glutamate receptor activation triggers a calcium-dependent and SNARE protein-dependent release of the gliotransmitter D-serine. Proc Natl Acad Sci U S A 102:5606–5611

Nabavi S, Kessels HW, Alfonso S et al (2013) Metabotropic NMDA receptor function is required for NMDA receptor-dependent long-term depression. Proc Natl Acad Sci U S A 110:4027–4032

Naur P, Hansen KB, Kristensen AS et al (2007) Ionotropic glutamate-like receptor δ2 binds D-serine and glycine. Proc Natl Acad Sci U S A 104:14116–14121
Newman EA (2003) New roles for astrocytes: regulation of synaptic transmission. Trends Neurosci 26:536–542
Oliet SH, Mothet JP (2009) Regulation of N-methyl-D-aspartate receptors by astrocytic D-serine. Neuroscience 158:275–283
Panatier A, Theodosis DT, Mothet JP et al (2006) Glia-derived D-serine controls NMDA receptor activity and synaptic memory. Cell 125:775–784
Rodriguez-Moreno A, Lerma J (1998) Kainate receptor modulation of GABA release involves a metabotropic function. Neuron 20:1211–1218
Schell MJ, Molliver ME, Snyder SH (1995) D-serine, an endogenous synaptic modulator: localization to astrocytes and glutamate-stimulated release. Proc Natl Acad Sci U S A 92:3948–3952
Schell MJ, Brady RO Jr, Molliver ME et al (1997) D-serine as a neuromodulator: regional and developmental localizations in rat brain glia resemble NMDA receptors. J Neurosci 17:1604–1615
Smith M, Spence MA, Flodman P (2009) Nuclear and mitochondrial genome defects in autisms. Ann N Y Acad Sci 1151:102–132
Traynelis SF, Wollmuth LP, McBain CJ et al (2010) Glutamate receptor ion channels: structure, regulation, and function. Pharmacol Rev 62:405–496
Treutlein J, Muhleisen TW, Frank J et al (2009) Dissection of phenotype reveals possible association between schizophrenia and Glutamate Receptor Delta 1 (GRID1) gene promoter. Schizophr Res 111:123–130
Wang LZ, Zhu XZ (2003) Spatiotemporal relationships among D-serine, serine racemase, and D-amino acid oxidase during mouse postnatal development. Acta Pharmacol Sin 24:965–974
Weimar WR, Neims AH (1977) The development of D-amino acid oxidase in rat cerebellum. J Neurochem 29:649–656
Williams SM, Diaz CM, Macnab LT et al (2006) Immunocytochemical analysis of D-serine distribution in the mammalian brain reveals novel anatomical compartmentalizations in glia and neurons. Glia 53:401–411
Wolosker H (2007) NMDA receptor regulation by D-serine: new findings and perspectives. Mol Neurobiol 36:152–164
Wolosker H, Blackshaw S, Snyder SH (1999a) Serine racemase: a glial enzyme synthesizing D-serine to regulate glutamate-N-methyl-D-aspartate neurotransmission. Proc Natl Acad Sci U S A 96:13409–13414
Wolosker H, Sheth KN, Takahashi M et al (1999b) Purification of serine racemase: biosynthesis of the neuromodulator D-serine. Proc Natl Acad Sci U S A 96:721–725
Yadav R, Rimerman R, Scofield MA et al (2011) Mutations in the transmembrane domain M3 generate spontaneously open orphan glutamate delta1 receptor. Brain Res 1382:1–8
Yadav R, Gupta SC, Hillman BG et al (2012) Deletion of glutamate delta-1 receptor in mouse leads to aberrant emotional and social behaviors. PLoS ONE 7, e32969
Yadav R, Hillman BG, Gupta SC et al (2013) Deletion of glutamate delta-1 receptor in mouse leads to enhanced working memory and deficit in fear conditioning. PLoS ONE 8, e60785
Yamamoto T, Nishizaki I, Nukada T et al (2004) Functional identification of ASCT1 neutral amino acid transporter as the predominant system for the uptake of L-serine in rat neurons in primary culture. Neurosci Res 49:101–111
Yao Y, Harrison CB, Freddolino PL et al (2008) Molecular mechanism of ligand recognition by NR3 subtype glutamate receptors. EMBO J 27:2158–2170
Yasumura M, Yoshida T, Lee SJ et al (2012) Glutamate receptor δ1 induces preferentially inhibitory presynaptic differentiation of cortical neurons by interacting with neurexins through cerebellin precursor protein subtypes. J Neurochem 121:705–716
Yoshikawa M, Takayasu N, Hashimoto A et al (2007) The serine racemase mRNA is predominantly expressed in rat brain neurons. Arch Histol Cytol 70:127–134

Yuzaki M (2003) The δ2 glutamate receptor: 10 years later. Neurosci Res 46:11–22

Yuzaki M (2005) Transgenic rescue for characterizing orphan receptors: a review of δ2 glutamate receptor. Transgenic Res 14:117–121

Yuzaki M (2013) Cerebellar LTD vs. motor learning-lessons learned from studying GluD2. Neural Netw 47:36–41

Zhang Z, Gong N, Wang W et al (2008) Bell-shaped D-serine actions on hippocampal long-term depression and spatial memory retrieval. Cereb Cortex 18:2391–2401

Chapter 6
D-Serine Signaling and Schizophrenia

Toru Nishikawa

Abstract It has been widely accepted that the hypofunction of the N-methyl-D-aspartate-type glutamate receptor (NMDAR) may be implicated in the pathophysiology of both positive- and negative-cognitive symptomatologies of schizophrenia because NMDAR antagonists, including phencyclidine (PCP) and anti-NMDAR antibodies, mimic these respective antipsychotic-responsive and antipsychotic-resistant symptoms. D-Serine and other agonists for the glycine modulatory site of the NMDAR, which facilitate the receptor function, are found to not only inhibit behavioral models of schizophrenia and hyperdopaminergic transmission caused by schizophrenomimetics, PCP, and amphetamines, in experimental animals, but also ameliorate the entire extent of the above schizophrenic symptoms. Moreover, D-serine has been revealed to be a brain-enriched endogenous substance displaying an NMDAR-like distribution. At least, in the forebrain areas, the NMDAR function levels are under control of the extracellular D-serine concentrations that are regulated in a different manner from that of classical neurotransmitters by neuronal and glial activities, the calcium-permeable AMPA receptor, the Asc-1 neutral amino acid transporter, and the neuronal serine racemase, a D-serine synthesizing enzyme. These findings raise the possibility that insufficient extracellular D-serine signaling could be a part of a key factor that leads to the presumed hypofunction of the NMDAR in schizophrenia. Further investigations on the molecular and cellular mechanisms of the D-serine metabolism and their alterations in schizophrenia may contribute to the elucidation of the pathophysiology of and development of a novel therapeutic approach to this intractable mental disorder.

Keywords N-Methyl-D-aspartate-type glutamate receptor • Dopaminergic transmission • Schizophrenia • D-Serine

---

T. Nishikawa (✉)
Department of Psychiatry and Behavioral Sciences, Graduate School of Medical and Dental Sciences, Tokyo Medical and Dental University, 1-5-45, Yushima, Bunkyo-ku, Tokyo 113-8519, Japan
e-mail: tnis.psyc@tmd.ac.jp

T. Yoshimura et al. (eds.), *D-Amino Acids*, DOI 10.1007/978-4-431-56077-7_6

6.1 Introduction: A Link Between D-Serine and Schizophrenia

Prominent stereoselective effects of D-serine in the central nervous system were first demonstrated on the sodium-independent [3H]glycine binding to the homogenate of the rat cerebral cortex (Kishimoto et al. 1981). At that time, the exact nature of the target substance of D-serine was unknown, while the L-isomer of serine had been described to be as equipotent as or more effective than the D-isomer during interactions with the strychnine-sensitive inhibitory glycine receptor (Kishimoto et al. 1981; Tokutomi et al. 1989; White et al. 1989). In 1988, the potent effects of D-serine were possibly explained by the findings of Kleckner and Dingledine (1988) that D-serine and D-alanine enhanced the ability of N-methyl-D-aspartate (NMDA) to activate the NMDA-type glutamate receptor in neurons or expressed in Xenopus Oocytes much stronger than their enantiomers. They also revealed that D-serine and D-alanine stimulated the strychnine-insensitive glycine site of the NMDAR that was shown to be stimulated by glycine, but different from the inhibitory glycine receptor (Johnson and Ascher 1987). The stimulation of the glycine site was required for activation of the NMDAR although an agonist for the site, such as glycine, D-serine, or D-alanine, alone could not activate the NMDAR. In the light of these unique actions, the three amino acids have been called "coagonists for the NMDAR."

The author noted the stereoselective facilitating effects of the D-amino acids on the NMDAR functions as a tool to develop a novel pharmacotherapy for intractable schizophrenia based on the glutamate hypothesis of schizophrenia. Thus, phencyclidine (PCP: 1-(1-phenyl cyclohexyl)piperidine), which generates an antipsychotic-resistant psychosis displaying both positive and negative symptomatologies indistinguishable from those of schizophrenia, was found to block the NMDAR in 1983 (Anis et al. 1983). These observations have suggested that diminished NMDAR-mediated glutamate neurotransmission might be involved in the pathophysiology of schizophrenia. This hypothesis is further supported by the facts that sub-psychotomimetic doses of the NMDAR antagonists for healthy volunteers produce schizophrenic symptoms in the patients with schizophrenia under remission (Javitt and Zulkin 1991; Petersen and Stillman 1978). Moreover, the S-isomer of ketamine, an NMDAR antagonist, has a higher affinity for the glutamate receptor and a greater schizophrenomimetic effect than the R-isomer. Recent observations that autoantibodies against the NMDAR subunits elicit a psychotic state including symptoms that resemble those of schizophrenia also favor the postulated NMDAR dysfunction in schizophrenia.

Therefore, the NMDAR hypofunction hypothesis has indicated that recovery of the glutamate receptor function by its agonists, allosteric agonists, or indirect positive modulators could be expected to ameliorate the antipsychotic-resistant negative symptoms and cognitive deficits in patients with schizophrenia. To test this possibility, the author started to study the effects of the D-serine and D-alanine on the drug-induced model of schizophrenia because (1) agonists for the glutamate site, but not the glycine site, of the NMDAR produce neuronal cell injury or death, (2) the selectivity of the actions of these D-amino acids at the glycine site can be verified by using their L-isomers, and (3) the D-amino acids had been believed not to be endogenous substances that would not be easily broken down due to the lack of their specific metabolic pathways. This plan appears to be one of the beginnings for the link between schizophrenia and D-serine and lead to the detection of intrinsic D-serine in the brain.

The purpose of this section is to summarize the behavioral and neurochemical analyses supporting the possible relationship between D-serine and schizophrenia and introduce the studies that clarify the molecular and cellular mechanisms underlying endogenous D-serine metabolism and functions that could be targets for investigation of the pathophysiology of schizophrenia and for the development of a novel pharmacotherapy for this mental disorder.

6.2 Ameliorating Effects of D-Serine on Pharmacological Models of Schizophrenia

6.2.1 *Phencyclidine Model*

In rodents, the acute or chronic administration of a schizophrenomimetic, PCP, induces abnormal behavior including hyperlocomotion, stereotypy, and ataxia and disturbances in preattentive information processing/gating mechanisms and cognitive and social functions (Javitt and Zukin 1991; Tanii et al. 1994). Because the single and repeated use of PCP causes a schizophrenia-like psychosis resistant to antipsychotic drugs and because the above behavioral changes are refractory to or only partially attenuated by antipsychotic drugs, it has been well accepted that PCP-treated animals provide an experimental and comprehensive model for schizophrenia (Javitt and Zukin 1991).

Research groups of the author (Tanii et al. 1990, 1991a, b, 1994) and Contreras (1990) first demonstrated that D-serine possesses an anti-PCP model property by showing the attenuating effects of the D-amino acid given intra-cerebroventricurally on the ability of the systemic administration of PCP to produce increased locomotion, stereotypy, and/or ataxia. These effects of D-serine have been considered to be mediated by the strychnine-insensitive glycine-binding site of the NMDAR on the basis of the following observations: (1) D-serine and D-alanine more potently inhibited the PCP-induced abnormal behavior than L-serine and L-alanine, respectively, coinciding the stereoselectivity of the serine and alanine isomers as agonists for the glycine site (Tanii et al. 1991a, 1994), (2) the decreasing effects of D-serine and D-alanine on PCP-induced hyperactivity were antagonized by intraventricular application of 7-chlorokynurenate and 5,7-dichlorokynurenate, respectively, which are selective antagonists of the glycine modulatory site (Tanii et al. 1994), and (3) D-serine improved the behavioral abnormalities after administration of another NMDAR antagonist, dizocilpine (MK-801) (Contreras 1990). The detailed mechanisms underlying the anti-PCP or anti-dizocilpine effects of D-serine and other glycine modulatory site agonists including glycine, glycine transporter inhibitor, and D-amino acid oxidase inhibitors that increase the brain concentrations of endogenous D-serine (see Chap. 19) are still unclear. These allosteric agonists could reduce the behavioral effects of noncompetitive antagonists for the NMDAR by increasing the open frequency of the NMDAR ion channel that facilitates liberation of these antagonists from the channel or by activation of the NMDAR that is not occupied by these ion channel blockers (Tanii et al. 1991a, 1994). Furthermore, the NMDAR antagonist induction of disturbances in information processing, sensorimotor gating, and/or attentional processes, which can be evaluated by the prepulse inhibition and lateral inhibition test, respectively, has been found to be ameliorated by direct and indirect agonists of the glycine modulatory site (Depoortere et al. 1999; Hashimoto et al. 2009; Kanahara et al. 2008; Lipina et al. 2005).

The previously mentioned data support the ideas that stimulating agents of the NMDAR glycine site may have a broader therapeutic efficacy for various symptoms of schizophrenia than current antipsychotic drugs and that behavioral disturbances evoked by the NMDAR antagonists could be useful indices for screening such agents.

6.2.2 Amphetamine Model

Another pharmacological model for schizophrenia is the amphetamine or dopamine agonist model that is presumed to mimic the pathophysiology of enhanced dopaminergic transmission, at least, the D2-type dopamine receptor underlying positive symptoms of the disorder. This presumption is based upon the facts that the potency of the therapeutic effects of antipsychotics on the hallucinatory and paranoid state of patients with schizophrenia parallels not only that of their inhibiting effects on hyperlocomotion and/or stereotyped behavior induced by acute treatment with a dopamine agonist such as amphetamines but also that of their antagonizing effects on and affinity for the D2-type dopamine receptor. Moreover, hallucinations and delusions easily relapse in remitted patients with schizophrenia following a sub-psychotomimetic dose of a dopamine agonist for healthy volunteers.

Intraventricular infusion of D-alanine and D-serine but not L-alanine (Hashimoto et al. 1991) also attenuated the methamphetamine production of hyperlocomotion in rats without affecting the stereotypy and catalepsy, suggesting that stimulation of the NMDAR glycine modulatory site could improve the positive symptoms of schizophrenia and produce minimal extrapyramidal symptoms. In line with these results, the attenuating influence of a large dose of D-serine given systemically on amphetamine-induced psychomotor activity was observed in the rats by Smith et al. (2009).

6.3 Dopaminergic Transmission and D-Serine

Acute application of schizophrenomimetic NMDAR antagonists, PCP or dizocilpine, given either systemically or locally has been demonstrated to increase the tissue dopamine metabolism and the extracellular dopamine concentrations in the discrete brain regions of experimental animals including the prefrontal cortex; the nucleus accumbens; the limbic forebrain area consisting of the septum, olfactory tubercle, and nucleus accumbens; and the striatum (Deutch et al. 1987; Kashiwa et al. 1995; Nishijima et al. 1996; Rao et al. 1989; Tanii et al. 1990; Umino et al. 1998). The magnitude of the enhancement of dopamine transmission after PCP or dizocilpine application was most prominent in the prefrontal cortical areas (Deutch et al. 1987; Nishijima et al. 1996; Rao et al. 1989; Umino et al. 1998). In the medial prefrontal cortex of the rat, the facilitating effects of PCP on the dopamine metabolism were attenuated by an intra-prefrontal infusion of D-alanine, but not L-alanine, in a dose-dependent fashion (Umino et al. 1998). Furthermore, a similar augmented dopamine metabolism was elicited by local injection of competitive antagonists for the NMDAR, but not for the non-NMDA ionotropic glutamate receptors, into the prefrontal cortex in an NMDA-reversible manner (Hata et al. 1990; Nishijima et al. 1994). These data indicate that the prefrontal NMDAR exerts a tonic facilitatory control over the terminals of dopamine neurons projecting to the prefrontal cortex from the ventral tegmental area.

While the accurate neuron setups of the NMDAR-dopamine interaction are still unclear, the over-liberation of the prefrontal extracellular dopamine following the NMDAR antagonists appears to result from a reduced inhibitory GABAergic tone that is provoked by the interruption of the tonic facilitation by the NMDAR expressed on the GABA interneurons. This possibility was substantiated by the following results obtained by in vivo dialysis experiments of Yonezawa et al. (1998): (1) systemic administration of PCP or dizocilpine leads to an increase and a decrease in the extracellular dopamine and GABA concentrations, respectively, (2) local infusion of a GABAA receptor agonist, muscimol, reversed the elevated levels of the extracellular dopamine, and (3) a GABAA receptor antagonist, bicuculline, augmented the extracellular dopamine contents. In agreement with this view, ketamine-induced acceleration of the frontal dopamine metabolism was normalized by the systemic administration of diazepam, an allosteric agonist of the GABAA receptor (Irifune et al. 1998).

It was also demonstrated by Balla et al. (2003, 2012) that repeated blockade of the NMDAR by the daily systemic administration of PCP caused augmentation of an increase in the extracellular dopamine release in response to a challenge dose of amphetamine in the prefrontal cortex and striatum. Such augmentation of the amphetamine-provoked dopamine release was seen in the caudate-putamen of the patients with schizophrenia as revealed by in vivo studies using positron-emission tomography measuring the rate of displacement of radioligand binding to the D2 dopamine receptor by endogenous dopamine (Abi-Dargham et al. 2009; Laruelle et al. 1996). Balla et al. (2003, 2012) further reported that the subchronic daily co-administration of an agonist for the glycine modulatory site with PCP lowered the hyperresponsiveness of the prefrontal dopamine release to an amphetamine challenge.

Finally, an in vivo dialysis study in freely moving rats (Smith et al. 2009) pointed out that an elevation of the extracellular dopamine levels in the nucleus accumbens by amphetamine was reduced by the systemic injection of a high dose of D-serine. The reducing effect could be related to the attenuation by D-alanine and D-serine of methamphetamine induction of hyperactivity in the rats (Hashimoto et al. 1991; see the previous Sect. 6.2).

6.4 Clinical Treatment Trials with D-Serine of Patients with Schizophrenia

As indicated in Fig. 6.1, the aforementioned studies of the PCP model or NMDAR dysfunction model and the amphetamine model or hyperdopaminergic transmission model extrapolate that, at least in a group of schizophrenia, the NMDAR hypofunction could elicit positive symptoms by hyperdopaminergic activity subsequent to the reduced GABAergic tone and negative symptoms and cognitive defects by disturbances of non-dopaminergic systems. Accelerated serotonergic transmission would lead to the part of the positive and/or negative symptoms by composing the pathophysiological consequences of the NMDAR dysfunction, because (1) the NMDAR antagonists produce an increase in the extracellular serotonin concentrations in the prefrontal cortex, hippocampus, nucleus accumbens, and striatum (López-Gil et al. 2007; Martin et al. 1998; Whitton et al. 1992; Yan et al. 1997) and (2) a selective agonist, dimethyltryptamine, provokes a type of formal thought disorder (Gouzoulis-Mayfrank et al. 2005) (Fig. 6.1). The presumed mechanisms underlying the schizophrenic symptoms suggest that facilitation of the NMDAR function by a direct or indirect agonist could ameliorate both antipsychotic-responsive dopamine-related symptoms and antipsychotic-resistant dopamine-unrelated symptoms (Fig. 6.1).

Allosteric agonists for the NMDAR acting at the glycine modulatory site including glycine, D-serine (Table 6.1), D-alanine, glycine transporter inhibitor sarcosine, and D-cycloserine have indeed been reported to improve the rating scores of the antipsychotic-refractory negative and cognitive symptoms of patients with

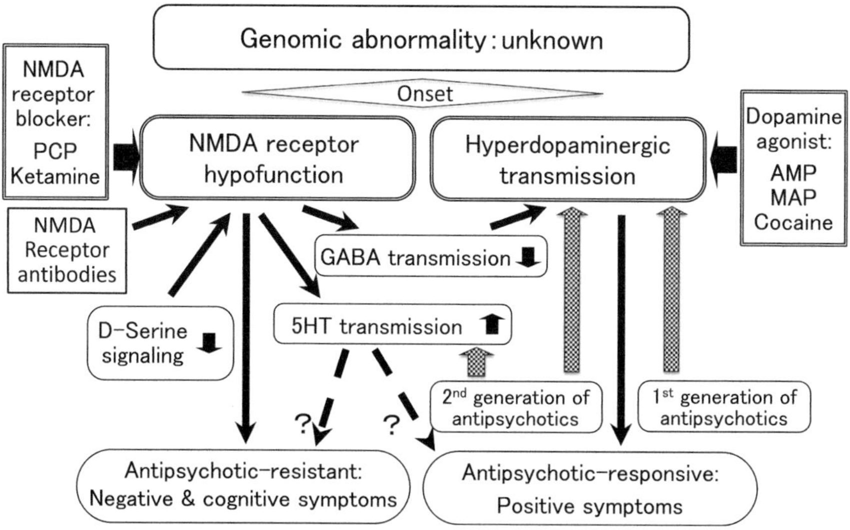

Fig. 6.1 Schematic representation of a pharmacological analysis of the molecular basis of schizophrenic symptoms. Mainly based on the clinical and experimental pharmacological data, the hypothetical pathological changes and their relationships among the neurotransmitter and neuromodulator systems in schizophrenia and substance-induced schizophrenia-like psychoses are depicted. Abbreviations: *AMP* amphetamine, *GABA* γ-aminobutyric acid, *5HT* serotonin, *MAP* methamphetamine, *PCP* phencyclidine

schizophrenia treated with non-clozapine antipsychotics but not with clozapine. The therapeutic efficacy of glycine, D-serine, and/or sarcosine has been confirmed by meta-analyses of their double-blind clinical trials (Singh and Singh 2011; Tsai and Lin 2010; Tuominen et al. 2006;). Recent dose-related investigations of D-serine (Kantrowitz et al. 2010; Weiser et al. 2012) revealed that significant improvement of rating scores of the negative and cognitive symptoms was found after treatment of D-serine at more than 60 mg/kg/day, but not at 30 mg/kg/day and 2 g/day. It is also noteworthy that the results of pilot double-blind studies showed that monotherapy by D-serine reduced the total and negative symptom PANSS scores of patients with schizophrenia (Ermilov et al. 2013) or negative symptoms of individuals at clinical high risk of schizophrenia as estimated by the scale of prodromal symptoms (Kantrowitz et al. 2015).

Consequently, the ameliorating efficacy of the exogenous application of D-serine and other glycine site agonists on schizophrenic symptoms is fitted with the hypofunction of the NMDAR in a group of schizophrenia. However, the effect size of these agents is rather small, which could be due to insufficiencies in their potency or selectivity for the NMDAR and/or their permeability of the blood-brain barrier (BBB).

Table 6.1 Clinical trials of D-serine treatment for patients with schizophrenia or individuals at high risk of schizophrenia

D-serine dosage	Period (weeks)	N	Method (control drug)	Combined antipsychotics	Positive symptoms	Negative symptoms	Cognitive disturbances	Year	Authors
30 mg/kg/day	6	31	Double blind (placebo)	Various (except clozapine)	Improved	Improved	Improved	1998	Tsai et al.
30 mg/kg/day	6	20	Double blind (placebo)	Clozapine	Unchanged	Unchanged	Unchanged	1999	Tsai et al.
30 mg/kg/day	6	39	Double blind, crossover (placebo)	Risperidone or olanzapine	Improved	Improved	Improved	2005	Heresco-Levy et al.
2 g/day (sarcosine or D-ser)	6	44	Double blind (placebo)	Various (except clozapine)	Improved	Improved	Improved	2005	Lane et al.
30, 60, 120 mg/kg/day	4	42	Open label	Various (except clozapine)	Improved	Improved	Improved (≥60 mg)	2010	Kantrowitz et al.
2g/day	6	40	Double blind (placebo)	Various (except clozapine)	Unchanged	Unchanged	Unchanged	2010	Lane et al.
2g/day	16	195	Double blind (placebo)	Various (except clozapine)	Unchanged	Unchanged	Unchanged	2012	Weiser et al.
1.5 (1 week) ~3.0 (9 weeks) g/day	10	8[a]	Double blind (olanzapine)	None	Unchanged[c]	Improved[c]	Unchanged[c]	2013	Ermilov et al.
60 mg/kg/day	16	44[b]	Double blind (placebo)	None	(–)	Improved	(–)	2015	Kantrowitz et al.

[a]Inpatients with a documented history of treatment resistance according to Kane et al. (1988)
[b]Individuals at high risk of schizophrenia
[c]As compared to the scores at the start of the trial but not after the high-dose olanzapine treatment

6.5 Endogenous D-Serine and the Pathophysiology of Schizophrenia

6.5.1 Detection of Brain Endogenous D-Serine

In the initial behavioral study for the development of a novel therapy for intractable schizophrenic symptoms, the author considered the use of D-serine and D-alanine to facilitate the NMDAR function (see the Sect. 6.1) together with a way to overcome their low permeability to the brain. In response to the request of the author, Dr. Hibino from the Nippon Oil & Fat Company synthesized N-myristoyl-D-serine and N-myristoyl-D-alanine as fatty acid derivatives of these amino acids with improved permeability across the BBB. These compounds could be predicted to decompose to their respective fatty acid and free D-amino acid that stimulate the glycine site of the NMDAR (Tanii et al. 1991b). As expected, similar to D-serine and D-alanine, N-myristoyl-D-serine and N-myristoyl-D-alanine inhibited the PCP-induced hyperactivity, stereotypy, and ataxia in a glycine site antagonist-reversible manner (Tanii et al. 1991b), suggesting their action at the NMDAR regulation site. The possible emergence of assumed "non-intrinsic compounds" of the two free D-amino acids in the rats treated by their N-myristic acid derivatives was tested by gas chromatography (GC) and GC-mass spectrometry under cooperation of Dr. Fujii (Tasukuba University, presently Kyoto University) and the late Dr. Hayashi (National Institute of Neuroscience), and endogenous free D-serine was detected at high concentrations in the forebrain, low in the kidney, and non-detectable levels in the liver and serum in the control rats without injections of the lipid-modified D-amino acids (Hashimoto et al. 1992a).

Endogenous D-serine shows a brain-preferring and NMDAR-associated distribution pattern in the rats, mice, and humans (Hashimoto et al. 1992a, b, 1993a, b, 1993c, 1995b; Kumashiro et al. 1995). The author's research group revealed the synthesis (Takahashi et al 1997), extracellular release (Hashimoto et al. 1995a), binding (Matsui et al. 1995; Matoba et al. 1997), uptake (Hayashi et al. 1997; Yamamoto et al. 2001), and breakdown (Hashimoto et al. 1993c) processes of D-serine. Serine racemase and D-amino acid oxidase are believed to catalyze the synthesis of D-serine from L-serine (see Sect. 6.5.1) and the degradation of D-serine to β-hydroxypyruvate and ammonia (see Sect. 6.5.1) in mammals, respectively. However, the molecular machineries specific to the uptake and extracellular release for D-serine are still unclear. In addition, the cell types that contain D-serine and each D-serine metabolic pathway need further elucidation (Nishikawa 2011).

Interestingly, the distribution pattern and the levels of the tissue D-serine concentrations in the discrete brain portions dramatically change during postnatal development, and these changes also resemble those of the GRIN2B (NR2B) subunit mRNA of the NMDAR (Hashimoto et al. 1993a, 1995b). The neuroanatomical features of D-serine suggest that D-serine may be an intrinsic coagonist for the NMDAR (Hashimoto et al. 1993a; Nishikawa 2011). Consistent with this view, the elimination of D-serine without reduction in the tissue and/or extracellular concentration of another NMDAR coagonist, glycine, from brain slice preparation (Kim et al. 2005; Mothet et al. 2000; Panatier et al. 2006) or in vivo (Ishiwata et al. 2015) by application of D-amino acid oxidase or neuron-selective deletion of the gene for a D-serine-synthesizing enzyme, serine racemase, respectively, produces a hypoactivity of the NMDAR.

Despite the D-serine action at a receptor site, regulation mechanisms of the extracellular levels of D-serine differ from those of classical neurotransmitters including glutamate, glycine, and dopamine in the medial frontal cortex of rodents: (1) depolarization by a potassium channel opener evokes a considerable elevation of the extracellular levels of glutamate and glycine but a significant diminution in those of D-serine (Hashimoto et al. 1995a), (2) cessation of nerve impulse traffic by a sodium channel blocker results in a almost complete loss of the extracellular dopamine (Nishijima et al. 1996) but an increase in the extracellular D-serine contents (Hashimoto et al. 1995a), and (3) chelation of the extracellular contents of calcium ion leads to a drastic decrease in those of dopamine, but rather to an upregulation of those of D-serine (Hashimoto et al. 1995a), although some researchers report that maintenance of the extracellular D-serine levels requires calcium ions (Mothet et al. 2005). No reduction in the extracellular D-serine contents by the calcium ion depletion appears to fit with the demonstration by Pan et al. (2015) that calcium ion-independent D-serine release occurs through the pannexin-1 hemichannel in the astroglia.

6.5.2 Possible Involvement of Disturbed D-Serine Signaling in the Pathophysiology of Schizophrenia

The accumulating lines of evidence indicating that D-serine plays a critical role as an endogenous coagonist for the tNMDAR display improving actions against PCP, and amphetamine models for schizophrenia and therapeutic effects on non-clozapine antipsychotic-refractory symptoms of schizophrenia agree with the assumption that aberrant D-serine signaling could be a causative mechanism of the hypofunction of the NMDAR in a group of schizophrenia. While there are so far no findings directly proving the disturbed D-serine metabolism in the pathogenesis or pathophysiology of schizophrenia, a lot of observations in both human and animal studies concur with the implication.

In the postmortem brain tissues, no differences in the D-serine concentrations have been detected between patients with schizophrenia and control subjects without neuropsychiatric disorders (Bendikov et al. 2007; Kumashiro et al. 1995). The results of case-control studies on the quantitative measurement of the D-serine levels in the blood and cerebrospinal fluid (see Brouwer et al. 2013 for a review) are controversial, and meta-analysis of these data indicates no significant changes in the D-amino acid levels in schizophrenia (Brouwer et al. 2013).

It should be noted that the density of [^{3}H]glycine binding to the glycine modulatory site of the NMDAR, which is a target of D-serine, was upregulated in the various cerebral cortical areas of the postmortem brains of patients with schizophrenia as compared to the controls (Ishimaru et al 1994). This upregulation could be explained by a compensatory phenomenon for diminished signaling of the endogenous glycine site agonists, glycine, and /or D-serine. The hypothetical reduced stimulation of the glycine site appears to accord with the in vivo study using single photon emission tomography demonstrating that the drug-naive patients with schizophrenia showed the loss of radioligand ([^{123}I]CNS-1261) binding to the PCP site within the ion channel of the NMDAR in the hippocampus (Pilowsky et al. 2006), because the binding loss could reflect decreased channel open frequencies due to insufficient signals to the NMDAR coagonist site.

These data could be related to the plausible disturbed D-serine metabolism in schizophrenia that is suggested by postmortem brain studies reporting the expressional changes in the transcripts or proteins of genes for D-serine synthesis and degradation enzymes, serine racemase, and D-amino acid oxidase, respectively, in the various brain portions (Madeira et al. 2008; Shinkai et al. 2007; Steffek et al. 2006; Verrall et al. 2007). The plasma levels of protein products of the D-amino acid oxidase activator that was identified from chromosome 13q34, a linkage region for schizophrenia, have been observed to be higher in patients with schizophrenia than in the healthy controls (Lin et al. 2014). In terms of the controversial data (Detera-Wadleigh and McMahon 2006; Li and He 2007; Yamada et al. 2005) or no ratification research, the reproducibility, specificity, and functional significance of these data are still open to argument.

Some results of genetic investigations also agree with the possible deficits in the D-serine signaling system. Single nucleotide polymorphisms and/or haplotypes of the genes for serine racemase (Goltsov et al. 2006; Morita et al. 2007), D-amino acid oxidase (Chumakov et al. 2002), and D-amino acid oxidase activator (G72) (Chumakov et al. 2002) have been found to be significantly associated with schizophrenia. Moreover, a meta-analysis (Shi et al. 2008) or genome-wide association study (Schizophrenia Working Group of the Psychiatric Genomics Consortium 2014) has confirmed the genetic association of serine racemase or G72 with schizophrenia, respectively.

In animal experiments, serine racemase gene knockout mice that deplete the brain tissue and extracellular D-serine concentrations to approximately 10 % compared to those of the wild-type controls (Balu et al. 2013; Horio et al. 2011; Labrie et al. 2009) exhibit hypofunction of the NMDAR and behavioral, molecular, and morphological abnormalities that are considered to be models of schizophrenia (Balu et al. 2013; Labrie et al. 2009; see Chap. 7 for details). These reports reinforce the notion drown from the abovementioned biochemical analysis of human brain tissues that a decrement in the synaptic D-serine concentrations could yield certain portions of schizophrenic symptomatologies, whereas no significant changes in the tissue contents of D-serine in the postmortem schizophrenic brains seem to deny a critical drop in the brain serine racemase activity.

6.5.3 Regulation Mechanisms of D-Serine Signaling in Mammalian Brains

From the viewpoint of the coagonist nature of D-serine for the NMDAR, the extracellular concentrations of D-serine require exceedingly precise control systems to maintain adequate levels for physiological activation of the glutamate receptor that is essential for higher-order brain functions. The lack of a depolarization-induced elevation and of a nerve impulse cessation-dependent diminishment of the extracellular D-serine concentrations indicates the possibility that glial cells could participate in their modulation. As expected, our in vivo dialysis experiments revealed that a local infusion of a reversible glia toxin, fluorocitrate, via the dialysis tubing caused a significant reduction in not only the extracellular concentrations of glutamine, a glial activity marker, but also those of D-serine in the medial frontal cortex of the rat (Kanematsu et al. 2006). Subsequent studies by Oliet group (Henneberger et al. 2010; Panatier et al. 2006) added the data denoting the substantial role of astroglia in the regulation of the extracellular release of D-serine.

The extracellular liberation of brain D-serine has been pointed out to be under the influence of the ionotropic glutamate receptors. Snyder's research group found that the kainate-type glutamate receptor facilitates the release of the [3H]D-serine preloaded to the primary culture of the astroglia from the rat cortex (Schell et al. 1995). The α-amino-3-hydroxy-5-methyl-4-isoxazolepropioinic acid (AMPA)-type glutamate receptor has also been described to play a phasic stimulatory role in the control of the extracellular release of preloaded radiolabeled or intrinsic D-serine from the in vitro preparations including the primary culture of the rat cortical astroglia (Mothet et al. 2005; Rosenberg et al. 2010; Schell et al. 1995) or neurons (Kartvelishvily et al. 2006; Rosenberg et al. 2010), C6 glioma cells (Mothet et al. 2005), and isolated retinal tissues (Sullivan and Miller 2010, 2012). In contrast, our in vivo studies have demonstrated that the calcium-permeable-type GRIA2 subunit-lacking AMPA receptor exerts a phasic inhibitory regulation of the extracellular levels of D-serine in the medial frontal cortex of the rat (Ishiwata et al. 2013a). This discrepancy could be attributed to the differences in the cellular interactions (i.e., neuron-glia and glia-glia) and neuronal circuits between the in vitro and in vivo experimental conditions. The calcium-permeable AMPA receptor-connected inhibitory influence is considered to be a specific interaction because an AMPA receptor agonist-induced decrease in the extracellular taurine contents is insensitive to a selective antagonist for the calcium-permeable AMPA receptor (Ishiwata et al. 2013a).

Neutral amino acid transporters containing sodium-dependent ASCT1 and ASCT2 (Gliddon et al. 2009; Hayashi et al. 1997; O'Brien et al. 2005; Ribeiro et al. 2002), sodium-independent Asc-1 (Fukasawa et al. 2000; Rutter et al. 2007), and proton-dependent PAT1 (Metzner et al. 2005) have been documented to provide an uptake capacity of D-serine with low to relatively high affinity (IC_{50} of approximately 10–50 μM for Asc-1) and to be expressed in neurons or astroglia in the brain tissues. The pharmacological and neuroanatomical features of these transporters suggest their plausible involvement in the physiological uptake of D-serine in mammalian brains. In fact, we have found using an in vivo dialysis technique that an Asc-1 inhibitor, S-methyl-L-cysteine, augments the extracellular D-serine levels in the medial frontal cortex of the rat (Ishiwata et al. 2013b).

A profound loss of the extracellular and tissue amounts of D-serine and disturbances in the NMDAR-mediated formation of the long-term potentiation (LTP) by genetic total depletion of serine racemase in the mice seems to endow the D-serine-synthesizing enzyme with a role in the setting of the extracellular levels of D-serine (Benneyworth et al. 2011). CamKIIα-expressing neuron-specific deficits in the serine racemase lead to an NMDAR hypofunction with a significant diminution in the extracellular D-serine contents in the hippocampus (Benneyworth et al. 2011; Ishiwata et al. 2015).

Although the data indicating the molecules that modulate the extracellular D-serine concentrations have been accumulated as already mentioned, the exact machinery that directly liberates D-serine from the intracellular site to the extracellular fluid and the precise intracellular storage site of the releasable D-serine are mostly unknown. The SNARE proteins (Mothet et al. 2005), the hemichannel of connexin-43 (Stehberg et al. 2012) or pannexin-1 (Pan et al. 2015), the Asc-1 transporter (Rosenberg et al. 2013), or Dsm-1 (PAPST-1) (Shimazu et al. 2006) have been proposed to be involved in the extracellular release of D-serine and need further elucidation regarding their detailed functions in the release process and cellular localization.

6.6 Conclusions

Experimental and clinical investigations have indicated that the antipsychotic-responsive and antipsychotic-resistant symptoms of schizophrenia and their pharmacological models are ameliorated by D-serine through facilitation of the NMDAR functions. Since D-serine has been demonstrated to be an endogenous coagonist for the NMDAR, the molecular cascades and cellular setups for the brain D-serine signaling may be suitable targets for analyses of the pathophysiology of the NMDAR hypofunction and generation of a novel pharmacotherapy of schizophrenia and related psychoses.

References

Abi-Dargham A, van de Giessen E, Slifstein M et al (2009) Baseline and amphetamine-stimulated dopamine activity are related in drug-naïve schizophrenic subjects. Biol Psychiatry 65 (12):1091–1093

Anis NA, Berry SC, Burton NR et al (1983) The dissociative anaesthetics, ketamine and phencyclidine, selectively reduce excitation of central mammalian neurons by N-methyl-aspartate. Br J Pharmacol 79:565–575

Balla A, Sershen H, Serra M (2003) Subchronic continuous phencyclidine administration potentiates amphetamine-induced frontal cortex dopamine release. Neuropsychopharmacology 28 (1):34–44

Balla A, Schneider S, Sershen H et al (2012) Effects of novel, high affinity glycine transport inhibitors on frontostriatal dopamine release in a rodent model of schizophrenia. Eur Neuropsychopharmacol 22(12):902–910

Balu DT, Li Y, Puhl MD et al (2013) Multiple risk pathways for schizophrenia converge in serine racemase knockout mice, a mouse model of NMDA receptor hypofunction. Proc Natl Acad Sci U S A 110(26):E2400–E2409

Bendikov I, Nadri C, Amar S et al (2007) A CSF and postmortem brain study of D-serine metabolic parameters in schizophrenia. Schizophr Res 90(1–3):41–51

Benneyworth MA, Basu AC, Coyle JT (2011) Discordant behavioral effects of psychotomimetic drugs in mice with altered NMDA receptor function. Psychopharmacology (Berlin) 213 (1):143–153

Brouwer A, Luykx JJ, van Boxmeer L et al (2013) NMDA-receptor coagonists in serum, plasma, and cerebrospinal fluid of schizophrenia patients: a meta-analysis of case-control studies. Neurosci Biobehav Rev 37(8):1587–1596

Chumakov I, Blumenfeld M, Guerassimenko O et al (2002) Genetic and physiological data implicating the new human gene G72 and the gene for D-amino acid oxidase in schizophrenia. Proc Natl Acad Sci U S A 99(21):13675–13680

Contreras PC (1990) D-serine antagonized phencyclidine- and MK-801-induced stereotyped behavior and ataxia. Neuropharmacology 29(3):291–293

Depoortere R, Perrault G, Sanger DJ (1999) Prepulse inhibition of the startle reflex in rats: effects of compounds acting at various sites on the NMDA receptor complex. Behav Pharmacol 10 (1):51–62

Detera-Wadleigh SD, McMahon FJ (2006) G72/G30 in schizophrenia and bipolar disorder: review and meta-analysis. Biol Psychiatry 60(2):106–114

Deutch AY, Tam SY, Freeman AS et al (1987) Mesolimbic and mesocortical dopamine activation induced by phencyclidine: contrasting pattern to striatal response. Eur J Pharmacol 134 (3):257–264

Ermilov M, Gelfin E, Levin R et al (2013) A pilot double-blind comparison of d-serine and high-dose olanzapine in treatment-resistant patients with schizophrenia. Schizophr Res 150 (2–3):604–605

Fukasawa Y, Segawa H, Kim JY et al (2000) Identification and characterization of a Na(+)-independent neutral amino acid transporter that associates with the 4F2 heavy chain and exhibits substrate selectivity for small neutral D- and L-amino acids. J Biol Chem 275 (13):9690–9698

Gliddon CM, Shao Z, LeMaistre JL et al (2009) Cellular distribution of the neutral amino acid transporter subtype ASCT2 in mouse brain. J Neurochem 108(2):372–383

Goltsov AY, Loseva JG, Andreeva TV et al (2006) Polymorphism in the 5′-promoter region of serine racemase gene in schizophrenia. Mol Psychiatry 11:325–326

Gouzoulis-Mayfrank E, Heekeren K, Neukirch A et al (2005) Psychological effects of (S)-ketamine and N, N-dimethyltryptamine (DMT): a double-blind, cross-over study in healthy volunteers. Pharmacopsychiatry 38(6):301–311

Hashimoto A, Nishikawa T, Oka T et al (1991) D-Alanine inhibits methamphetamine-induced hyperactivity in rats. Eur J Pharmacol 202:105–107

Hashimoto A, Nishikawa T, Hayashi T et al (1992a) The presence of free D-serine in rat brain. FEBS Lett 296:33–36

Hashimoto A, Nishikawa T, Oka T et al (1992b) Determination of free amino acid enantiomers in rat brain and serum by high performance liquid chromatography after derivatization with N-tert.-butyloxycarbonyl-L-cysteine and o-phthaldialdehyde. J Chromatogr 582:41–48

Hashimoto A, Nishikawa T, Oka T et al (1993a) Endogenous D-serine in rat brain: N-Methyl-D-aspartate receptor-related distribution and aging. J Neurochem 60:783–786

Hashimoto A, Kumashiro S, Nishikawa T et al (1993b) Embryonic development and postnatal changes in free D-aspartate and D-serine in the human prefrontal cortex. J Neurochem 61:348–351

Hashimoto A, Nishikawa T, Konno R et al (1993c) Free D-serine, D-aspartate and D-alanine in central nervous system and serum in mutant mice lacking D-amino acid oxidase. Neurosci Lett 152:33–36

Hashimoto A, Oka T, Nishikawa T (1995a) Extracellular concentration of endogenous free D-serine in the rat brain as revealed by in vivo microdialysis. Neuroscience 66(3):635–643

Hashimoto A, Oka T, Nishikawa T (1995b) Anatomical distribution and postnatal changes in endogenous free D-aspartate and D-serine in rat brain and periphery. Eur J Neurosci 7:1657–1663

Hashimoto K, Fujita Y, Horio M et al (2009) Co-administration of a D-amino acid oxidase inhibitor potentiates the efficacy of D-serine in attenuating prepulse inhibition deficits after administration of dizocilpine. Biol Psychiatry 65(12):1103–1106

Hata N, Nishikawa T, Umino A et al (1990) Evidence for involvement of N-methyl-D-aspartate receptor in tonic inhibitory control of dopaminergic transmission in rat medial frontal cortex. Neurosci Lett 120(1):101–104

Hayashi F, Takahashi K, Nishikawa T (1997) Uptake of L- and D-serine in C6 glioma cells. Neurosci Lett 239:85–88

Henneberger C, Papouin T, Oliet SH et al (2010) Long-term potentiation depends on release of D-serine from astrocytes. Nature 463(7278):232–236

Heresco-Levy U, Javitt DC, Ebstein R et al (2005) D-serine efficacy as add-on pharmacotherapy to risperidone and olanzapine for treatment-refractory schizophrenia. Biol Psychiatry 57 (6):577–585

Horio M, Kohno M, Fujita Y et al (2011) Levels of D-serine in the brain and peripheral organs of serine racemase (Srr) knock-out mice. Neurochem Int 59(6):853–859

Irifune M, Sato T, Kamata Y et al (1998) Inhibition by diazepam of ketamine-induced hyperlocomotion and dopamine turnover in mice. Can J Anaesth 45(5 Pt 1):471–478

Ishimaru M, Kurumaji A, Toru M (1994) Increases in strychnine-insensitive glycine binding sites in cerebral cortex of chronic schizophrenics: evidence for glutamate hypothesis. Biol Psychiatry 35:84–95

Ishiwata S, Ogata S, Umino A et al (2013a) Increasing effects of S-methyl-L-cysteine on the extracellular D-serine concentrations in the rat medial frontal cortex. Amino Acids 44 (5):1391–1395

Ishiwata S, Umino A, Umino M et al (2013b) Modulation of extracellular d-serine content by calcium permeable AMPA receptors in rat medial prefrontal cortex as revealed by in vivo microdialysis. Int J Neuropsychopharmacol 16(6):1395–1406

Ishiwata S, Umino A, Balu DT et al (2015) Neuronal serine racemase regulates extracellular D-serine levels in the adult mouse hippocampus. J Neural Transm 122(8):1099–1103

Javitt DC, Zukin SR (1991) Recent advances in the phencyclidine model of schizophrenia. Am J Psychiatry 148(10):1301–1308

Johnson JW, Ascher P (1987) Glycine potentiates the NMDA response in cultured mouse brain neurons. Nature 325(6104):529–531

Kanahara N, Shimizu E, Ohgake S et al (2008) Glycine and D: −serine, but not D: −cycloserine, attenuate prepulse inhibition deficits induced by NMDA receptor antagonist MK-801. Psychopharmacology (Berlin) 198(3):363–374

Kane JM, Honigfeld G, Singer J et al (1988) Clozapine in treatment-resistant schizophrenics. Psychopharmacol Bull 24(1):62–67

Kanematsu S, Ishii S, Umino A et al (2006) Evidence for involvement of glial cell activity in the control of extracellular D-serine contents in the rat brain. J Neural Transm 113:1717–1721
Kantrowitz JT, Malhotra AK, Cornblatt B et al (2010) High dose D-serine in the treatment of schizophrenia. Schizophr Res 121(1–3):125–130
Kantrowitz JT, Woods SW, Petkova E et al (2015) D-serine for the treatment of negative symptoms in individuals at clinical high risk of schizophrenia: a pilot, double-blind, placebo-controlled, randomised parallel group mechanistic proof-of-concept trial. Lancet Psychiatry 2(5):403–412
Kartvelishvily E, Schleper M, Balan L et al (2006) Neuron-derived D-serine release provides a novel means to activate N-Methyl-D-aspartate receptors. J Biol Chem 281:14151–14162
Kashiwa A, Nishikawa T, Nishijima K et al (1995) Dizocilpine (MK-801) elicits a tetrodotoxin-sensitive increase in extracellular release of dopamine in rat medial frontal cortex. Neurochem Int 26(3):269–279
Kim PM, Aizawa H, Kim PS et al (2005) Serine racemase: activation by glutamate neurotransmission via glutamate receptor interacting protein and mediation of neuronal migration. Proc Natl Acad Sci U S A 102(6):2105–2110
Kishimoto H, Simon JR, Aprison MH (1981) Determination of the equilibrium dissociation constants and number of glycine binding sites in several areas of the rat central nervous system, using a sodium-independent system. J Neurochem 37(4):1015–1024
Kleckner NW, Dingledine R (1988) Requirement for glycine in activation of NMDA-receptors expressed in Xenopus oocytes. Science 241(4867):835–837
Kumashiro S, Hashimoto A, Nishikawa T (1995) Free D-serine in post-mortem brains and spinal cords of individuals with and without neuropsychiatric diseases. Brain Res 681:117–125
Labrie V, Fukumura R, Rastogi A et al (2009) Serine racemase is associated with schizophrenia susceptibility in humans and in a mouse model. Hum Mol Genet 18(17):3227–3243
Lane HY, Chang YC, Liu YC et al (2005) Sarcosine or D-serine add-on treatment for acute exacerbation of schizophrenia: a randomized, double-blind, placebo-controlled study. Arch Gen Psychiatry 62(11):1196–1204
Lane HY, Lin CH, Huang YJ et al (2010) A randomized, double-blind, placebo-controlled comparison study of sarcosine (N-methylglycine) and D-serine add-on treatment for schizophrenia. Int J Neuropsychopharmacol 13(4):451–460
Laruelle M, Abi-Dargham A, van Dyck CH et al (1996) Single photon emission computerized tomography imaging of amphetamine-induced dopamine release in drug-free schizophrenic subjects. Proc Natl Acad Sci U S A 93(17):9235–9240
Li D, He L (2007) G72/G30 genes and schizophrenia: a systematic meta-analysis of association studies. Genetics 175(2):917–922
Lin CH, Chang HT, Chen YJ et al (2014) Distinctively higher plasma G72 protein levels in patients with schizophrenia than in healthy individuals. Mol Psychiatry 19(6):636–637
Lipina T, Labrie V, Weiner I et al (2005) Modulators of the glycine site on NMDA receptors, D-serine and ALX 5407, display similar beneficial effects to clozapine in mouse models of schizophrenia. Psychopharmacology (Berlin) 179(1):54–67
López-Gil X, Babot Z, Amargós-Bosch M et al (2007) Clozapine and haloperidol differently suppress the MK-801-increased glutamatergic and serotonergic transmission in the medial prefrontal cortex of the rat. Neuropsychopharmacology 32(10):2087–2097
Madeira C, Freitas ME, Vargas-Lopes C et al (2008) Increased brain D-amino acid oxidase (DAAO) activity in schizophrenia. Schizophr Res 101(1–3):76–83
Martin P, Carlsson ML, Hjorth S (1998) Systemic PCP treatment elevates brain extracellular 5-HT: a microdialysis study in awake rats. Neuroreport 9(13):2985–2988
Matoba M, Tomita U, Nishikawa T (1997) Characterization of 5, 7-dichlorokynurenate-insensitive [3H] D-serine binding to synaptosomal fraction isolated from rat brain tissues. J Neurochem 69:399–405
Matsui T, Sekiguchi M, Hasimoto A et al (1995) Functional comparison of D-serine and glycine in rodents: the effects on cloned NMDA receptors and the extracellular concentration. J Neurochem 65:454–458

Metzner L, Kottra G, Neubert K et al (2005) Serotonin, L-tryptophan, and tryptamine are effective inhibitors of the amino acid transport system PAT1. FASEB J 19(11):1468–1473

Morita Y, Ujike H, Tanaka Y et al (2007) A genetic variant of the serine racemase gene is associated with schizophrenia. Biol Psychiatry 61(10):1200–1203

Mothet JP, Parent AT, Wolosker H et al (2000) D-serine is an endogenous ligand for the glycine site of the N-methyl-D-aspartate receptor. Proc Natl Acad Sci U S A 97(9):4926–4931

Mothet JP, Pollegioni L, Ouanounou G et al (2005) Glutamate receptor activation triggers a calcium-dependent and SNARE protein-dependent release of the gliotransmitter D-serine. Proc Natl Acad Sci U S A 102(15):5606–5611

Nishijima K, Kashiwa A, Nishikawa T (1994) Preferential stimulation of extracellular release of dopamine in rat frontal cortex to striatum following competitive inhibition of the N-methyl-D-aspartate receptor. J Neurochem 63(1):375–378

Nishijima K, Kashiwa A, Hashimoto A et al (1996) Differential effects of phencyclidine and methamphetamine on dopamine metabolism in rat frontal cortex and striatum as revealed by in vivo dialysis. Synapse 22(4):304–312

Nishikawa T (2011) Analysis of free D-serine in mammals and its biological relevance. J Chromatogr B Anal Technol Biomed Life Sci 879:3169–3183

O'Brien KB, Miller RF, Bowser MT (2005) D-Serine uptake by isolated retinas is consistent with ASCT-mediated transport. Neurosci Lett 385(1):58–63

Pan HC, Chou YC, Sun SH (2015) P2X7 R-mediated Ca(2+) -independent d-serine release via pannexin-1 of the P2X7 R-pannexin-1 complex in astrocytes. Glia 63(5):877–893

Panatier A, Theodosis DT, Mothet JP et al (2006) Glia-derived D-serine controls NMDA receptor activity and synaptic memory. Cell 125(4):775–784

Petersen RC, Stillman RC (1978) Phencyclidine (PCP) abuse: an appraisal. In: Petersen RC, Stillman RC (eds) NIDA research monograph 21. Superintendent of Documents, US Government Printing Office, Washington, DC, pp 1–17

Pilowsky LS, Bressan RA, Stone JM et al (2006) First in vivo evidence of an NMDA receptor deficit in medication-free schizophrenic patients. Mol Psychiatry 11(2):118–119

Rao TS, Kim HS, Lehmann J et al (1989) Differential effects of phencyclidine (PCP) and ketamine on mesocortical and mesostriatal dopamine release in vivo. Life Sci 45(12):1065–1072

Ribeiro CS, Reis M, Panizzutti R et al (2002) Glial transport of the neuromodulator D-serine. Brain Res 929(2):202–209

Rosenberg D, Kartvelishvily E, Shleper M et al (2010) Neuronal release of d-serine: a physiological pathway controlling extracellular d-serine concentration. FASEB J 24(8):2951–2961

Rosenberg D, Artoul S, Segal AC et al (2013) Neuronal D-serine and glycine release via the Asc-1 transporter regulates NMDA receptor-dependent synaptic activity. J Neurosci 33 (8):3533–3544

Rutter AR, Fradley RL, Garrett EM et al (2007) Evidence from gene knockout studies implicates Asc-1 as the primary transporter mediating d-serine reuptake in the mouse CNS. Eur J Neurosci 25(6):1757–1766

Schell MJ, Molliver ME, Snyder SH (1995) D-serine, an endogenous synaptic modulator: localization to astrocytes and glutamate-stimulated release. Proc Natl Acad Sci U S A 92 (9):3948–3952

Schizophrenia Working Group of the Psychiatric Genomics Consortium (2014) Biological insights from 108 schizophrenia-associated genetic loci. Nature 511(7510):421–427

Shi J, Badner JA, Gershon ES (2008) Allelic association of G72/G30 with schizophrenia and bipolar disorder: a comprehensive meta-analysis. Schizophr Res 98(1–3):89–97

Shimazu D, Yamamoto N, Umino A et al (2006) Inhibition of D-serine accumulation to the Xenopus Oocyte by expression of the rat ortholog of human 3′-phosphoadenosine 5′-phosphosulfate transporter gene isolated from the neocortex as D-serine modulator-1. J Neurochem 96:30–42

Shinkai T, De Luca V, Hwang R (2007) Association analyses of the DAOA/G30 and D-amino-acid oxidase genes in schizophrenia: further evidence for a role in schizophrenia. Neuromolecular 9(2):169–177

Singh SP, Singh V (2011) Meta-analysis of the efficacy of adjunctive NMDA receptor modulators in chronic schizophrenia. CNS Drugs 25(10):859–885

Smith SM, Uslaner JM, Yao L et al (2009) The behavioral and neurochemical effects of a novel D-amino acid oxidase inhibitor compound 8 [4H-thieno [3,2-b]pyrrole-5-carboxylic acid] and D-serine. J Pharmacol Exp Ther 328(3):921–930

Steffek AE, Haroutunian V, Meador-Woodruff JH (2006) Serine racemase protein expression in cortex and hippocampus in schizophrenia. Neuroreport 17(11):1181–1185

Stehberg J, Moraga-Amaro R et al (2012) Release of gliotransmitters through astroglial connexin 43 hemichannels is necessary for fear memory consolidation in the basolateral amygdala. FASEB J 26(9):3649–3657

Sullivan SJ, Miller RF (2010) AMPA receptor mediated D-serine release from retinal glial cells. J Neurochem 115(6):1681–1689

Sullivan SJ, Miller RF (2012) AMPA receptor-dependent, light-evoked D-serine release acts on retinal ganglion cell NMDA receptors. J Neurophysiol 108(4):1044–1051

Takahashi K, Hayashi F, Nishikawa T (1997) In vivo evidence for the link between L- and D-serine metabolism in rat cerebral cortex. J Neurochem 69:1286–1290

Tanii Y, Nishikawa T, Umino A et al (1990) Phencyclidine increases extracellular dopamine metabolites in rat medial frontal cortex as measured by in vivo dialysis. Neurosci Lett 112 (2–3):318–323

Tanii Y, Nishikawa T, Hashimoto A et al (1991a) Stereoselective inhibition by D- and L-alanine of phencyclidineinduced locomotor stimulation in the rat. Brain Res 563:281–284

Tanii Y, Nishikawa T, Hibino H et al (1991b) Effects of allosteric agonists for the N-methyl D-aspartate receptor and their derivatives on phencyclidine-induced abnormal behavior in the rat. Brain Sci Ment Disord 2:497–502

Tanii Y, Nishikawa T, Hashimoto A et al (1994) Stereoselective antagonism by enantiomers of alanine and serine of phencyclidine-induced hyperactivity, stereotypy and ataxia in the rat. J Pharmacol Exp Ther 269:1040–1048

Tokutomi N, Kaneda M, Akaike N (1989) What confers specificity on glycine for its receptor site? Br J Pharmacol 97(2):353–360

Tsai GE, Lin PY (2010) Strategies to enhance N-methyl-D-aspartate receptor-mediated neurotransmission in schizophrenia, a critical review and meta-analysis. Curr Pharm Des 16 (5):522–537

Tsai G, Yang P, Chung LC et al (1998) D-serine added to antipsychotics for the treatment of schizophrenia. Biol Psychiatry 44(11):1081–1089

Tsai GE, Yang P, Chung LC (1999) D-serine added to clozapine for the treatment of schizophrenia. Am J Psychiatry 156(11):1822–1825

Tuominen HJ, Tiihonen J, Wahlbeck K (2006) Glutamatergic drugs for schizophrenia. Cochrane Database Syst Rev 19(2):CD003730

Umino A, Takahashi K, Nishikawa T (1998) Characterization of phencyclidine-induced increase in prefrontal cortical dopamine metabolism in the rat. Br J Pharmacol 124:377–385

Verrall L, Walker M, Rawlings N et al (2007) d-Amino acid oxidase and serine racemase in human brain: normal distribution and altered expression in schizophrenia. Eur J Neurosci 26 (6):1657–1669

Weiser M, Heresco-Levy U, Davidson M et al (2012) A multicenter, add-on randomized controlled trial of low-dose d-serine for negative and cognitive symptoms of schizophrenia. J Clin Psychiatry 73(6):e728–e734

White WF, Brown KL, Frank DM (1989) Glycine binding to rat cortex and spinal cord: binding characteristics and pharmacology reveal distinct populations of sites. J Neurochem 53 (2):503–512

Whitton PS, Biggs CS, Pearce BR (1992) MK-801 increases extracellular 5-hydroxytryptamine in rat hippocampus and striatum in vivo. J Neurochem 58(4):1573–1575

Yamada K, Ohnishi T, Hashimoto K et al (2005) Identification of multiple serine racemase (SRR) mRNA isoforms and genetic analyses of SRR and DAO in schizophrenia and D-serine levels. Biol Psychiatry 57(12):1493–1503

Yamamoto N, Tomita U, Umino A et al (2001) Uptake of D-serine by synaptosomal P2 fraction isolated from rat brain. Synapse 42:84–86

Yan QS, Reith ME, Jobe PC et al (1997) Dizocilpine (MK-801) increases not only dopamine but also serotonin and norepinephrine transmissions in the nucleus accumbens as measured by microdialysis in freely moving rats. Brain Res 765(1):149–158

Yonezawa Y, Kuroki T, Kawahara T (1998) Involvement of gamma-aminobutyric acid neurotransmission in phencyclidine-induced dopamine release in the medial prefrontal cortex. Eur J Pharmacol 341(1):45–56

Chapter 7
D-Serine and the Pathophysiology of Schizophrenia

Joseph T. Coyle

Abstract Schizophrenia is a chronic, disabling psychiatric disorder characterized by psychosis, cognitive impairments, and negative symptoms such as social withdrawal and anhedonia. Pathologic changes in the brain include cortical atrophy, loss of pyramidal cell dendritic complexity and spine density, ventricular enlargement, and reduction in intracellular mediators of neuronal plasticity. Acute blockade of N-methyl-D-aspartate receptors (NMDARs) with ketamine reproduces the full range of symptoms of schizophrenia in normal volunteers. A recent genome-wide association study revealed that many of the risk genes for schizophrenia interact directly with the NMDAR such as serine racemase (SR), which synthesizes its co-agonist, D-serine, or downstream mediators of NMDAR activity. Consistent with the genetic findings, alterations in the levels of SR, D-serine, and D-amino acid oxidase have been found in schizophrenia. Genetically silencing the SR gene (SR−/−) in mice causes an 85 % reduction of D-serine in the brain and reproduces the synaptic neuropathology of schizophrenia as well as the cognitive deficits and anhedonia. Treatment of adult SR−/− mice with D-serine to restore brain levels reverses most of the neuropathology and the cognitive deficits. Placebo-controlled clinical trials in schizophrenic patients on antipsychotic drugs with D-serine or agents that act at its site on the NMDAR have shown significant reductions in symptoms including psychosis. These findings support the hypothesis that drugs that enhance NMDAR function might be effective treatments for schizophrenia, especially the cognitive and negative symptoms, which are unaffected by current treatments.

Keywords D-serine • N-methyl-D-aspartate receptor (NMDAR) • Neuroplasticity: D-serine • Serine racemase • Schizophrenia

---

J.T. Coyle (✉)
Eben S. Draper Professor of Psychiatry and Neuroscience, Harvard Medical School, McLean Hospital, Belmont, MA 02478, USA
e-mail: Joseph_coyle@hms.harvard.edu

T. Yoshimura et al. (eds.), *D-Amino Acids*, DOI 10.1007/978-4-431-56077-7_7

7.1 Introduction

Schizophrenia is a chronic debilitating psychiatric disorder, which afflicts approximately 1 % of the population worldwide. Schizophrenia affects multiple domains: positive symptoms including psychosis and thought disorder, cognitive impairments affecting memory and executive functions, and negative symptoms with anhedonia, alogia, and poverty of thought (Goff et al. 2001). Prospective studies with youth at high risk for schizophrenia because of having an affected first-degree relative and prepsychotic symptoms (Bang et al. 2015) and the so-called retrospective-prospective studies that exploit the longitudinal videotapes taken by families of their children's development (Schiffman et al. 2004) indicate that most youth, who go on to develop schizophrenia, exhibit reduced intelligence, academic underachievement, social and physical awkwardness, and abnormal thinking. However, these symptoms are poor predictors of schizophrenia as they affect nearly 10 % of youth, and less than half of genetically high-risk adolescents convert to active schizophrenia (Horton et al. 2015). These observations support the notion that schizophrenia is a developmental disorder; however, the factors that precipitate the conversion to psychosis remain still poorly understood (Nelson et al. 2013).

After the symptomatic onset of schizophrenia in late adolescence and early adulthood, most individuals with schizophrenia experience lifelong unemployment, social isolation, and substance abuse in spite of treatment with first- or second-generation antipsychotic medications (Goff et al. 2001). In the United States, schizophrenia is the seventh most costly medical condition to Society because of the costs of care and loss of productivity. Yet the basic approach to treatment remains largely unchanged over the last 40 years. (Tune et al. 1982; Goff et al. 2001). The mainstay for treatment is the dopamine D2 receptor antagonists, the antipsychotic drugs, which target positive symptoms but have negligible effects on cognitive impairments and negative symptoms, the major contributors to disability (Goff et al. 2001).

This review will summarize recent research addressing the etiology of schizophrenia that indicates that impaired function of the N-methyl-D-aspartate receptors (NMDARs) plays a central role in its pathophysiology and the associated reduced synaptic connectivity. The consequent cortical atrophy correlates with cognitive impairments and negative symptoms (Keefe et al. 2006), suggesting that the former are core features of the disorder with psychosis being a downstream consequence of the cortical pathology (Coyle et al. 2010). The review will summarize the evidence showing that enhancing NMDAR function directly with D-serine or indirectly appears to be an important strategy for not only reducing symptoms but also for restoring synaptic plasticity and synaptic reconnection.

7.2 Brain Imaging

Early in the twentieth century, pneumoencephalographic studies, in which cerebral spinal fluid is drained from the ventricles and replaced with air to permit their visualization with x-rays, provided the first evidence of ventricular enlargement in schizophrenia (Jacoby and Winkle 1927) as a result of cortical atrophy and (Haug 1982). However, it was with the development of modern brain imaging techniques – computer-assisted tomography (CAT) scans and magnetic resonance imaging (MRI) – that the evidence of reduced volume of cortico-limbic structures and increased ventricular volume in schizophrenia became incontrovertible. Although the frontal and temporal cortices are strongly affected, no region of the cortex appears to be spared from atrophy (Kuperberg et al. 2003). Cortical atrophy is present at the first episode of psychosis in schizophrenia and progresses over the next decade of illness (DeLisi et al. 2006). Of concern, antipsychotic drugs, the mainstay of treatment of schizophrenia, appear to exacerbate cortical atrophy in schizophrenia (Ho et al. 2011). Default mode functional MRI in schizophrenia shows intracortical and thalamocortical disconnection (Tu et al. 2015; Chang et al. 2015).

7.3 Neuropathology

This progressive cortical atrophy revealed by brain imaging studies suggested that schizophrenia might be a neurodegenerative disorder like other dementias, consistent with Kraepelin's appellation dementia praecox. However, studies using stereologic cell counting methods have not revealed loss of cortical neurons in schizophrenia commensurate with the degree of atrophy (Selemon et al. 1995; 1998). Rather, the density of neurons was increased, and there was a significant reduction in the neuropil. This reduction in cortical neuropil but not in neuronal numbers stood in contrast to findings in Huntington's disease, in which there is reduced neuronal density and increased glial density (Selemon et al. 1998).

To understand the loss of neuropil, which is composed of axons, dendrites, and synapses, several studies have exploited Golgi staining to examine the structure of cortical pyramidal neurons in postmortem frontal and temporal cortices from patients with schizophrenia and suitable controls (Glantz and Lewis 2000; Sweet et al. 2009; Konopaske et al. 2014). Golgi stain results in random silver impregnation of entire neurons, revealing the three-dimensional structure of their perikarya, axons, and dendritic processes including synaptic spines (Mancuso et al. 2013). Conditions are optimized to stain 1–2 % of the neurons; with greater neuronal impregnation, individual neurons cannot be visualized for analysis because they are obscured by the densely stained neuropil. These studies have consistently demonstrated reduced dendritic complexity (primary, secondary, and tertiary branching), reduced total dendritic length, and reduced spine density in schizophrenia.

The total number of glutamatergic synapses on apical and basilar dendrites, which receive their inputs from different brain regions, can be estimated by multiplying the total dendritic length by spine density (Sheng and Hoogenraad 2007). The estimates vary but range from 20 to nearly a 40 % reduction in the number of glutamatergic synapses on cortical pyramidal neurons, the principal projecting neurons, in schizophrenia. Thus, the significant but modest 3–4 % loss of cortical volume due shrinkage of the neuropil in schizophrenia obscures the substantial loss of glutamatergic synaptic connections, consistent with Goldman-Rakic's description of schizophrenia as a "disconnection syndrome" (Selemon and Goldman-Rakic 1999). A recent study has shown comparable losses of glutamatergic synapses on pyramidal neurons in layer III of the dorsolateral prefrontal cortex in bipolar disorder as in schizophrenia (Konopaske et al. 2014). However, brain imaging studies indicate that the cortical atrophy is more circumscribed in bipolar disorder than in schizophrenia (De Peri et al. 2012).

7.4 Neurochemistry

Nearly 40 years ago, postmortem neurochemical studies revealed reduction in the activity of glutamic acid decarboxylase (GAD), the enzyme that synthesizes γ-aminobutyric acid (GABA), the major inhibitory neurotransmitter, in the cortex in schizophrenia (Bird et al. 1977). Results of subsequent studies disputed this conclusion, suggesting that the reduction in GAD was an artifact of the slow death in the schizophrenic subjects (Bird et al. 1978). However, the last 10 years have witnessed an accumulation of compelling evidence that there is a downregulation of the fast-firing parvalbumin-positive (PV+) GABAergic interneurons in the intermediate layers of the cortex in schizophrenia. The evidence includes immunocytochemical staining, neurochemical assays, in situ hybridization, and DNA chip array measurements of mRNA levels from the brain material obtained from several different brain banks from hundreds of subjects (for review, Lewis 2014). Cortical GABAergic neurons also express two other calcium-binding proteins, calbindin and calretinin, besides PV; however, these other GABAergic neuronal cell types appear to be relatively spared in schizophrenia (Fung et al. 2010). In addition to PV and GAD67, GABA transporter (GAT1) expression, another presynaptic marker, is reduced (Hoftman et al. 2015). This subpopulation of GABAergic neurons, known as chandelier cells, synapse on the proximal hillock of the axon of the pyramidal neuron, thus having the final "say" over the firing of an action potential (Lewis 2014).

The functional impact of this downregulation of the presynaptic components of the PV+–GABAergic interneurons is supported by evidence of postsynaptic $GABA_A$ receptor upregulation consistent with supersensitivity (Benes et al. 1996). Early studies by Hanada et al. (1987) revealed a 40 % increase in the

B_{max} but no change in affinity in the specific binding of [^{3}H] muscimol in the prefrontal cortex of patients with schizophrenia. Subsequent studies demonstrated significant increases in [^{3}H] muscimol binding to the $GABA_A$ receptor in subfields of the hippocampal formation, the anterior cingulate cortex and the prefrontal cortex (Benes et al. 1996; Benes et al. 1992). Consistent with the selective vulnerability of chandelier cells, Volk et al. (2002) reported a 100 % increase of immunoreactive alpha 2 subunit of the $GABA_A$ receptor on the pyramidal neuron initial axon segments. Impagnatiello et al. (1996) also found increased alpha 1 and alpha 5 subunits of the $GABA_A$ receptor in the prefrontal cortex.

What could account for this disparate neuropathology affecting a subpopulation of cortical GABAergic neurons and the dendrites of the principal output neurons, the glutamatergic pyramidal neurons? The pathology does not comport with the innervation pattern of the dopaminergic terminal fields in the cortex. Given the reciprocal synaptic relationship between the pyramidal neurons and the chandelier GABAergic neurons, a primary dysfunction of glutamatergic neurotransmission would be the plausible explanation (Lisman et al. 2008).

7.5 NMDA Receptor Antagonists

The potential involvement of impaired glutamatergic neurotransmission in the pathophysiology of schizophrenia was initially supported by the clinical observations of the psychotomimetic effects of dissociative anesthetics, first reported over 50 years ago (Luby et al. 1959; Itil et al. 1967). The demonstration that these drugs act as use-dependent noncompetitive inhibitors of NMDA receptors at the concentration associated with their psychotomimetic properties pointed to inhibition of NMDA receptors as a potential pathologic mechanism (Anis et al. 1983). However, as the psychiatric manifestations of dissociative anesthetics were generally based on anecdotal clinical observations of intoxicated patients presenting in emergency rooms, there were concerns that the "psychosis" could be due to co-occurring abuse of other substances such as stimulants, which were known to cause psychosis, or due to latent schizophrenia (Bowers and Swigar 1983).

The growing appreciation of the importance of negative symptoms and cognitive impairments as critical components of the endophenotype of schizophrenia prompted a more careful analysis of the psychiatric effects of ketamine, another NMDA receptor antagonist. When normal volunteers were studied under laboratory conditions (as opposed to substance abusers observed in emergency rooms) and administered low doses of intravenous ketamine that did not cause impairments in consciousness, they developed negative symptoms (blunted affect, withdrawal) and the rather selective impairments in memory and cognition that are specifically associated with schizophrenia (Krystal et al. 1994). Positive symptoms with auditory hallucinations and fully formed delusions do not typically occur after acute administration of ketamine to normal subjects but are seen in neuroleptic-free schizophrenics (Lahti et al. 2001). In addition, low-dose ketamine produces several physiologic abnormalities associated with schizophrenia including abnormal eye tracking, enhanced subcortical dopamine release (Kegeles et al. 2000), impaired prepulse inhibition in experimental animals (Braff et al. 2001), hypofrontality, and abnormal cortical event-related potentials (Umbricht et al. 2000).

In an "experiment of nature," individuals with NMDAR autoimmune encephalitis, which results from the development of antibodies against the extracellular epitopes of the NR1A subunit, initially present with psychiatric symptoms similar to schizophrenia (Hughes et al. 2010). Many are young women with occult teratomas that precipitate the autoimmune reaction; the syndrome is resolved by removing the teratoma. It is believed that these antibodies cause NMDAR internalization, resulting in NMDAR hypofunction. A recent study also detected NMDAR antibodies, particularly IgG-NR1A, that are different from the antibodies detected in NMDAR encephalitis, in acutely ill schizophrenic patients (Steiner et al. 2013).

7.6 D-Serine and Glycine

The NMDAR is a heterotetrameric, cation channel receptor primarily composed of two GluN1 channel subunits and two GluN2 subunits (for review, Glasgow et al. 2015). It is widely distributed throughout the brain and is the primary mediator of activity-dependent postsynaptic plasticity. What makes the NMDAR unique is that in addition to the binding of its agonist glutamate to the GluN2 subunit, the neuron must also be contemporaneously depolarized, which relieves the Mg2+ blockade of the channel, and either glycine or D-serine must be bound at the glycine modulatory site (GMS) on the GluN1 subunit to open the cation channel (McBain et al. 1989). Although the concentration of glycine in the cerebrospinal fluid is high, the GMS of the NMDAR is not saturated in vivo because the glycine transporter-1 (GlyT1), which is expressed primarily on astrocytes in the adult brain, maintains subsaturating concentrations of glycine in the synapse (Bergeron et al. 1998; Berger et al. 1998). The NMDAR ion channel is highly permeable to Ca^{2+}, which upon entry into the neuron, triggers a cascade of intracellular events that mediate local, acute functional synaptic plasticity but also changes in gene expression that influence long-term neural structural plasticity (Bading 2013).

Although the presence of small amounts of D-serine in mammalian tissues was long known (Nagata et al. 1989), Hashimoto et al. (1992) were the first to demonstrate the remarkably high free levels of D-serine in mammalian brain using gas chromatography-mass spectroscopy. In their comprehensive review, Hashimoto and Oka (1997) summarize the compelling findings that D-serine is concentrated in the forebrain across several mammalian species including humans. The synthetic enzyme for D-serine, SR, was purified to homogeneity by Wolosker et al. (1999a) and subsequently cloned and shown to be expressed in the forebrain of the adult rat (Wolosker et al. 1999b). Immunocytochemical staining for both SR and D-serine suggested that they were co-localized to astrocytes consistent with tissue culture results showing expression of SR in type 2 astrocytes (Schell et al. 1995; Wolosker et al. 1999a). Given that D-serine was threefold more potent agonist at the GMS on the NMDAR (Matsui et al. 1995), it would appear that D-serine is the preferred GMS agonist in the forebrain, whereas glycine would play this role in the midbrain, brainstem, and cerebellum where the glycine transporter, GlyT1, is highly expressed and D-serine levels are negligible (Schell et al. 1995). Recent neurophysiologic studies show that D-serine appears to be the primary co-agonist at CA1-Schaffer collateral and thalamic-lateral amygdala synapse, whereas glycine appears to act at extra-junctional NMDARs and to activate synaptic NMDARs when there is a high presynaptic activity (Papouin et al. 2012; Li et al. 2013).

As noted above, SR and D-serine were originally thought to be localized to astrocytes. However, Kartvelishvily et al. (2006), having developed more selective antibodies against SR, reported substantial SR immunostaining in cultured forebrain neurons and in forebrain tissue sections. Furthermore, they reported a neuronal localization of D-serine, which was released by stimulating ionotropic glutamate receptors but via an alanine-serine-cysteine-1 (Asc-1) transporter and not through a vesicular exocytotic mechanism (Rosenberg et al. 2013). More recent studies, which used SR−/− mice to optimize immunostaining by providing a true "control" devoid of SR and D-serine to establish immuno-specificity for both SR and D-serine, describe SR and D-serine as having a predominant neuronal localization in the rodent forebrain (Miya et al. 2008; Balu et al. 2014). Notably, under immunostaining conditions used in prior studies with regard to the blocking concentration of L-serine, astrocytes exhibited robust staining in SR−/− mice (Balu et al. 2014).

To resolve the confusion over the relative level of expression of SR and D-serine in neurons versus glia, Benneyworth et al. (2012) used mice with cell-specific expression of Cre-recombinase crossed with mice with the *srr*'s first coding exon flanked by lox-P sites to suppress SR expression in astrocytes (GFAP promoter) or glutamatergic neurons (CamKIIα promoter). They demonstrated that ~65 % of SR was localized to cortical glutamatergic neurons, whereas there was a quite modest to negligible expression of SR in forebrain astrocytes. Immunostaining for SR and conditional inactivation of SR with Cre-recombinase driven by a GAD65 promoter indicate that the remaining SR was expressed in forebrain GABAergic neurons, especially those co-expressing PV and somatostatin (Takagi et al. in preparation). Consistent with this localization of SR, D-serine immunostaining was localized to glutamatergic and GABAergic neurons in the murine cortex. Using conditions for optimal specific immunostaining established in mice, Balu et al. (2014) showed that SR was expressed exclusively in pyramidal neurons and GABAergic neurons but not astrocytes in the human cortex.

Serum/plasma glycine and serine levels have been investigated as biological markers for schizophrenia. Plasma total serine and glycine levels were found to be significantly higher in schizophrenic patients than in controls (Baruah et al. 1991). An inverse association between plasma glycine levels and negative symptoms in schizophrenic patients has also been reported (Neeman et al. 2005;

Sumiyoshi et al. 2005). Ketamine impairs prepulse inhibition (PPI) in rodents, but heightens it in humans. Thus, NMDA receptor blockade in experimental animals may be related to a prepulse inhibition deficit in schizophrenic patients (Braff et al. 2001). In this regard, Heresco-Levy et al. (2007) found an inverse relationship between plasma glycine levels and PPI in schizophrenic subjects. Decreased SR and increased D-amino acid oxidase (DAAO), which mediates the degradation of D-serine, were reported in brains of schizophrenic subjects (Bendikov et al. 2007), consistent with reduced D-serine availability in schizophrenic brains.

Although a postmortem study did not reveal a significant difference in the D-serine levels in the brains of individuals diagnosed with schizophrenia and controls, the means were lower for the schizophrenic subjects but the number of subjects was quite small and underpowered so as to preclude statistical analysis (Kumashiro et al. 1995). Subsequent studies revealed significantly lower levels of D-serine in serum (Hashimoto et al. 2003) and in cerebrospinal fluid (CSF) in drug-naive patients with schizophrenia as compared to suitable controls (Hashimoto et al. 2005). Consistent with this finding, Yamamori et al. (2014) found that the plasma D-serine to L-serine ratio was lower in subjects with schizophrenia and that treatment with clozapine, the antipsychotic with the greatest effects on negative symptoms, normalized the plasma D-serine to L-serine ratio and increased the plasma glycine levels and the glycine to L-serine ratio. In a genome-wide association study (GWAS) of human subjects from the general population, Luykx et al. (2015) found an association between DAAO gene and the D-serine plasma-CSF ratio and L-serine to D-serine ratio in CSF with an allelic variant of *SRR*.

7.7 Genetics and the NMDA Receptor

Over the last 50 years, twin and adoption studies have provided compelling evidence of the heritability of schizophrenia (Kety 1959; Kendler and Diehl 1993). The risk of schizophrenia in a first-degree relatives of an affected proband is 15-fold greater than in the general population similar the risk in fraternal twins. Concordance in identical twins is approximately fourfold higher than in fraternal twins. Nevertheless, environmental factors including fetal infections, dietary deficiencies during pregnancy, and perinatal insults account for approximately a third of the risk. The pattern of heritable risk indicates that schizophrenia is due to complex genetics wherein multiple risk alleles of modest effect interacting with environmental factors produce the phenotype (Kendler and Diehl 1993).

Most of the early studies of the molecular genetics of schizophrenia focused on putative risk genes identified by postmortem neurochemical findings or pathways related to the dominant hypothesis of pathophysiology based on D2-dopamine receptor blocking action of antipsychotic drugs. For example, using the case control method, several studies involving at the most hundreds of subjects implicated dopamine receptors D2, D3, and D4 as risk genes for schizophrenia while others contradicted these results (Crocq et al. 1992; Arinami et al. 1994; Weiss et al. 1996;

Sommer et al. 1993). The contradictory results emanating from these early studies stemmed from multiple problems: biased selection of genes, underpowered studies leading to type 1 errors, and publication bias where negative findings were either not submitted to journals or were denied publication (Farrell et al. 2015).

With the completion of the sequencing of the human genome in 2001, the ability to carry out molecular genetic studies to identify genes that confer risk for schizophrenia became feasible (Williams and Hayward 2001). On rare occasions highly penetrant mutations, often de novo, were associated with schizophrenia (Kirov et al. 2012). These mutations are the result of deletion or replication of kilobases of DNA including multiple genes and have been designated as "copy number variants" (CNV). The same CNV has been associated with autism and schizophrenia as indicated by studies of carriers of the 22q11.2 deletion (Vorstman et al. 2013). Kirov et al. (2012) sequenced the CNVs involving nearly eight thousand subjects with schizophrenia. They found that the CNVs were highly significantly enriched with gene-encoding proteins localized to the glutamatergic synapse and its postsynaptic density.

Geneticists interested in psychiatric disorders have advocated for sufficiently powered genome-wide association studies so that risk genes could achieve the high statistical threshold required: 5×10^{-8} (Farrell et al. 2015). The Schizophrenia Working Group of the Psychiatric Genomics Consortium carried out a genome-wide association study (GWAS) comprised of nearly 37,000 cases and over 110,000 controls and identified 108 loci that met genome-wide significance (Schizophrenic Working Group 2014). Associations were enriched with genes expressed primarily in the brain, especially those associated with NMDA receptor neurotransmission or its downstream mediators. In fact, nearly a dozen risk genes were within 2° of separation from the NMDA receptor including SR. Thus, the findings from the GWAS are consistent with the gene analysis of the CNVs associated with schizophrenia.

7.8 Serine Racemase and the Pathology of Schizophrenia

As reviewed in this volume (Balu 2015), genetically silencing in mice, the gene-encoding serine racemase (SR), a risk gene identified in the large GWAS (Schizophrenic Working Group 2014), results in a phenotype that closely replicates many aspects of schizophrenia including reduced dendritic length, reduced spine density, ventricular enlargement, cortical atrophy, downregulation of the cortical fast-firing parvalbumin-positive GABAergic interneurons, and cognitive impairments (Basu et al. 2009; DeVito et al. 2011; Balu et al. 2012). In mice with *SRR* constitutively silenced, sub-chronic treatment of adult SR−/− mice with doses of D-serine that normalize their cortico-hippocampal D-serine levels restores trophic factor markers, and transcriptional downstream to the NMDA receptor corrects cognitive deficits and partially restores spine density (Balu et al. 2013). The dendritic deficits are also observed in mice in which SR is conditionally silenced in glutamatergic neurons in the forebrain in late adolescence (Balu and Coyle 2012). Thus, NMDA receptor hypofunction in cortico-limbic brain regions caused by the lack of D-serine replicates those aspects of schizophrenia associated with negative symptoms and cognitive impairments (Puhl et al. 2014). Furthermore, these seemingly irreversible structural brain changes due to congenital NMDA receptor hypofunction may be corrected with restoration NMDA receptor function in adulthood (Balu and Coyle 2014).

SR stands at the apex of a family of risk genes for schizophrenia that are within two degrees of separation from the NMDA receptor. For example, the kinase, Akt3, is a major regulator in the class I PI3 kinase pathway and is activated in neurons by phosphorylation as a consequence of NMDA receptor activity. Silencing Akt3 expression in neurons adversely affects neuronal survival and axon extension (Diez et al. 2012). Notably, the activation of Akt, a risk gene for schizophrenia, is reduced in SR knockout mice, pointing to the potential of epistatic interactions of two risk genes (Balu et al. 2013).

7.9 Clinical Trials to Increase NMDA Receptor Function

Intervention studies to test the hypothesis that enhancement of NMDA receptor function would attenuate symptoms of schizophrenia were initiated quite early. Deutsch's group carried out an open-label study in which they examined the efficacy of glycine (10.8 g/day) in chronically psychotic patients as an adjunct to conventional antipsychotic treatment (Rosse et al. 1989) and had inconclusive results. In a subsequent open-label study, they examined the effects of milacemide, a prodrug for glycine, and found no significant therapeutic effects. Javitt and his colleagues initiated a series of placebo-controlled clinical trials with glycine in patients stabilized on antipsychotics (Javitt et al. 1994; Lederman et al. 1996; Heresco-Levy et al. 1996). Because of the poor penetration of glycine through the blood-brain barrier, they studied high daily doses ranging from 0.4 to 0.8 g/kg/day. The high-dose intervention increased plasma levels of glycine by sixfold and resulted in significant reductions in negative symptoms, the very symptoms resistant to antipsychotic treatment, and improved cognition.

At the same time, Goff et al. (1995a, b) examined the effects of D-cycloserine (DCS), drug used to treat tuberculosis, which is a partial agonist at the glycine modulatory site and crosses the blood-brain barrier better than glycine. Preclinical studies indicated that DCS exhibited cognitive enhancing effects (Schuster and Schmidt 1992). Clinically, DCS exhibited a U-shaped dose-response curve with the optimal dose, 50 mg/day, significantly reducing negative symptoms and improving performance on a cognitive task in patients with chronic schizophrenia stabilized on antipsychotics. A subsequent placebo-controlled, parallel clinical trial with DCS replicated the finding of a significant reduction in negative symptoms but not positive symptoms (Goff et al. 1999). In contrast, the addition of DCS to clozapine resulted in a significant exacerbation of negative symptoms (Goff et al. 1996), a counterintuitive finding later interpreted as the result of the antagonistic effect of the partial agonist at a fully occupied glycine modulatory site caused by clozapine (Goff and Coyle 2001). A functional imaging study in schizophrenic patients receiving antipsychotics demonstrated that the addition of DCS significantly enhanced performance on a memory task and increased activation of the left temporal superior gyrus required for performance and that this effect correlated inversely with negative symptom change (Yurgelun-Todd et al. to, 2005). A troubling aspect of the therapeutic effects DCS is that its efficacy decreased with increasing duration of treatment, resulting in variable effects across studies (Goff et al. 2005).

A multicenter, placebo-controlled trial of add-on of DCS and glycine to a stable antipsychotic regimen in patients with schizophrenia reported that neither were more effective than placebo (Buchanan et al. 2007). However, there was a highly significant difference in site-by-site response, indicating a "failed" study. Post hoc analysis by site revealed significant effects on negative symptoms by DCS and glycine.

Double-blind, placebo-controlled trials of D-serine on schizophrenic patients receiving stable doses of antipsychotic treatment have similarly shown effects on negative symptoms and cognition (Tsai et al. 1998; Lane et al. 2005). Notably, D-serine treatment of schizophrenic patients receiving clozapine did not affect negative symptoms or cognition, consistent with the hypothesis that clozapine's effects on negative symptoms are mediated by full occupancy of the NMDAR GMS (Tsai et al. 1999). In a large multicenter, placebo-controlled trial of D-serine add-on treatment to antipsychotics in patients with schizophrenia or schizoaffective disorder, D-serine did not separate from placebo on Scale the for Assessment of Negative Symptoms (SANS) and Measurement and Treatment Research to Improve Cognition in Schizophrenia (MATRICS); however, placebo response was unusually robust (Weiser et al. 2012). A blinded comparison of high-dose D-serine as monotherapy to olanzapine revealed that D-serine was significantly less effective than olanzapine on Positive and Negative Syndrome Scale but nevertheless significantly improved negative symptoms without affecting psychotic symptoms (Ermilov et al. 2013).

Between 1995 and 2010, 26 double-blind, placebo-controlled trials had been carried out with agents that directly or indirectly (sarcosine) act at the NMDAR GMS in patients with chronic schizophrenia, who were receiving antipsychotic medications. These include glycine, D-serine, alanine, and sarcosine. The effects on negative and depressive symptoms are substantial (effect size = 0.4) and highly significant, whereas the effects on cognition, positive symptoms, and general psychopathology are more modest but still highly significant (Tsai and Lin 2010). Because of problems with crossing the blood-brain barrier, metabolism, and intrinsic potency, these natural products are unlikely to be useful treatments, but they have provided proof of principle that enhancing NMDA receptor function can significantly reduce symptoms in schizophrenia, especially those least responsive to existing antipsychotic medications (Coyle et al. 2002).

7.10 Conclusion

Twenty years after the first proposals that hypofunction of NMDARs might be a core pathophysiologic feature of schizophrenia, especially with regard to the negative symptoms and the cognitive impairments (Krystal et al. 1994; Tsai et al. 1995), a sufficiently powered GWAS has revealed that several gene-encoding proteins involved with glutamatergic neurotransmission and neuroplasticity are risk genes for schizophrenia (Schizophrenic Working Group 2014). Notably, SR, located presynaptically in forebrain glutamatergic neurons and the co-agonist at forebrain NMDARs, sits at the apex of a pathway of risk genes critical to synaptogenesis and synaptic plasticity (Balu and Coyle 2015). Genetically silencing *srr* in mice reproduces the cortical atrophy, reduced glutamatergic synaptic connectivity, and the cognitive deficit characteristic of schizophrenia (Balu et al. 2012). Remarkably, this "schizophrenic" neuropathology in the SR−/− mice can be reversed by treating them with D-serine in adulthood (Balu et al. 2013). This finding raises the possibility that pharmacologically enhancing NMDAR function in adult individuals with schizophrenia could restore synaptic plasticity, thereby attenuating or even eliminating the cognitive impairments and negative symptoms currently responsible for the persistent disability.

References

Anis NA, Berry SC, Burton NR, Lodge D (1983) The dissociative anaesthetics, ketamine and phencyclidine, selectively reduce excitation of central mammalian neurones by N-methyl-aspartate. Br J Pharmacol 79:565–575

Arinami T, Itokawa M, Enguchi H, Tagaya H, Yano S, Shimizu H, Hamaguchi H, Toru M (1994) Association of dopamine D2 receptor molecular variant with schizophrenia. Lancet 343:703–704

Bading H (2013) Nuclear calcium signalling in the regulation of brain function. Nat Rev Neurosci 14(9):593–608

Balu DT, Coyle JT (2012) Neuronal D-serine regulates dendritic architecture in the somatosensory cortex. Neurosci Lett 517(2):77–81

Balu DT, Coyle JT (2014) Chronic D-serine reverses arc expression and partially rescues dendritic abnormalities in a mouse model of NMDA receptor hypofunction. Neurochem Int 75:76–78

Balu DT, Coyle JT (2015) The NMDA receptor 'glycine modulatory site' in schizophrenia: D-serine, glycine, and beyond. Curr Opin Pharmacol 20:109–115

Balu DT, Basu AC, Corradi JP, Cacace AM, Coyle JT (2012) The NMDA receptor co-agonists, D-serine and glycine, regulate neuronal dendritic architecture in the somatosensory cortex. Neurobiol Dis 45(2):671–682

Balu DT, Li Y, Puhl MD, Benneyworth MA, Basu AC, Takagi S, Bolshakov VY, Coyle JT (2013) Multiple risk pathways for schizophrenia converge in serine racemase knockout mice, a mouse model of NMDA receptor hypofunction. Proc Natl Acad Sci U S A 110(26):E2400–E2409

Balu DT, Takagi S, Puhl MD, Benneyworth MA, Coyle JT (2014) D-serine and serine racemase are localized to neurons in the adult mouse and human forebrain. Cell Mol Neurobiol 34 (3):419–435

Bang M, Kim KR, Song YY, Baek S, Lee E, An SK (2015) Neurocognitive impairments in individuals at ultra-high risk for psychosis: who will really convert? Aust N Z J Psychiatry 49:462–470

Baruah S, Waziri R, Hegwood TS, Mallis LM (1991) Plasma serine in schizophrenics and controls measured by gas chromatography-mass spectrometry. Psychiatry Res 37(3):261–270

Basu AC, Tsai GE, Ma CL, Ehmsen JT, Mustafa AK, Han L, Jiang ZI, Benneyworth MA, Froimowitz MP, Lange N, Snyder SH, Bergeron R, Coyle JT (2009) Targeted disruption of serine racemase affects glutamatergic neurotransmission and behavior. Mol Psychiatry 14 (7):719–727

Bendikov I, Nadri C, Amar S, Panizzutti R, De Miranda J, Wolosker H, Agam G (2007) A CSF and postmortem brain study of D-serine metabolic parameters in schizophrenia. Schizophr Res 90 (1–3):41–51

Benes FM, Vincent SL, Alsterberg G, Bird ED, SanGiovanni JP (1992) Increased GABAA receptor binding in superficial layers of cingulate cortex in schizophrenics. J Neurosci 12 (3):924–929

Benes FM, Vincent SL, Marie A, Khan Y (1996) Up-regulation of GABAA receptor binding on neurons of the prefrontal cortex in schizophrenic subjects. Neuroscience 75(4):1021–1031

Benneyworth MA, Li Y, Basu AC, Bolshakov VY, Coyle JT (2012) Cell selective conditional null mutations of serine racemase demonstrate a predominate localization in cortical glutamatergic neurons. Cell Mol Neurobiol 32(4):613–624

Berger AJ, Dieudonné S, Ascher P (1998) Glycine uptake governs glycine site occupancy at NMDA receptors of excitatory synapses. J Neurophysiol 80(6):3336–3340

Bergeron R, Meyer TM, Coyle JT, Greene RW (1998) Modulation of N-methyl-D-aspartate receptor function by glycine transport. Proc Natl Acad Sci U S A 95(26):15730–15734

Bird ED, Spokes EG, Barnes J, MacKay AV, Iversen LL, Shepherd M (1977) Increased brain dopamine and reduced glutamic acid decarboxylase and choline acetyl transferase activity in schizophrenia and related psychoses. Lancet 2(8049):1157–1158

Bird ED, Spokes EG, Barnes J, Mackay AV, Iversen LL, Shepherd M (1978) Glutamic-acid decarboxylase in schizophrenia. Lancet 1(8056):156

Bowers MB Jr, Swigar ME (1983) Vulnerability to psychosis associated with hallucinogen use. Psychiatry Res 9(2):91–97

Braff DL, Geyer MA, Swerdlow NR (2001) Human studies of prepulse inhibition of startle: normal subjects, patient groups, and pharmacological studies. Psychopharmacology (Berl) 156:234–258

Buchanan RW, Javitt DC, Marder SR, Schooler NR, Gold JM, McMahon RP, Heresco-Levy U, Carpenter WT (2007) The Cognitive and Negative Symptoms in Schizophrenia Trial (CONSIST): the efficacy of glutamatergic agents for negative symptoms and cognitive impairments. Am J Psychiatry 164:1593–1602

Chang X, Xi YB, Cui LB, Wang HN, Sun JB, Zhu YQ, Huang P, Collin G, Liu K, Xi M, Qi S, Tan QR, Miao DM, Yin H (2015) Distinct inter-hemispheric dysconnectivity in schizophrenia patients with and without auditory verbal hallucinations. Sci Rep 5:11218

Coyle JT, Tsai G, Goff DC (2002) Ionotropic glutamate receptors as therapeutic targets in schizophrenia. Curr Drug Targets CNS Neurol Disord 1(2):183–189

Coyle JT, Balu D, Benneyworth MA, Basu AC, Roseman A (2010) Beyond the dopamine receptor: novel therapeutic targets for treating schizophrenia. Dialogues Clin Neurosci 12:233–270

Crocq MA, Mant R, Asherson P et al (1992) Association between schizophrenia and homozygosity at the dopamine D3 receptor gene. J Med Genet 29:858–860

De Peri L, Crescini A, Deste G, Fusar-Poli P, Sacchetti E, Vita A (2012) Brain structural abnormalities at the onset of schizophrenia and bipolar disorder: a meta-analysis of controlled magnetic resonance imaging studies. Curr Pharm Des 18:486–494

DeLisi LE, Szulc KU, Bertisch HC, Majcher M, Brown K (2006) Understanding structural brain changes in schizophrenia. Dialogues Clin Neurosci 8:71–78

DeVito LM, Balu DT, Kanter BR, Lykken C, Basu AC, Coyle JT, Eichenbaum H (2011) Serine racemase deletion disrupts memory for order and alters cortical dendritic morphology. Genes Brain Behav 10:210–222
Diez H, Garrido JJ, Wandosell F (2012) Specific roles of Akt iso forms in apoptosis and axon growth regulation in neurons. PLoS One 7:e32715
Ermilov M, Gelfin E, Levin R, Lichtenberg P, Hashimoto K, Javitt DC, Heresco-Levy U (2013) A pilot double-blind comparison of d-serine and high-dose olanzapine in treatment-resistant patients with schizophrenia. Schizophr Res 150:604–605
Farrell MS, Werge T, Sklar P, Owen MJ, Ophoff RA, O'Donovan MC, Corvin A, Cichon S, Sullivan PF (2015) Evaluating historical candidate genes for schizophrenia. Mol Psychiatry 20:555–562
Fung SJ, Webster MJ, Sivagnanasundaram S, Duncan C, Elashoff M, Weickert CS (2010) Expression of interneuron markers in the dorsolateral prefrontal cortex of the developing human and in schizophrenia. Am J Psychiatry 167:1479–1488
Glantz LA, Lewis DA (2000) Decreased dendritic spine density on prefrontal cortical pyramidal neurons in schizophrenia. Arch Gen Psychiatry 57:65–73
Glasgow NG, Siegler Retchless B, Johnson JW (2015) Molecular bases of NMDA receptor subtype-dependent properties. J Physiol 593:83–95
Goff DC, Coyle JT (2001) The emerging role of glutamate in the pathophysiology and treatment of schizophrenia. Am J Psychiatry 158:1367–1377
Goff DC, Tsai G, Beal MF, Coyle JT (1995a) Tardive dyskinesia and substrates of energy metabolism in CSF. Am J Psychiatry 152:1730–1736
Goff DC, Tsai G, Manoach DS, Coyle JT (1995b) Dose-finding trial of D-cycloserine added to neuroleptics for negative symptoms in schizophrenia. Am J Psychiatry 152:1213–1215
Goff DC, Tsai G, Levitt J, Amico E, Manoach D, Schoenfeld DA, Hayden DL, McCarley R, Coyle JT (1999) A placebo-controlled trial of D-cycloserine added to conventional neuroleptics in patients with schizophrenia. Arch Gen Psychiatry 56:21–27
Goff DC, Heckers S, Freudenreich O (2001) Schizophrenia. Med Clin North Am 85:663–689
Goff DC, Herz L, Posever T, Shih V, Tsai G, Henderson DC, Freudenreich O, Evins AE, Yovel I, Zhang H, Schoenfeld D (2005) A six-month, placebo-controlled trial of D-cycloserine co-administered with conventional antipsychotics in schizophrenia patients. Psychopharmacology (Berl) 179:144–150
Hanada S, Mita T, Nishino N, Tanaka C (1987) [3H]muscimol binding sites increased in autopsied brains of chronic schizophrenics. Life Sci 40:259–266
Hashimoto A, Oka T (1997) Free D-aspartate and D-serine in the mammalian brain and periphery. Prog Neurobiol 52:325–353
Hashimoto A, Nishikawa T, Hayashi T, Fujii N, Harada K, Oka T, Takahashi K (1992) The presence of free D-serine in rat brain. FEBS Lett 296:33–36
Hashimoto K, Fukushima T, Shimizu E, Komatsu N, Watanabe H, Shinoda N, Nakazato M, Kumakiri C, Okada S, Hasegawa H, Imai K, Iyo M (2003) Decreased serum levels of D-serine in patients with schizophrenia: evidence in support of the N-methyl-D-aspartate receptor hypofunction hypothesis of schizophrenia. Arch Gen Psychiatry 60:572–576
Hashimoto K, Engberg G, Shimizu E, Nordin C, Lindström LH, Iyo M (2005) Reduced D-serine to total serine ratio in the cerebrospinal fluid of drug naive schizophrenic patients. Prog Neuropsychopharmacol Biol Psychiatry 29:767–769
Haug JO (1982) Pneumoencephalographic evidence of brain atrophy in acute and chronic schizophrenic patients. Acta Psychiatr Scand 66:374–378
Heresco-Levy U, Silipo G, Javitt DC (1996) Glycinergic augmentation of NMDA receptor-mediated neurotransmission in the treatment of schizophrenia. Psychopharmacol Bull 32:731–740
Heresco-Levy U, Bar G, Levin R, Ermilov M, Ebstein RP, Javitt DC (2007) High glycine levels are associated with prepulse inhibition deficits in chronic schizophrenia patients. Schizophr Res 91:14–21

Ho BC, Andreasen NC et al (2011) Long-term antipsychotic treatment and brain volumes: a longitudinal study of first-episode schizophrenia. Arch Gen Psychiatry 68:128–137

Hoftman GD, Volk DW, Bazmi HH, Li S, Sampson AR, Lewis DA (2015) Altered cortical expression of GABA-related genes in schizophrenia: illness progression vs developmental disturbance. Schizophr Bull 41:180–191

Horton LE, Tarbox SI, Olino TM, Haas GL (2015) Trajectories of premorbid childhood and adolescent functioning in schizophrenia-spectrum psychoses: a first-episode study. Psychiatry Res 227:339–346

Hughes EG, Peng X, Gleichman AJ, Lai M, Zhou L, Tsou R, Parsons TD, Lynch DR, Dalmau J, Balice-Gordon RJ (2010) Cellular and synaptic mechanisms of anti-NMDA receptor encephalitis. J Neurosci 30:5866–5875

Impagnatiello F, Pesold C, Longone P, Caruncho H, Fritschy JM, Costa E, Guidotti A (1996) Modifications of gamma-aminobutyric acidA receptor subunit expression in rat neocortex during tolerance to diazepam. Mol Pharmacol 49:822–831

Itil T, Keskiner A, Kiremitci N, Holden JM (1967) Effect of phencyclidine in chronic schizophrenics. Can Psychiatr Assoc J 12:209–212

Jacoby A, Winkler H (1927) Encephalographischen studien an schizophrenen. Arch Psycho 84:208–226

Javitt DC, Zylberman I, Zukin SR, Heresco-Levy U, Lindenmayer JP (1994) Amelioration of negative symptoms in schizophrenia by glycine. Am J Psychiatry 151:1234–1236

Kartvelishvily E, Shleper M, Balan L, Dumin E, Wolosker H (2006) Neuron-derived D-serine release provides a novel means to activate N-methyl-D-aspartate receptors. J Biol Chem 281:14151–14162

Keefe RS, Bilder RM, Harvey PD et al (2006) Baseline neurocognitive deficits in the CATIE schizophrenia trial. Neuropsychopharmacology 31:2033–2046

Kegeles LS, Abi-Dargham A, Zea-Ponce Y, Rodenhiser-Hill J, Mann JJ, Van Heertum RL, Cooper TB, Carlsson A, Laruelle M (2000) Modulation of amphetamine-induced striatal dopamine release by ketamine in humans: implications for schizophrenia. Biol Psychiatry 48:627–640

Kendler KS, Diehl SR (1993) The genetics of schizophrenia: a current, genetic-epidemiologic perspective. Schizophr Bull 19:261–285

Kety SS (1959) Biochemical theories of schizophrenia. II. Science 129:1590–1596

Kirov G, Pocklington AJ, Holmans P et al (2012) De novo CNV analysis implicates specific abnormalities of postsynaptic signalling complexes in the pathogenesis of schizophrenia. Mol Psychiatry 17:142–153

Konopaske GT, Lange N, Coyle JT, Benes FM (2014) Prefrontal cortical dendritic spine pathology in schizophrenia and bipolar disorder. JAMA Psychiatry 71:1323–1331

Krystal JH, Karper LP, Seibyl JP, Freeman GK, Delaney R, Bremner JD et al (1994) Subanesthetic effects of the noncompetitive NMDA antagonist, ketamine, in humans. Psychotomimetic, perceptual, cognitive, and neuroendocrine responses. Arch Gen Psychiatry 51:199–214

Kumashiro S, Hashimoto A, Nishikawa T (1995) Free D-serine in post-mortem brains and spinal cords of individuals with and without neuropsychiatric diseases. Brain Res 681:117–125

Kuperberg GR, Broome MR, McGuire PK et al (2003) Regionally localized thinning of the cerebral cortex in schizophrenia. Arch Gen Psychiatry 60:878–888

Lahti AC, Weiler MA, TamaraMichaelidis BA, Parwani A, Tamminga CA (2001) Effects of ketamine in normal and schizophrenic volunteers. Neuropsychopharmacology 25:455–467

Lane HY, Chang YC, Liu YC, Chiu CC, Tsai GE (2005) Sarcosine or D-serine add-on treatment for acute exacerbation of schizophrenia: a randomized, double-blind, placebo-controlled study. Arch Gen Psychiatry 62:1196–1204

Leiderman E, Zylberman I, Zukin SR, Cooper TB, Javitt DC (1996) Preliminary investigation of high-dose oral glycine on serum levels and negative symptoms in schizophrenia: an open-label trial. Biol Psychiatry 39:213–215

Lewis DA (2014) Inhibitory neurons in human cortical circuits: substrate for cognitive dysfunction in schizophrenia. Curr Opin Neurobiol 26:22–26

Li Y, Sacchi S, Pollegioni L, Basu AC, Coyle JT, Bolshakov VY (2013) Identity of endogenous NMDAR glycine site agonist in amygdala is determined by synaptic activity level. Nat Commun 4:1760

Lisman JE, Coyle JT, Green RW, Javitt DC, Benes FM, Heckers S, Grace AA (2008) Circuit-based framework for understanding neurotransmitter and risk gene interactions in schizophrenia. Trends Neurosci 31:234–242

Luby ED, Cohen BD, Rosenbaum G, Gottlieb JS, Kelley R (1959) Study of a new schizophrenomimetic drug; sernyl. AMA Arch Neurol Psychiatry 81:363–369

Luykx JJ, Bakker SC, Visser WF et al (2015) Genome-wide association study of NMDA receptor coagonists in human cerebrospinal fluid and plasma. Mol Psychiatry 20:1557–1564.

Mancuso JJ, Chen Y, Li X, Xue Z, Wong ST (2013) Methods of dendritic spine detection: from Golgi to high-resolution optical imaging. Neuroscience 251:129–140

Matsui T, Sekiguchi M, Hashimoto A, Tomita U, Nishikawa T, Wada K (1995) Functional comparison of D-serine and glycine in rodents: the effect on cloned NMDA receptors and the extracellular concentration. J Neurochem 65:454–458

McBain CJ, Kleckner NW, Wyrick S, Dingledine R (1989) Structural requirements for activation of the glycine coagonist site of N-methyl-D-aspartate receptors expressed in Xenopus oocytes. Mol Pharmacol 36:556–565

Miya K, Inoue R, Takata Y, Abe M, Natsume R, Sakimura K, Hongou K, Miyawaki T, Mori H (2008) Serine racemase is predominantly localized in neurons in mouse brain. J Comp Neurol 510:641–654

Nagata Y, Akino T, Ohno K (1989) The presence of free D-amino acids in mouse tissues. Experientia 45:330–332

Neeman G, Blanaru M, Bloch B, Kremer I, Ermilov M, Javitt DC, Heresco-Levy U (2005) Relation of plasma glycine, serine, and homocysteine levels to schizophrenia symptoms and medication type. Am J Psychiatry 162:1738–1740

Nelson B, Yuen HP, Wood SJ, Lin A et al (2013) Long-term follow-up of a group at ultra high risk ("prodromal") for psychosis: the PACE 400 study. JAMA Psychiatry 70:793–802

Papouin T, Ladépêche L, Ruel J, Sacchi S, Labasque M, Hanini M, Groc L, Pollegioni L, Mothet JP, Oliet SH (2012) Synaptic and extrasynaptic NMDA receptors are gated by different endogenous coagonists. Cell 15:633–646

Puhl MD, Mintzopoulos D, Jensen JE, Gillis TE, Konopaske GT, Kaufman MJ, Coyle JT (2014) In vivo magnetic resonance studies reveal neuroanatomical and neurochemical abnormalities in the serine racemase knockout mouse model of schizophrenia. Neurobiol Dis 73C:269–274

Rosenberg D, Artoul S, Segal AC, Kolodney G, Radzishevsky I, Dikopoltsev E, Foltyn VN, Inoue R, Mori H, Billard JM, Wolosker H (2013) Neuronal D-serine and glycine release via the Asc-1 transporter regulates NMDA receptor-dependent synaptic activity. J Neurosci 33:3533–3544

Rosse RB, Theut SK, Banay-Schwartz M, Leighton M, Scarcella E, Cohen CG, Deutsch SI (1989) Glycine adjuvant therapy to conventional neuroleptic treatment in schizophrenia: an open-label, pilot study. Clin Neuropharmacol 12:416–424

Schell MJ, Molliver ME, Snyder SH (1995) D-serine, an endogenous synaptic modulator: localization to astrocytes and glutamate-stimulated release. Proc Natl Acad Sci U S A 92:3948–3952

Schiffman J, Walker E, Ekstrom M, Schulsinger F et al (2004) Childhood videotaped social and neuromotor precursors of schizophrenia: a prospective investigation. Am J Psychiatry 161:2021–2027

Schizophrenia Working Group of the Psychiatric Genomics Consortium (2014) Biological insights from 108 schizophrenia-associated genetic loci. Nature 511:421–427

Schuster GM, Schmidt WJ (1992) D-cycloserine reverses the working memory impairment of hippocampal-lesioned rats in a spatial learning task. Eur J Pharmacol 224:97–98

Selemon LD, Goldman-Rakic PS (1999) The reduced neuropil hypothesis: a circuit based model of schizophrenia. Biol Psychiatry 45:17–25

Selemon LD, Rajkowska G, Goldman-Rakic PS (1995) Abnormally high neuronal density in the schizophrenic cortex. A morphometric analysis of prefrontal area 9 and occipital area 17. Arch Gen Psychiatry 52:805–818

Selemon LD, Rajkowska G, Goldman-Rakic PS (1998) Elevated neuronal density in prefrontal area 46 in brains from schizophrenic patients: application of a three-dimensional, stereologic counting method. J Comp Neurol 392:402–412

Sheng M, Hoogenraad CC (2007) The postsynaptic architecture of excitatory synapses: a more quantitative view. Annu Rev Biochem 76:823–847

Sommer SS, Lind TJ, Heston LL, Sobell JL (1993) Dopamine D4 receptor variants in unrelated schizophrenic cases and controls. Am J Med Genet 48:90–93

Steiner J, Walter M, Glanz W et al (2013) Increased prevalence of diverse N-methyl-D-aspartate glutamate receptor antibodies in patients with an initial diagnosis of schizophrenia: specific relevance of IgG NR1a antibodies for distinction from N-methyl-D-aspartate glutamate receptor encephalitis. JAMA Psychiatry 70:271–278

Sumiyoshi T, Jin D, Jayathilake K, Lee M, Meltzer HY (2005) Prediction of the ability of clozapine to treat negative symptoms from plasma glycine and serine levels in schizophrenia. Int J Neuropsychopharmacol 8:451–455

Sweet RA, Henteleff RA, Zhang W, Sampson AR, Lewis DA (2009) Reduced dendritic spine density in auditory cortex of subjects with schizophrenia. Neuropsychopharmacology 34:374–389

Tsai GE, Lin PY (2010) Strategies to enhance N-methyl-D-aspartate receptor-mediated neurotransmission in schizophrenia, a critical review and meta-analysis. Curr Pharm Des 16:522–537

Tsai G, Passani LA, Slusher BS, Carter R, Baer L, Kleinman JE, Coyle JT (1995) Abnormal excitatory neurotransmitter metabolism in schizophrenic brains. Arch Gen Psychiatry 52:829–836

Tsai G, Yang P, Chung LC, Lange N, Coyle JT (1998) D-serine added to antipsychotics for the treatment of schizophrenia. Biol Psychiatry 44:1081–1089

Tsai GE, Yang P, Chung LC, Tsai IC, Tsai CW, Coyle JT (1999) D-serine added to clozapine for the treatment of schizophrenia. Am J Psychiatry 156:1822–1825

Tu PC, Lee YC, Chen YS, Hsu JW, Li CT, Su TP (2015) Network-specific cortico-thalamic disconnection in schizophrenia revealed by intrinsic functional connectivity analyses. Schizophr Res 166:137–143

Tune LE, McHugh PR, Coyle JT (1982) Drug management in chronic schizophrenia. Johns Hopkins Med J 150:45–48

Umbricht D, Schmid L, Koller R, Vollenweider FX, Hell D, Javitt DC (2000) Ketamine-induced deficits in auditory and visual context-dependent processing in healthy volunteers: implications for models of cognitive deficits in schizophrenia. Arch Gen Psychiatry 57:1139–1147

Volk DW, Pierri JN, Fritschy JM, Auh S, Sampson AR, Lewis DA (2002) Reciprocal alterations in pre- and postsynaptic inhibitory markers at chandelier cell inputs to pyramidal neurons in schizophrenia. Cereb Cortex 12:1063–1070

Vorstman JA, Breetvelt EJ, Thode KI, Chow EW, Bassett AS (2013) Expression of autism spectrum and schizophrenia in patients with a 22q11.2 deletion. Schizophr Res 143:55–59

Weiser M, Heresco-Levy U, Davidson M, Javitt DC, Werbeloff N, Gershon AA, Abramovich Y, Amital D, Doron A, Konas S, Levkovitz Y, Liba D, Teitelbaum A, Mashiach M, Zimmerman Y (2012) A multicenter, add-on randomized controlled trial of low-dose d-serine for negative and cognitive symptoms of schizophrenia. J Clin Psychiatry 73:e728–e734

Weiss J, Mägert HJ, Cieslak A, Forssmann WG (1996) Association between different psychotic disorders and the DRD4 polymorphism, but no differences in the main ligand binding region of the DRD4 receptor protein compared to controls. Eur J Med Res 1(9):439–445

Williams SJ, Hayward NK (2001) The impact of the Human Genome Project on medical genetics. Trends Mol Med 7:229–231

Wolosker H, Blackshaw S, Snyder SH (1999a) Serine racemase: a glial enzyme synthesizing D-serine to regulate glutamate-N-methyl-D-aspartate neurotransmission. Proc Natl Acad Sci U S A 96:13409–13414

Wolosker H, Sheth KN, Takahashi M, Mothet JP, Brady RO Jr, Ferris CD, Snyder SH (1999b) Purification of serine racemase: biosynthesis of the neuromodulator D-serine. Proc Natl Acad Sci U S A 96:721–725

Yamamori H, Hashimoto R, Fujita Y, Numata S, Yasuda Y, Fujimoto M, Ohi K, Umeda-Yano S, Ito A, Ohmori T, Hashimoto K, Takeda M (2014) Changes in plasma D-serine, L-serine, and glycine levels in treatment-resistant schizophrenia before and after clozapine treatment. Neurosci Lett 582:93–98

Yurgelun-Todd DA, Coyle JT, Gruber SA, Renshaw PF, Silveri MM, Amico E, Cohen B, Goff DC (2005) Functional magnetic resonance imaging studies of schizophrenic patients during word production: effects of D-cycloserine. Psychiatry Res 138:23–31

Chapter 8
Serine Racemase Knockout Mice: Neurotoxicity, Epilepsy, and Schizophrenia

Ran Inoue and Hisashi Mori

Abstract The mammalian brain contains high levels of D-serine, which acts as a coagonist at the glycine site of the *N*-methyl D-aspartate (NMDA)-type glutamate receptor (NMDAR). The synthesis of D-serine from L-serine is catalyzed by serine racemase (SR). To date, several SR knockout (KO) mouse strains have been established to elucidate the role of the SR-D-serine pathway in the regulation of NMDAR activity under both physiological and pathological conditions. Here, we will review the phenotypes of these SR-KO mice used as animal models of NMDAR-mediated neurotoxicity, epilepsy, and schizophrenia and discuss the mechanistic involvement of the SR-D-serine pathway in these neurological and psychiatric disorders.

Keywords Serine racemase knockout • D-Serine • Neurotoxicity • Epilepsy • Schizophrenia

8.1 Introduction

D-Amino acids were initially identified in bacteria and many invertebrate species and found to be used for cellular functions. It has long been considered that higher organisms were confined to the use of L-amino acids until the discovery of significant amounts of D-serine in the rodent brain (Hashimoto et al. 1992). The distribution pattern of D-serine is similar to that of *N*-methyl-D-aspartate (NMDA)-type glutamate receptor (NMDAR) (Schell et al. 1997). The NMDAR plays key roles in excitatory synaptic transmission, plasticity, learning and memory, neural network formation during development, and various neurological and pathological disorders (Bliss and Collingridge 1993; Komuro and Rakic 1993; Lancelot and Beal 1998; Coyle et al. 2003; Wang and Zhang 2005). D-Serine was found to be 100 times more potent than glycine in enhancing the NMDAR component of miniature synaptic currents recorded in rat hypoglossal motoneurons (Berger et al. 1998), and selective

R. Inoue (✉) • H. Mori
Department of Molecular Neuroscience, Graduate School of Medicine and Pharmaceutical Sciences, University of Toyama, Toyama, Japan
e-mail: eiran@med.u-toyama.ac.jp

T. Yoshimura et al. (eds.), *D-Amino Acids*, DOI 10.1007/978-4-431-56077-7_8

destruction of endogenous D-serine by D-amino acid oxidase (DAAO), which depletes endogenous D-serine but not glycine, greatly reduces NMDAR activation (Mothet et al. 2000). Therefore, D-serine is widely recognized as a major endogenous ligand for the glycine site of NMDARs.

The origin of brain D-serine had merely been speculated before serine racemase (SR) was discovered in the mammalian brain (Wolosker et al. 1999a, b). Using a conventional biochemical purification method, Wolosker et al. first isolated SR and cloned SR cDNA from the rat brain. SR not only catalyzes the racemization of L-serine to synthesize D-serine but also catalyzes the α,β-elimination of water from L- or D-serine to form pyruvate and NH_4 (Foltyn et al. 2005).

As mentioned above, NMDAR dysfunctions are involved in various neurological and psychiatric disorders, including disorders resulting from acute excitotoxic insults (e.g., ischemic stroke and traumatic brain injury), diseases due to chronic neurodegeneration [e.g., Alzheimer's disease (AD), Parkinson's disease (PD), Huntington's disease (HD), and amyotrophic lateral sclerosis (ALS)], disorders arising from sensitization of neurons (e.g., epilepsy and neuropathic pain), and neurodevelopmental disorders associated with NMDAR hypofunction (e.g., schizophrenia) (Lau and Tymianski 2010; Ghasemi and Schachter 2011; Lakhan et al. 2013). Many in vitro and in vivo studies have focused on the pharmacological inhibition of the NMDAR as a therapeutic strategy for some of these diseases, and a number of blockers of the NMDAR were proven to have beneficial effects. However, undesirable side effects due to the inhibition of a wide range of NMDAR function have limited their clinical applications (Chen and Lipton 2006). Therefore, alternative methods of modulating NMDAR function need to be determined and developed. In contrast to the wide distribution of the NMDAR, SR is localized mainly in the forebrain region (Miya et al. 2008). Recently, D-serine and serine racemase have emerged as new potential targets for the regulation of NMDAR activity in forebrain regions under pathological conditions (Labrie and Roder 2010).

To fully elucidate the role of the SR-D-serine pathway in the regulation of NMDAR activity under physiological conditions, and the mechanisms underlying neurological and psychiatric disorders associated with NMDAR hyper- and hypofunctions, several SR knockout (KO) mouse strains have been established. Here, we will review the phenotypes of these SR-KO mice used as animal models of NMDAR-mediated neurotoxicity, epilepsy, and schizophrenia, and discuss the mechanistic involvement of the SR-D-serine pathway in these neurological and psychiatric disorders, and conclusively recommend SR as a target for developing new therapeutic strategies.

8.2 Modulation of NMDAR by D-Serine

The NMDAR is a heteromeric protein complex composed of at least one GluN1 subunit in combination with GluN2 and/or GluN3 subunits (Cull-Candy et al. 2001). Unlike other neurotransmitter receptors, the activation of the NMDAR requires, besides the binding of glutamate to the GluN2 subunit, the binding of glycine or D-serine to the glycine site on the GluN1 subunit (Dingledine et al. 1999). The mammalian forebrain contains high levels of D-serine. D-Serine is efficacious in potentiating the activity of the NMDAR (Fadda et al. 1988; Matsui et al. 1995), and its deletion by treatment with DAAO was demonstrated to markedly decrease NMDAR activity (Mothet et al. 2000).

NMDARs are located in neuronal cell membranes at synaptic and extrasynaptic locations, where they are believed to mediate distinct physiological and pathological processes. Recently, Papouin et al. have demonstrated that synaptic and extrasynaptic NMDARs are gated by distinct endogenous coagonists (Papouin et al. 2012). Whereas D-serine is the coagonist at synaptic NMDARs, glycine acts at extrasynaptic NMDARs. The observation that D-serine but not glycine gates synaptic NMDARs could be the consequence of the preferential affinity of synaptic NMDARs for D-serine. GluN2A-containing NMDARs, which exhibit a stronger affinity for D-serine over glycine, appear to compose 70–75 % of the NMDAR population at hippocampal CA3-CA1 synapses in adult rats. Regulation of synaptic NMDARs by D-serine has been shown to be essential for long-term potentiation (LTP) induction and NMDA-induced neurotoxicity.

8.3 Serine Racemase and Modulation of D-Serine Synthesis

SR is a fold-type II of the pyridoxal 5’-phosphate (PLP)-dependent enzyme that catalyzes the racemization of L-serine to synthesize D-serine and does not use glycine as a substrate (Wolosker et al. 1999a, b). SR also catalyzes the α,-β-elimination of water from L- or D-serine to form pyruvate and NH_4 (Foltyn et al. 2005). PLP is the main cofactor that stimulates the activity of SR (Wolosker et al. 1999a). In addition to PLP, other compounds such as Mg^{2+} and ATP are capable of stimulating the synthesis of D-serine by increasing SR activity (De Miranda et al. 2002).

SR activity can be regulated by protein-protein interactions. SR binds to the glutamate receptor interacting protein (GRIP), which also binds to the α-amino-3-hydroxy-5-methyl-4-isoxazolepropionic acid (AMPA) receptor (AMPAR). Upon neuronal depolarization, GRIP dissociates from the AMPAR and binds to SR to increase its activity and D-serine production (Kim et al. 2005). SR also binds to the protein interacting with C-kinase (Pick-1). Neonatal Pick-1-deficient mice display 30 % lower D-serine levels in the forebrain than wild-type (WT) mice, and HEK293 cells transfected with both SR and Pick-1 show an increase in D-serine levels

(Hikida et al. 2008). SR is also modulated by Golga-3, which stabilizes SR levels through the inhibition of its ubiquitination, which leads to slower degradation by the ubiquitin-proteasome system (Dumin et al. 2006).

In addition to the regulation by protein interactions, SR activity is dynamically regulated by glutamate receptors. The activation of metabotropic glutamate receptors leads to the cleavage of phosphatidylinositol (4,5)-bisphosphate (PIP2) by phospholipase C (PLC), thereby diminishing the inhibition by PIP2 of SR (Mustafa et al. 2009). On the other hand, activation of NMDAR inhibits SR activity by promoting SR translocation to the membrane (Balan et al. 2009) and by enhancing SR *S*-nitrosylation (Mustafa et al. 2007).

The capability of SR to produce D-serine and to simultaneously degrade both D-serine and L-serine makes it a unique enzyme. As mentioned above, a number of regulatory mechanisms are proposed to play roles in modulating SR activity, but very little is known in terms of how they regulate SR physiologically, let alone the effects they may have on racemase versus eliminase activity. Further study is needed to elucidate the role of SR in the dynamic regulation of D-serine under both physiological and pathological conditions in vivo.

8.4 SR-KO Mouse Strains

To date, four SR-KO mouse strains have been established and used for phenotype analysis. The first SR-KO mouse strain, which we refer to as SR-KO^{HM}, was the insertional mutation-type SR-KO mouse strain with a pure C57BL/6 genetic background (Miya et al. 2008). The second, the SR-KO^{JTC} mouse strain, was generated by targeted deletion of the exon encoding initiation methionine of the SR gene. SR-KO^{JTC} mice were developed from 129Sv-derived embryonic stem (ES) cells and were backcrossed with the C57BL/6 strain for at least seven generations (Basu et al. 2009). The third, Srr^{Y269*}, was identified from the F1 progeny of ethylnitrosourea (ENU)-mutagenized C57BL/6 male mice and DBA/2 female mice (Labrie et al. 2009). Srr^{Y269*}mice show a complete loss of the SR protein owing to a nonsense point mutation at the start of exon 9. Srr^{Y269*}mice contained >99.6 % of the C57BL/6 genome. The fourth, the SR-KO^{KH} mouse strain, was generated by the disruption of the third exon of the SR gene containing the initiation codon (Miyoshi et al. 2012). The number of backcrossings of SR-KO^{KH} mice (129Sv genetic background) with C57BL/6 mice was not mentioned in the report. All these mutant mice exhibit an 80–90 % reduction in D-serine level in the forebrain. The level of L-serine in the forebrain of SR-KO^{HM}, SR-KO^{JTC}, and SR-KO^{KH} mice are is equivalent to those of WT controls (Inoue et al. 2008; Basu et al. 2009; Miyoshi et al. 2012), whereas Srr^{Y269*} mice have elevated L-serine level in the frontal cortex (Labrie et al. 2009). These findings indicate that D-serine production in the mouse brain is predominantly catalyzed by SR.

The origin of the residual 10–20 % of D-serine is not known. Possible alternative pathways for D-serine synthesis include the glycine cleavage system (Iwama et al. 1997), the hydrolysis of phosphoserine by phosphoserine phosphatase (Wood et al. 1996), and exogenous D-serine from intestinal bacterial flora. In addition to the brain, D-serine may also be produced in peripheral organs such as the liver (Benneyworth et al. 2012) and the skin (Inoue et al. 2014). The increase in the SR expression level in the liver of neuron-specific SR-KO^{JTC} (nSR-KO^{JTC}) mice suggests that there may be a cross talk between the brain and the liver, possibly to maintain brain and peripheral D-serine levels (Benneyworth et al. 2012).

8.5 Analysis of the Localization of SR Using SR-KO Mice

Early data on the cellular distribution of SR indicated that the enzyme is expressed in astrocytes (Wolosker et al. 1999a). However, later studies by immunohistochemistry and in situ hybridization demonstrated that SR (Kartvelishvily et al. 2006) and its mRNA (Yoshikawa et al. 2007) are present primarily in neurons. More definitive data on SR distribution was obtained using SR-KO^{HM} mice as negative controls (Miya et al. 2008). This approach ensured the specificity of the antibodies and revealed that SR is predominantly localized in pyramidal neurons of the cerebral cortex and hippocampus. Double-immunofluorescence experiments revealed that SR colocalizes with the neuron-specific nuclear protein (NeuN), but not with astrocytic markers, namely, the glial fibrillary acidic protein (GFAP) and 3-phosphoglycerate dehydrogenase (Phgdh). In the striatum, SR is expressed by γ-aminobutyric acid (GABA)ergic medium spiny neurons. In the adult cerebellum, weak but significant SR signals are detected in GABAergic Purkinje cells. Taken together, these findings indicate that SR is expressed by the main neuronal populations in most brain regions, irrespective of the excitatory or inhibitory nature of the cells. These findings, however, cannot exclude the possibility that SR is expressed in glial cells in vivo below the levels that could not be detected by the available antibodies. Recently, the notion that SR is predominantly neuronal has gathered additional support from the analysis of nSR-KO^{JTC} mice. These mice exhibit low D-serine and SR expression levels in the brain, especially in cortical glutamatergic neurons (Benneyworth et al. 2012). In contrast, astrocyte-specific SR-KO mice (aSR-KO^{JTC}) exhibit no change in D-serine levels. Therefore, neurons are likely the main source of D-serine.

8.6 Altered NMDAR-Mediated Neurotransmission in SR-KO Mice

The contribution of endogenous D-serine to the activation of NMDAR has been assessed in SR-KO mouse brain slices. Analyses of neurotransmission between Schaffer collaterals and hippocampal CA1 by whole-cell patch-clamp recording from brain slices of juvenile (postnatal days 21–28) SR-KO^{JTC} mice revealed slower decay kinetics of NMDAR-mediated excitatory postsynaptic currents (EPSCs), suggesting an increased contribution of the GluN2B-containing NMDAR in SR-KO mice (Basu et al. 2009). The increase in GluN2B protein expression level was later detected in juvenile SR-KO^{JTC} mouse brains (Balu and Coyle 2011). The juvenile SR-KO^{JTC} mice also exhibited impaired induction of NMDAR-dependent long-term potentiation (LTP) of synaptic transmission, as shown using a pairing stimulation protocol (Basu et al. 2009). The impaired LTP induction can be rescued by D-serine bath application. In contrast to juvenile SR-KO mice, adult SR-KO^{HM} mice did not show impairment of glutamatergic neurotransmission in hippocampal CA1 synapses (Rosenberg et al. 2013).

The contribution of endogenous D-serine to NMDAR-mediated neurotransmission was further assessed in the dentate gyrus (DG) using SR-KO^{JTC} mice (Balu et al. 2013). The amplitude of NMDAR-mediated miniature excitatory postsynaptic currents (mEPSCs) was found to decrease significantly in the mutant mice. Exogenously applied D-serine potentiated the amplitude and prolonged the decay time of NMDAR-mediated mEPSCs in both mutant and WT mice. SR-KO^{JTC} mice also displayed diminished magnitude of LTP at the medial perforant pathway to DG synapses.

To differentiate between the roles of D-serine produced by neurons and those produced by astrocytes, in the induction of NMDAR-dependent LTP, Benneyworth et al. (2012) stimulated Schaffer collaterals and recorded field excitatory postsynaptic potentials (fEPSPs) from the stratum radiatum of the CA1 of the hippocampus in slices from aSR-KO^{JTC} or nSR-KO^{JTC} mice. The magnitude of LTP induced by a weak single-train tetanus protocol was significantly reduced in slices from nSR-KO^{JTC} mice but unaffected in slices from aSR-KO^{JTC} mice. Consistent with the results obtained from SR-KO^{JTC} mice, nSR-KO^{JTC} mice also displayed reduced mEPSCs in CA1 pyramidal neurons. Taken together, these findings demonstrate the principal role of neuronal SR in the induction of NMDAR-dependent LTP.

8.7 Diminished Neurotoxicity in SR-KO Mice

Glutamate excitotoxicity, which occurs mainly through excessive activation of the NMDAR and subsequent massive calcium influx into a cell, has been linked to acute central nervous system (CNS) insults, such as ischemia and traumatic brain injury, and chronic neurodegenerative disorders, such as AD, ALS, and PD (Lau and Tymianski 2010). The role of brain D-serine in NMDAR-mediated neurotoxicity was first demonstrated in studies using in vitro culture systems. Shleper et al. proposed that D-serine is a predominant endogenous glycine site agonist for the induction of NMDA neurotoxicity in hippocampal slice cultures (Shleper et al. 2005). In this slice culture system, despite the high glycine levels in the medium, removal of endogenous D-serine by D-serine deaminase completely abolishes NMDA neurotoxicity. However, another study demonstrated that the neurotoxicity of NMDA in rat cerebrocortical slice cultures is potentiated by the addition of both glycine and D-serine (Katsuki et al. 2004). These findings stimulated the interest in the study of the importance of the two coagonists, D-serine and glycine, in NMDAR-mediated neurotoxicity in vivo.

Injection of NMDA into the cerebral cortex may induce excitotoxic neuronal death. SR-KO^{HM} mice displayed significantly attenuated NMDA-induced neurotoxicity, and this phenotype was abolished by the coinjection of D-serine and NMDA (Inoue et al. 2008). The levels of glycine in the forebrain were not altered in SR-KO^{HM} mice (Inoue et al. 2008), indicating that endogenous D-serine is the predominant and necessary coagonist for NMDAR activation in NMDA-induced neurotoxicity in vivo. This notion was supported by a recent study demonstrating that synaptic NMDARs, which use D-serine but not glycine as the coagonist, mainly contribute to NMDA-mediated excitotoxicity (Papouin et al. 2012). In contrast to the findings obtained from SR-KO^{HM} mice, Mustafa et al. reported a 50 % increase in NMDA-induced lesion volume in the striatum of SR-KO^{JTC} mice (Mustafa et al. 2010). Although the expression levels of NMDAR GluN1 remained unchanged in the forebrains of SR-KO^{HM} mice, it increased fourfold in the striatum of SR-KO^{JTC} mice. The difference in genetic background may account for the variations in GluN1 subunit expression level and for the development of NMDA-induced lesion volume in these two SR-KO mouse strains.

SR and D-serine were also found to play roles in the development of stroke. In comparison with WT mice, the stroke model SR-KO^{JTC} mice exhibited a 50–60 % reduction in infarct volume in the cerebral cortex and caudate-putamen region following middle cerebral artery occlusion (Mustafa et al. 2010). Activation of the NMDAR by glutamate/D-serine leads to the formation of nitric oxide (NO), which promotes oxidative neuronal damage by forming reactive oxygen species and nitrosylation of diverse proteins (Mustafa et al. 2007). Cerebral cortical cultures derived from SR-KO^{JTC} mice displayed a ~50 % reduction in NO generation and markedly diminished oxygen-glucose deprivation neurotoxicity (Mustafa et al. 2010).

AD is a major neurodegenerative disorder in which the excitotoxic effect of D-serine may be involved. The amyloid beta (Aβ) peptide is a main constituent of amyloid plaques found in the brains of patients with AD and proposed as the main pathological factor for AD. Aβ is reported to induce the release of D-serine and glutamate from cultured microglia (Wu et al. 2004). Aβ also enhances microglial SR transcription through the JNK-dependent recruitment of the AP-1 complex to the SR promoter (Wu and Barger 2004). In addition, the hippocampus of AD patients shows increased SR activity (Wu et al. 2004). Supporting the roles of D-serine and SR in AD, the neuronal injury induced by the direct injection of $A\beta_{1\text{-}42}$ into the hippocampus is shown to be attenuated in SR-KO^{HM} mice and in WT mice pretreated with MK-801, an antagonist of the NMDAR, prior to $A\beta_{1\text{-}42}$ injection (Inoue et al. 2008).

ALS is a neurodegenerative disorder characterized by dysfunction and death of both upper and lower motoneurons, leading to fatal paralysis. The mechanism underlying selective motoneuronal death is still unclear, but excitotoxicity has been proposed to play a principal role (Bruijn et al. 2004; Spreux-Varoquaux et al. 2002). In the G93A mutant of superoxide dismutase (SOD1) transgenic mice, a frequently used model of ALS, SR, and D-serine levels in the spinal cord gradually increased with disease progression. In addition, in the spinal cords of patients with the familial and sporadic forms of ALS, the levels of SR and D-serine also markedly increased (Sasabe et al. 2007). These findings offered the first evidence of a link between mutated SOD1 and D-serine. The role of SR and D-serine in the pathogenesis of ALS has recently been studied using the SR-deficient (Srr^{Y269*}) G93A mutant of SOD1 mice (Thompson et al. 2012). The SR-deficient G93A mutant of SOD1 mice showed an earlier symptom onset, but decreasing rate of disease progression and longer life span in a gene-dose-dependent manner. SR deletion also decreased spinal cord levels of D-serine. These findings suggest that SR and D-serine are fundamentally involved in both the presymptomatic and progression phases of ALS.

In conclusion, the diminished neurotoxicity in SR-KO mice in various models of acute and chronic neurodegenerative diseases suggest that SR and D-serine can be the targets for the development of new therapies for the treatment of these diseases through the regulation of NMDARs.

8.8 Decreased Susceptibility to Epileptic Seizures in SR-KO Mice

Epilepsy is a serious neurological disorder in human beings, which is characterized by neuronal death in brains and spontaneous recurrent seizures. It is well established that alterations in central inhibitory [e.g., γ-aminobutyric acid (GABA)] and excitatory (e.g., glutamate) neurotransmission play a crucial role in the etiology of epilepsy (Cloix and Hevor 2009; McCormick and Contreras 2001). Animal models of epilepsy and clinical studies have demonstrated that NMDAR activity and expression can be altered in association with epilepsy and particularly with some specific seizure types (Ghasemi and Schachter 2011). Therefore, many studies have focused on the pharmacological inhibition of the NMDAR as a therapeutic strategy for epilepsy, and a number of antagonists of the NMDAR were proven to have a potent anticonvulsant effect in a wide range of animal epileptic models (Brandt et al. 2003; Clifford et al. 1990; Croucher et al. 1982). However, blocking NMDARs leads to severe adverse effects, which limited their clinical applications (Kohl and Dannhardt 2001).

Growing evidence has shown that D-serine may be involved in the pathogenesis and progression of epilepsy. Upon systemic injection of pilocarpine, a mouse model for epilepsy, D-serine expression was upregulated in GABAergic neurons, which underwent degenerating death in the cerebral cortex and hippocampus (Liu et al. 2009). In addition, following status epilepticus, D-serine and SR immunoreactivities were found to increase in astrocytes (Ryu et al. 2010). These findings suggest that abnormal D-serine generation may be involved in the epileptogenesis and recurrent seizure development.

The effect of D-serine deficiency on seizures induced by a single injection of pentylenetetrazol (PTZ), a $GABA_A$ receptor blocker, was examined in $SR\text{-}KO^{HM}$ mice (Harai et al. 2012). $SR\text{-}KO^{HM}$ mice showed the attenuation of seizure expression in terms of a significantly shortened duration of generalized seizures and resistance to generalized clonic-tonic seizures. Consistent with this finding, immunohistochemical analysis of c-Fos, a marker of cellular activation, demonstrated that the numbers of cells expressing c-Fos induced by high-dose PTZ in the cerebral cortex, hippocampal CA1, hippocampal CA3, and the basolateral nucleus of the amygdala in WT mice were significantly higher than those in SR-KO mice. The excessive release of glutamate has been implicated in the generation and maintenance of epileptic seizures (Geula et al. 1988; Li et al. 2000; Ueda and Tsuru 1994). PTZ induced an increase in extracellular glutamate level in the DG of WT mice, but this change was completely suppressed in SR-KO mice.

8.9 Schizophrenia-Like Behavioral Abnormalities in SR-KO Mice

Schizophrenia is a severe mental illness affecting approximately 1 % of the population worldwide and it ranks as one of the leading causes of chronic disability. Symptoms of schizophrenia are generally divided into three main classes: positive symptoms such as hallucinations, delusions, and thought disorder; negative symptoms, such as social withdrawal and blunted affect; and cognitive symptoms, which involve profound deficits in attention, learning, and memory (Ross et al. 2006; Lewis and Gonzalez-Burgos 2006). Current treatments for this disease have limited efficacy, particularly in ameliorating the negative and cognitive symptoms, and have significant side effects that often lead to poor compliance.

NMDAR hypofunction has been hypothesized to contribute to the pathophysiology of schizophrenia. This hypothesis was based on studies demonstrating that noncompetitive NMDAR antagonists, such as phencyclidine or ketamine, can produce transient schizophrenia-like symptoms (Javitt and Zukin 1991; Krystal et al. 1994). Transgenic mice with reduced NMDAR activity also display phenotypes relevant to schizophrenia, particularly to the negative and cognitive symptoms (Mohn et al. 1999). Several studies have shown that D-serine may also be involved in the pathophysiology of schizophrenia. Significant reductions in D-serine level have been found in the cerebrospinal fluid (CSF) of drug-naïve patients with schizophrenia (Hashimoto et al. 2005; Bendikov et al. 2007). D-Serine concentrations were also decreased in the serum of patients with schizophrenia (Hashimoto et al. 2003; Yamada et al. 2005). Treatment with D-serine was shown to attenuate positive, negative, and cognitive symptoms in patients with schizophrenia

(Tsai et al. 1998; Heresco-Levy et al. 2005), suggesting that D-serine metabolism may be altered in schizophrenia.

Several genetic association studies have implicated that a single-nucleotide polymorphism variant of SR is associated with schizophrenia (Goltsov et al. 2006; Morita et al. 2007; Labrie et al. 2009). Moreover, neonatal administration of Met-Phen, an SR inhibitor, was shown to cause schizophrenia-like behavioral abnormalities in juvenile and adult mice (Hagiwara et al. 2013).

Recently, in a limited number of studies, the role of SR in schizophrenia has been investigated using SR-KO mice. SR-KO^{JTC} mice exhibited behavioral abnormalities such as hyperactivity, impaired spatial memory, and elevated anxiety, which are all relevant to schizophrenia (Basu et al. 2009). Schizophrenia patients exhibit reduced hippocampal dendritic spine density and volume and altered neuroplasticity. These abnormalities have been recapitulated in SR-KO^{JTC} mice (Balu et al. 2012). Pyramidal neurons in the primary somatosensory cortex of SR-KO^{JTC} mice show reductions in the complexity, total length, and spine density of apical and basal dendrites. In accordance with reduced cortical neuropil, SR-KO ^{JTC} mice show a smaller cortical volume than WT mice. SR-KO^{JTC} mice also show alteration in the signaling pathways, including that of BDNF/TrkB, the Akt/mammalian target of rapamycin (mTOR), and microRNA (miR)-312, which have been found to be genetically associated with or perturbed in schizophrenia (Balu et al. 2013).

Prepulse inhibition (PPI) of the acoustic startle response (ASR) is one indication of the sensory-motor gating process, which is attenuated by the pharmacological suppression of NMDARs. PPI attenuation is also observed in patients with schizophrenia (Cadenhead et al. 2000). PPI deficits were not observed in SR-KO^{HM} (Mori and Inoue 2010) and SR-KO^{JTC} (Basu et al. 2009) mice. In contrast, Srr^{Y269*} mice showed diminished PPI (Labrie et al. 2009). The discrepancies between these genetic models may be related to the genetic background of SR-KO mouse strains used for the analysis of PPI. It has been demonstrated that inbred mouse strains differ in the regulation of ASR and PPI of ASR (Bullock et al. 1997).

Social dysfunction is an important component of schizophrenia negative symptoms, often present in the prodromal stages and persisting throughout life (Ross et al. 2006; Ellenbroek and Cools 2000). Srr^{Y269*} mice displayed social approach deficit in social affiliation task, and the application of D-serine was shown to reverse the reduced sociability in mutant mice (Labrie et al. 2009).

Cognitive impairments are recognized to be a primary and enduring core deficit in schizophrenia. As mentioned above, D-serine deficiency in SR-KO mice leads to alteration in NMDAR-dependent neurotransmission and impaired LTP. To assess the effect of D-serine deficiency on cognitive function, several memory tasks have been conducted in SR-KO mice. SR-KO^{JTC} and SrrY269* male mice showed an impaired spatial reference memory, indicating a role of D-serine in spatial discrimination (Basu et al. 2009; Labrie et al. 2009). In a trace fear conditioning task that is dependent on both the hippocampus and amygdala, SR-KO^{JTC} mice exhibited impairment in contextual fear memory, and this deficit can be reversed by the chronic administration of D-serine (Balu et al. 2013). SR-KO^{JTC} mice also exhibit disrupted representation of the order associated with events in distinct experiences monitored by object recognition and the odor sequence test (DeVito et al. 2011). These findings strongly suggest that D-serine deficiency is associated with cognitive dysfunctions in schizophrenia.

8.10 Conclusions

Studies using SR-KO mice have provided compelling evidence that D-serine produced by neuronal SR constitutes the majority of forebrain D-serine and that an appropriate level of D-serine is required for normal NMDAR function. These studies also revealed the crucial roles of D-serine and SR in pathologies associated with NMDAR dysfunction, such as neurotoxicity, epilepsy, and schizophrenia (Table 8.1). Because SR-KO mice are viable and do not exhibit any gross physiological abnormalities, the pharmacological inhibition of SR may serve as a new strategy for the treatment of a number of neurological disorders associated with NMDAR overactivation, such as acute and chronic neurodegenerative disorders and epileptic seizures.

Alterations in D-serine level appear to be central to many neurological and psychiatric disorders that we have mentioned above. Nevertheless, the mechanisms underlying the abnormal increase or decrease in D-serine level in various pathological states remain largely unknown. Therefore, to fully understand the mechanisms involved in D-serine production, degradation, uptake, and release should be the first priority. In addition to using conventional pharmacological and genetic approaches, the development of an optimized probe for visualizing D-serine at synapse may facilitate the elucidation of D-serine dynamics under physiological and pathological conditions.

Table 8.1 Phenotypes of four established SR-KO mouse strains

Strains	SR-KO^{HM} Miya et al. (2008), Inoue et al. (2008), Rosenberg et al. (2013)	SR-KO^{JTC} Basu et al. (2009), DeVito et al. (2011), Balu et al. (2012), Balu et al. (2013)	Srr^{Y269*} Labrie et al. (2009), Thompson et al. (2012)	SR-KO^{KH} Miyoshi et al. (2012)
Expression of SR protein	Not detected	Not detected	Not detected	Not detected
D-Serine level	90 % reduction	90 % reduction	95 % reduction	90 % reduction
L-Serine level	No change	No change	Increased in the frontal cortex	No change
Glycine level	No change	NA	No change	NA
NMDAR-mediated neurotransmission in the hippocampal slices	Unaffected	Slower decay kinetics of EPSCs in hippocampal CA1 (P21–P28) Decreased amplitude of mEPSCs in dentate gyrus (adult) Impaired LTP	NA	NA
Neurotoxicity				
NMDA-induced neurotoxicity	Reduced (cerebral cortex)	Increased (striatum)	NA	NA
$A\beta_{1\text{-}42}$-induced neurotoxicity	Reduced (hippocampus)	NA	NA	NA
Neurotoxicity				
Stroke model	NA	Reduced	NA	NA
ALS model	NA	NA	Early disease onset and slow disease progression	NA
Epileptic seizures	Resistant to generalized clonic-tonic seizures	NA	NA	NA
Phenotypes associated with schizophrenia				
Neuronal morphology	NA	Dendritic dysplasia and reduced cortical volume	NA	NA
Locomotion	Unaffected	Hyperlocomotion	Unaffected	NA

(continued)

Table 8.1 (continued)

	SR-KO^{HM}	SR-KO^{JTC}	Srr^{Y269*}	SR-KO^{KH}
Strains	Miya et al. (2008), Inoue et al. (2008), Rosenberg et al. (2013)	Basu et al. (2009), DeVito et al. (2011), Balu et al. (2012), Balu et al. (2013)	Labrie et al. (2009), Thompson et al. (2012)	Miyoshi et al. (2012)
PPI	Unaffected	Unaffected	Reduced	NA
Sociability	NA	Unaffected	Decreased	NA
Cognitive functions	NA	Impaired spatial learning and memory; impaired contextual fear memory; impaired memory for the order of events	Impaired spatial learning and memory; impaired memory for order of events	

NA not analyzed

References

Balan L, Foltyn VN, Zehl M, Dumin E, Dikopoltsev E, Knoh D, Ohno Y, Kihara A, Jensen ON, Radzishevsky IS, Wolosker H (2009) Feedback inactivation of D-serine synthesis by NMDA receptor-elicited translocation of serine racemase to the membrane. Proc Natl Acad Sci U S A 106(18):7589–7594. doi:10.1073/pnas.0809442106

Balu DT, Coyle JT (2011) Glutamate receptor composition of the post-synaptic density is altered in genetic mouse models of NMDA receptor hypo- and hyperfunction. Brain Res 1392:1–7. doi:10.1016/j.brainres.2011.03.051

Balu DT, Basu AC, Corradi JP, Cacace AM, Coyle JT (2012) The NMDA receptor co-agonists, D-serine and glycine, regulate neuronal dendritic architecture in the somatosensory cortex. Neurobiol Dis 45(2):671–682. doi:10.1016/j.nbd.2011.10.006

Balu DT, Li Y, Puhl MD, Benneyworth MA, Basu AC, Takagi S, Bolshakov VY, Coyle JT (2013) Multiple risk pathways for schizophrenia converge in serine racemase knockout mice, a mouse model of NMDA receptor hypofunction. Proc Natl Acad Sci U S A 110(26):E2400–E2409. doi:10.1073/pnas.1304308110

Basu AC, Tsai GE, Ma CL, Ehmsen JT, Mustafa AK, Han L, Jiang ZI, Benneyworth MA, Froimowitz MP, Lange N, Snyder SH, Bergeron R, Coyle JT (2009) Targeted disruption of serine racemase affects glutamatergic neurotransmission and behavior. Mol Psychiatry 14 (7):719–727. doi:10.1038/mp.2008.130

Bendikov I, Nadri C, Amar S, Panizzutti R, De Miranda J, Wolosker H, Agam G (2007) A CSF and postmortem brain study of D-serine metabolic parameters in schizophrenia. Schizophr Res 90 (1-3):41–51. doi:10.1016/j.schres.2006.10.010

Benneyworth MA, Li Y, Basu AC, Bolshakov VY, Coyle JT (2012) Cell selective conditional null mutations of serine racemase demonstrate a predominate localization in cortical glutamatergic neurons. Cell Mol Neurobiol 32(4):613–624. doi:10.1007/s10571-012-9808-4

Berger AJ, Dieudonne S, Ascher P (1998) Glycine uptake governs glycine site occupancy at NMDA receptors of excitatory synapses. J Neurophysiol 80(6):3336–3340

Bliss TV, Collingridge GL (1993) A synaptic model of memory: long-term potentiation in the hippocampus. Nature 361(6407):31–39. doi:10.1038/361031a0

Brandt C, Potschka H, Loscher W, Ebert U (2003) *N*-methyl-D-aspartate receptor blockade after status epilepticus protects against limbic brain damage but not against epilepsy in the kainate model of temporal lobe epilepsy. Neuroscience 118(3):727–740

Bruijn LI, Miller TM, Cleveland DW (2004) Unraveling the mechanisms involved in motor neuron degeneration in ALS. Annu Rev Neurosci 27:723–749. doi:10.1146/annurev.neuro.27.070203.144244

Bullock AE, Slobe BS, Vazquez V, Collins AC (1997) Inbred mouse strains differ in the regulation of startle and prepulse inhibition of the startle response. Behav Neurosci 111(6):1353–1360

Cadenhead KS, Swerdlow NR, Shafer KM, Diaz M, Braff DL (2000) Modulation of the startle response and startle laterality in relatives of schizophrenic patients and in subjects with schizotypal personality disorder: evidence of inhibitory deficits. Am J Psychiatry 157 (10):1660–1668

Chen HS, Lipton SA (2006) The chemical biology of clinically tolerated NMDA receptor antagonists. J Neurochem 97(6):1611–1626. doi:10.1111/j.1471-4159.2006.03991.x

Clifford DB, Olney JW, Benz AM, Fuller TA, Zorumski CF (1990) Ketamine, phencyclidine, and MK-801 protect against kainic acid-induced seizure-related brain damage. Epilepsia 31 (4):382–390

Cloix JF, Hevor T (2009) Epilepsy, regulation of brain energy metabolism and neurotransmission. Curr Med Chem 16(7):841–853

Coyle JT, Tsai G, Goff D (2003) Converging evidence of NMDA receptor hypofunction in the pathophysiology of schizophrenia. Ann N Y Acad Sci 1003:318–327

Croucher MJ, Collins JF, Meldrum BS (1982) Anticonvulsant action of excitatory amino acid antagonists. Science 216(4548):899–901

Cull-Candy S, Brickley S, Farrant M (2001) NMDA receptor subunits: diversity, development and disease. Curr Opin Neurobiol 11(3):327–335

De Miranda J, Panizzutti R, Foltyn VN, Wolosker H (2002) Cofactors of serine racemase that physiologically stimulate the synthesis of the *N*-methyl-D-aspartate (NMDA) receptor coagonist D-serine. Proc Natl Acad Sci U S A 99(22):14542–14547. doi:10.1073/pnas.222421299

DeVito LM, Balu DT, Kanter BR, Lykken C, Basu AC, Coyle JT, Eichenbaum H (2011) Serine racemase deletion disrupts memory for order and alters cortical dendritic morphology. Genes Brain Behav 10(2):210–222. doi:10.1111/j.1601-183X.2010.00656.x

Dingledine R, Borges K, Bowie D, Traynelis SF (1999) The glutamate receptor ion channels. Pharmacol Rev 51(1):7–61

Dumin E, Bendikov I, Foltyn VN, Misumi Y, Ikehara Y, Kartvelishvily E, Wolosker H (2006) Modulation of D-serine levels via ubiquitin-dependent proteasomal degradation of serine racemase. J Biol Chem 281(29):20291–20302. doi:10.1074/jbc.M601971200

Ellenbroek BA, Cools AR (2000) Animal models for the negative symptoms of schizophrenia. Behav Pharmacol 11(3-4):223–233

Fadda E, Danysz W, Wroblewski JT, Costa E (1988) Glycine and D-serine increase the affinity of *N*-methyl-D-aspartate sensitive glutamate binding sites in rat brain synaptic membranes. Neuropharmacology 27(11):1183–1185

Foltyn VN, Bendikov I, De Miranda J, Panizzutti R, Dumin E, Shleper M, Li P, Toney MD, Kartvelishvily E, Wolosker H (2005) Serine racemase modulates intracellular D-serine levels through an alpha, beta-elimination activity. J Biol Chem 280(3):1754–1763. doi:10.1074/jbc.M405726200

Geula C, Jarvie PA, Logan TC, Slevin JT (1988) Long-term enhancement of K^+-evoked release of L-glutamate in entorhinal kindled rats. Brain Res 442(2):368–372

Ghasemi M, Schachter SC (2011) The NMDA receptor complex as a therapeutic target in epilepsy: a review. Epilepsy Behav 22(4):617–640. doi:10.1016/j.yebeh.2011.07.024

Goltsov AY, Loseva JG, Andreeva TV, Grigorenko AP, Abramova LI, Kaleda VG, Orlova VA, Moliaka YK, Rogaev EI (2006) Polymorphism in the 5'-promoter region of serine racemase gene in schizophrenia. Mol Psychiatry 11(4):325–326. doi:10.1038/sj.mp.4001801

Hagiwara H, Iyo M, Hashimoto K (2013) Neonatal disruption of serine racemase causes schizophrenia-like behavioral abnormalities in adulthood: clinical rescue by D-serine. PLoS One 8(4), e62438. doi:10.1371/journal.pone.0062438

Harai T, Inoue R, Fujita Y, Tanaka A, Horio M, Hashimoto K, Hongou K, Miyawaki T, Mori H (2012) Decreased susceptibility to seizures induced by pentylenetetrazole in serine racemase knockout mice. Epilepsy Res 102(3):180–187. doi:10.1016/j.eplepsyres.2012.06.001

Hashimoto A, Nishikawa T, Hayashi T, Fujii N, Harada K, Oka T, Takahashi K (1992) The presence of free D-serine in rat brain. FEBS Lett 296(1):33–36

Hashimoto K, Fukushima T, Shimizu E, Komatsu N, Watanabe H, Shinoda N, Nakazato M, Kumakiri C, Okada S, Hasegawa H, Imai K, Iyo M (2003) Decreased serum levels of D-serine in patients with schizophrenia: evidence in support of the *N*-methyl-D-aspartate receptor hypofunction hypothesis of schizophrenia. Arch Gen Psychiatry 60(6):572–576. doi:10.1001/archpsyc.60.6.572

Hashimoto K, Engberg G, Shimizu E, Nordin C, Lindstrom LH, Iyo M (2005) Reduced D-serine to total serine ratio in the cerebrospinal fluid of drug naive schizophrenic patients. Prog Neuropsychopharmacol Biol Psychiatry 29(5):767–769. doi:10.1016/j.pnpbp.2005.04.023

Heresco-Levy U, Javitt DC, Ebstein R, Vass A, Lichtenberg P, Bar G, Catinari S, Ermilov M (2005) D-Serine efficacy as add-on pharmacotherapy to risperidone and olanzapine for treatment-refractory schizophrenia. Biol Psychiatry 57(6):577–585. doi:10.1016/j.biopsych.2004.12.037

Hikida T, Mustafa AK, Maeda K, Fujii K, Barrow RK, Saleh M, Huganir RL, Snyder SH, Hashimoto K, Sawa A (2008) Modulation of D-serine levels in brains of mice lacking PICK1. Biol Psychiatry 63(10):997–1000. doi:10.1016/j.biopsych.2007.09.025

Inoue R, Hashimoto K, Harai T, Mori H (2008) NMDA- and beta-amyloid$_{1-42}$-induced neurotoxicity is attenuated in serine racemase knock-out mice. J Neurosci 28(53):14486–14491. doi:10.1523/JNEUROSCI.5034-08.2008

Inoue R, Yoshihisa Y, Tojo Y, Okamura C, Yoshida Y, Kishimoto J, Luan X, Watanabe M, Mizuguchi M, Nabeshima Y, Hamase K, Matsunaga K, Shimizu T, Mori H (2014) Localization of serine racemase and its role in the skin. J Investig Dermatol 134(6):1618–1626. doi:10.1038/jid.2014.22

Iwama H, Takahashi K, Kure S, Hayashi F, Narisawa K, Tada K, Mizoguchi M, Takashima S, Tomita U, Nishikawa T (1997) Depletion of cerebral D-serine in non-ketotic hyperglycinemia: possible involvement of glycine cleavage system in control of endogenous D-serine. Biochem Biophys Res Commun 231(3):793–796. doi:10.1006/bbrc.1997.6184

Javitt DC, Zukin SR (1991) Recent advances in the phencyclidine model of schizophrenia. Am J Psychiatry 148(10):1301–1308

Kartvelishvily E, Shleper M, Balan L, Dumin E, Wolosker H (2006) Neuron-derived D-serine release provides a novel means to activate *N*-methyl-D-aspartate receptors. J Biol Chem 281(20):14151–14162. doi:10.1074/jbc.M512927200

Katsuki H, Nonaka M, Shirakawa H, Kume T, Akaike A (2004) Endogenous D-serine is involved in induction of neuronal death by *N*-methyl-D-aspartate and simulated ischemia in rat cerebrocortical slices. J Pharmacol Exp Ther 311(2):836–844. doi:10.1124/jpet.104.070912

Kim PM, Aizawa H, Kim PS, Huang AS, Wickramasinghe SR, Kashani AH, Barrow RK, Huganir RL, Ghosh A, Snyder SH (2005) Serine racemase: activation by glutamate neurotransmission via glutamate receptor interacting protein and mediation of neuronal migration. Proc Natl Acad Sci U S A 102(6):2105–2110. doi:10.1073/pnas.0409723102

Kohl BK, Dannhardt G (2001) The NMDA receptor complex: a promising target for novel antiepileptic strategies. Curr Med Chem 8(11):1275–1289

Komuro H, Rakic P (1993) Modulation of neuronal migration by NMDA receptors. Science 260(5104):95–97

Krystal JH, Karper LP, Seibyl JP, Freeman GK, Delaney R, Bremner JD, Heninger GR, Bowers MB Jr, Charney DS (1994) Subanesthetic effects of the noncompetitive NMDA antagonist, ketamine, in humans. Psychotomimetic, perceptual, cognitive, and neuroendocrine responses. Arch Gen Psychiatry 51(3):199–214

Labrie V, Roder JC (2010) The involvement of the NMDA receptor D-serine/glycine site in the pathophysiology and treatment of schizophrenia. Neurosci Biobehav Rev 34(3):351–372. doi:10.1016/j.neubiorev.2009.08.002

Labrie V, Fukumura R, Rastogi A, Fick LJ, Wang W, Boutros PC, Kennedy JL, Semeralul MO, Lee FH, Baker GB, Belsham DD, Barger SW, Gondo Y, Wong AH, Roder JC (2009) Serine racemase is associated with schizophrenia susceptibility in humans and in a mouse model. Hum Mol Genet 18(17):3227–3243. doi:10.1093/hmg/ddp261

Lakhan SE, Caro M, Hadzimichalis N (2013) NMDA receptor activity in neuropsychiatric disorders. Front Psychiatry 4:52. doi:10.3389/fpsyt.2013.00052

Lancelot E, Beal MF (1998) Glutamate toxicity in chronic neurodegenerative disease. Prog Brain Res 116:331–347

Lau A, Tymianski M (2010) Glutamate receptors, neurotoxicity and neurodegeneration. Pflugers Arch 460(2):525–542. doi:10.1007/s00424-010-0809-1

Lewis DA, Gonzalez-Burgos G (2006) Pathophysiologically based treatment interventions in schizophrenia. Nat Med 12(9):1016–1022. doi:10.1038/nm1478

Li Z, Yamamoto Y, Morimoto T, Ono J, Okada S, Yamatodani A (2000) The effect of pentylenetetrazole-kindling on the extracellular glutamate and taurine levels in the frontal cortex of rats. Neurosci Lett 282(1-2):117–119

Liu YH, Wang L, Wei LC, Huang YG, Chen LW (2009) Up-regulation of D-serine might induce GABAergic neuronal degeneration in the cerebral cortex and hippocampus in the mouse pilocarpine model of epilepsy. Neurochem Res 34(7):1209–1218. doi:10.1007/s11064-008-9897-0

Matsui T, Sekiguchi M, Hashimoto A, Tomita U, Nishikawa T, Wada K (1995) Functional comparison of D-serine and glycine in rodents: the effect on cloned NMDA receptors and the extracellular concentration. J Neurochem 65(1):454–458

McCormick DA, Contreras D (2001) On the cellular and network bases of epileptic seizures. Annu Rev Physiol 63:815–846. doi:10.1146/annurev.physiol.63.1.815

Miya K, Inoue R, Takata Y, Abe M, Natsume R, Sakimura K, Hongou K, Miyawaki T, Mori H (2008) Serine racemase is predominantly localized in neurons in mouse brain. J Comp Neurol 510(6):641–654. doi:10.1002/cne.21822

Miyoshi Y, Konno R, Sasabe J, Ueno K, Tojo Y, Mita M, Aiso S, Hamase K (2012) Alteration of intrinsic amounts of D-serine in the mice lacking serine racemase and D-amino acid oxidase. Amino Acids 43(5):1919–1931. doi:10.1007/s00726-012-1398-4

Mohn AR, Gainetdinov RR, Caron MG, Koller BH (1999) Mice with reduced NMDA receptor expression display behaviors related to schizophrenia. Cell 98(4):427–436

Mori H, Inoue R (2010) Serine racemase knockout mice. Chem Biodivers 7(6):1573–1578. doi:10.1002/cbdv.200900293

Morita Y, Ujike H, Tanaka Y, Otani K, Kishimoto M, Morio A, Kotaka T, Okahisa Y, Matsushita M, Morikawa A, Hamase K, Zaitsu K, Kuroda S (2007) A genetic variant of the serine racemase gene is associated with schizophrenia. Biol Psychiatry 61(10):1200–1203. doi:10.1016/j.biopsych.2006.07.025

Mothet JP, Parent AT, Wolosker H, Brady RO Jr, Linden DJ, Ferris CD, Rogawski MA, Snyder SH (2000) D-Serine is an endogenous ligand for the glycine site of the *N*-methyl-D-aspartate receptor. Proc Natl Acad Sci U S A 97(9):4926–4931

Mustafa AK, Kumar M, Selvakumar B, Ho GP, Ehmsen JT, Barrow RK, Amzel LM, Snyder SH (2007) Nitric oxide S-nitrosylates serine racemase, mediating feedback inhibition of D-serine formation. Proc Natl Acad Sci U S A 104(8):2950–2955. doi:10.1073/pnas.0611620104

Mustafa AK, van Rossum DB, Patterson RL, Maag D, Ehmsen JT, Gazi SK, Chakraborty A, Barrow RK, Amzel LM, Snyder SH (2009) Glutamatergic regulation of serine racemase via

reversal of PIP2 inhibition. Proc Natl Acad Sci U S A 106(8):2921–2926. doi:10.1073/pnas.0813105106
Mustafa AK, Ahmad AS, Zeynalov E, Gazi SK, Sikka G, Ehmsen JT, Barrow RK, Coyle JT, Snyder SH, Dore S (2010) Serine racemase deletion protects against cerebral ischemia and excitotoxicity. J Neurosci 30(4):1413–1416. doi:10.1523/JNEUROSCI.4297-09.2010
Papouin T, Ladepeche L, Ruel J, Sacchi S, Labasque M, Hanini M, Groc L, Pollegioni L, Mothet JP, Oliet SH (2012) Synaptic and extrasynaptic NMDA receptors are gated by different endogenous coagonists. Cell 150(3):633–646. doi:10.1016/j.cell.2012.06.029
Rosenberg D, Artoul S, Segal AC, Kolodney G, Radzishevsky I, Dikopoltsev E, Foltyn VN, Inoue R, Mori H, Billard JM, Wolosker H (2013) Neuronal D-serine and glycine release via the Asc-1 transporter regulates NMDA receptor-dependent synaptic activity. J Neurosci 33(8):3533–3544. doi:10.1523/JNEUROSCI.3836-12.2013
Ross CA, Margolis RL, Reading SA, Pletnikov M, Coyle JT (2006) Neurobiology of schizophrenia. Neuron 52(1):139–153. doi:10.1016/j.neuron.2006.09.015
Ryu HJ, Kim JE, Yeo SI, Kim DS, Kwon OS, Choi SY, Kang TC (2010) Potential roles of D-serine and serine racemase in experimental temporal lobe epilepsy. J Neurosci Res 88(11):2469–2482. doi:10.1002/jnr.22415
Sasabe J, Chiba T, Yamada M, Okamoto K, Nishimoto I, Matsuoka M, Aiso S (2007) D-serine is a key determinant of glutamate toxicity in amyotrophic lateral sclerosis. EMBO J 26(18):4149–4159. doi:10.1038/sj.emboj.7601840
Schell MJ, Brady RO Jr, Molliver ME, Snyder SH (1997) D-Serine as a neuromodulator: regional and developmental localizations in rat brain glia resemble NMDA receptors. J Neurosci 17(5):1604–1615
Shleper M, Kartvelishvily E, Wolosker H (2005) D-Serine is the dominant endogenous coagonist for NMDA receptor neurotoxicity in organotypic hippocampal slices. J Neurosci 25(41):9413–9417. doi:10.1523/JNEUROSCI.3190-05.2005
Spreux-Varoquaux O, Bensimon G, Lacomblez L, Salachas F, Pradat PF, Le Forestier N, Marouan A, Dib M, Meininger V (2002) Glutamate levels in cerebrospinal fluid in amyotrophic lateral sclerosis: a reappraisal using a new HPLC method with coulometric detection in a large cohort of patients. J Neurol Sci 193(2):73–78
Thompson M, Marecki JC, Marinesco S, Labrie V, Roder JC, Barger SW, Crow JP (2012) Paradoxical roles of serine racemase and D-serine in the G93A mSOD1 mouse model of amyotrophic lateral sclerosis. J Neurochem 120(4):598–610. doi:10.1111/j.1471-4159.2011.07601.x
Tsai G, Yang P, Chung LC, Lange N, Coyle JT (1998) D-Serine added to antipsychotics for the treatment of schizophrenia. Biol Psychiatry 44(11):1081–1089
Ueda Y, Tsuru N (1994) Bilateral seizure-related changes of extracellular glutamate concentration in hippocampi during development of amygdaloid kindling. Epilepsy Res 18(1):85–88
Wang R, Zhang D (2005) Memantine prolongs survival in an amyotrophic lateral sclerosis mouse model. Eur J Neurosci 22(9):2376–2380. doi:10.1111/j.1460-9568.2005.04431.x
Wolosker H, Blackshaw S, Snyder SH (1999a) Serine racemase: a glial enzyme synthesizing D-serine to regulate glutamate-*N*-methyl-D-aspartate neurotransmission. Proc Natl Acad Sci U S A 96(23):13409–13414
Wolosker H, Sheth KN, Takahashi M, Mothet JP, Brady RO Jr, Ferris CD, Snyder SH (1999b) Purification of serine racemase: biosynthesis of the neuromodulator D-serine. Proc Natl Acad Sci U S A 96(2):721–725
Wood PL, Hawkinson JE, Goodnough DB (1996) Formation of D-serine from L-phosphoserine in brain synaptosomes. J Neurochem 67(4):1485–1490
Wu SZ, Barger SW (2004) Induction of serine racemase by inflammatory stimuli is dependent on AP-1. Ann N Y Acad Sci 1035:133–146. doi:10.1196/annals.1332.009
Wu SZ, Bodles AM, Porter MM, Griffin WS, Basile AS, Barger SW (2004) Induction of serine racemase expression and D-serine release from microglia by amyloid beta-peptide. J Neuroinflammation 1(1):2. doi:10.1186/1742-2094-1-2

Yamada K, Ohnishi T, Hashimoto K, Ohba H, Iwayama-Shigeno Y, Toyoshima M, Okuno A, Takao H, Toyota T, Minabe Y, Nakamura K, Shimizu E, Itokawa M, Mori N, Iyo M, Yoshikawa T (2005) Identification of multiple serine racemase (SRR) mRNA isoforms and genetic analyses of SRR and DAO in schizophrenia and D-serine levels. Biol Psychiatry 57 (12):1493–1503. doi:10.1016/j.biopsych.2005.03.018

Yoshikawa M, Takayasu N, Hashimoto A, Sato Y, Tamaki R, Tsukamoto H, Kobayashi H, Noda S (2007) The serine racemase mRNA is predominantly expressed in rat brain neurons. Arch Histol Cytol 70(2):127–134

Chapter 9
Abnormal D-Serine Metabolism in Amyotrophic Lateral Sclerosis

Jumpei Sasabe and Sadakazu Aiso

Abstract Amyotrophic lateral sclerosis (ALS) is a progressive neurodegenerative disease, whose pathology is characterized by death of upper and lower motor neurons, inclusion bodies in remaining neurons and glia, and glial activation around the neurons. The pathophysiological mechanisms underlying ALS are multifactorial and remain to be fully elucidated. Here, we review D-serine-related findings in ALS and delineate how D-serine metabolism is disturbed and why such D-serine derangement could potentially be toxic to motoneurons.

Motoneurons are vulnerable to glutamate excitotoxicity. D-Serine is an endogenous coagonist with glutamate for stimulation of *N*-methyl-D-aspartate (NMDA) subtype glutamate receptors. D-Serine accumulates progressively in the spinal cord of a mouse model of familial ALS, in which human superoxide dismutase 1 (SOD1) with a mutation G93A is overexpressed. Such accumulation was also reported in a few patients with sporadic ALS or familial ALS with A4T-SOD1. The D-serine accumulation in the mouse model is explained by a combination of increased D-serine-producing enzyme and decreased D-serine-degrading enzyme, D-amino acid oxidase (DAO), which are both found in activated glial cells. Importantly, a dominant negative mutation D199W in *DAO* has been reported in patients with familial ALS that exhibits classical motor symptoms of ALS. The mutant D199W-DAO increases autophagy in motor neurons through activation of NMDA receptors by D-serine, which results in motoneuronal apoptosis. Furthermore, a null mutation G181R in *DAO* significantly increases D-serine level in the spinal cord with mild motoneuronal degeneration in mice. Collectively, aberrant metabolism of D-serine in glial cells may trigger motoneuronal degeneration, which sheds light on a unique aspect of ALS pathophysiology.

Keywords D-Serine • Amyotrophic lateral sclerosis • Motoneuron • D-amino acid oxidase • Serine racemase

---

J. Sasabe (✉) • S. Aiso
Department of Anatomy, Keio University School of Medicine, 35 Shinanomachi, Shinjuku-ku, Tokyo 160-8582, Japan
e-mail: sasabe@a8.keio.jp

T. Yoshimura et al. (eds.), *D-Amino Acids*, DOI 10.1007/978-4-431-56077-7_9

9.1 Role of D-Serine in the Etiology of ALS

9.1.1 Overview of Amyotrophic Lateral Sclerosis

Amyotrophic lateral sclerosis (ALS), also called as Lou Gehrig's disease, is the most common motor neuron disease. "Amyotrophic" means muscular atrophy, and "lateral sclerosis" describes the scarring tissues in the lateral funiculus of the spinal cord, where the corticospinal tract of motor pathways is found. ALS exhibits a diverse and complex phenotype, which is clinically characterized by coexistence of upper and lower motoneuronal signs with progressive neurological deterioration (Kiernan et al. 2011). In addition to motor symptoms, subtle cognitive abnormalities are evident in up to 50 % of ALS patients, and frontotemporal dementia (FTD) may develop in up to 15 % (Phukan et al. 2012; Lomen-Hoerth et al. 2002, 2003). In most cases, disease onset is during late-adulthood, and fatal event due to respiratory paralysis occurs usually within 5 years after the onset. The mean incidence of ALS is 1.8/100,000 and the prevalence is 3.4/100,000 in North America (Chio et al. 2013), and these parameters are largely constant across the globe.

Whereas majority of ALS cases are sporadic, about 10 % are inherited in, mostly, a dominant manner. The identifications of mutations in the superoxide dismutase 1 (*SOD1*) gene (Rosen et al. 1993) followed by those in the genes of TAR DNA-binding protein 43 (*TDP-43*) (Kabashi et al. 2008; Sreedharan et al. 2008), fused in sarcoma/translocated in liposarcoma (*FUS/TLS*) (Kwiatkowski et al. 2009; Vance et al. 2009), chromosome 9 open reading frame 72 (*C9orf72*) (Renton et al. 2011; DeJesus-Hernandez et al. 2011), and D-amino acid oxidase (*DAO*) (Mitchell et al. 2010) in familial ALS have significantly moved forward in understanding biological mechanisms of ALS, although the mechanisms remain to be fully elucidated. The following are proposed principal hypotheses of motoneuronal degenerative processes (Sreedharan and Brown 2013; Vucic et al. 2014): (1) conformational instability of SOD1 triggers downstream neurotoxic processes; (2) perturbations of RNA processing related to TDP-43/FUS can be toxic to motoneurons; (3) glutamate excitotoxicity damages motoneurons; (4) neighboring glial cells mediate motoneuronal death; and (5) other mechanisms, such as impairment of axonal transport and oxidative/endoplasmic reticulum stresses, contribute to motoneuronal death.

Although a long list of investigational agents based on these hypotheses has been tested in clinical trials, there is only one agent that has been established as effective neuroprotective therapy for ALS. The agent with antiglutamatergic effect, riluzole, prolongs survival of patients for 3–6 months (Bensimon et al. 1994; Lacomblez et al. 1996), but the therapeutic effect is not radical. Therefore, the management of ALS remains best supportive therapy to maintain patient quality of life, and the therapeutic strategy that can control the progression of the disease is eagerly anticipated.

9.1.2 D-Serine-Related Pathology in ALS

9.1.2.1 Role of D-Serine in the Central Nervous System

D-Serine is a coagonist at D-serine-/glycine-binding site in NR1 subunit of *N*-methyl-D-aspartate (NMDA) glutamate receptors (Mothet et al. 2000; Panatier et al. 2006; Basu et al. 2009; Papouin et al. 2012). Activation of the NMDA receptors essentially requires bindings of both glutamate and a coagonist (Johnson and Ascher 1987; McBain et al. 1989). Although glycine also works as a coagonist, the affinity of D-serine to the site is threefold stronger than that of glycine due to additional hydrogen bonds (Furukawa and Gouaux 2003; Matsui et al. 1995). Accumulating evidence support that D-serine is an endogenous coagonist of NMDA receptors, and more recently Papouin et al. demonstrated that D-serine gates synaptic NMDA receptors, whereas glycine works on extrasynaptic ones (Papouin et al. 2012). NMDA receptors principally control synaptic plasticity and physiologically contribute to memory formation and learning through long-term potentiation (LTP). In line with the knowledge, depletion of a D-serine-producing enzyme, serine racemase (SR), in mice almost nullifies induction of LTP with 90 % reduction of D-serine in the hippocampus (Basu et al. 2009). On the other hand, overstimulation of NMDA receptors evokes excitotoxicity to neurons. Degradation of D-serine by externally added D-serine deaminase, a bacterial D-serine-degrading enzyme, in slice culture settings (Shleper et al. 2005) or SR knockout mice strikingly alleviates NMDA receptor-mediated excitotoxicity (Inoue et al. 2008), implying that regulation of D-serine is crucial for neurons to retain their physiological level of excitability.

In rodents, the physiological level of D-serine is highest (≈300 μM) (Hashimoto et al. 1993; Nagata et al. 1994; Morikawa et al. 2001) in the forebrain due to abundance of SR and lack of a D-serine-degrading enzyme, D-amino acid oxidase (DAO). In the cerebellum and spinal cord, the level of D-serine is maintained at a low level (≈ 5 μM) (Hashimoto et al. 1993; Nagata et al. 1994; Morikawa et al. 2001) since DAO activity is relatively high in the region. Miyoshi et al. confirmed the contributions of SR and DAO in D-serine metabolism by creating SR knockout mice without DAO activity (Miyoshi et al. 2012). In mouse cerebral cortex, SR knockout dramatically reduces D-serine down to 1/10 level of wild-type animals, and the D-serine levels stay unchanged in DAO-lacking SR knockout animals, showing that SR principally controls D-serine, while DAO has almost no roles in regulating cortical D-serine. In contrast, in mouse cerebellum and spinal cord, D-serine levels become ten-fold higher in DAO-lacking mice than in wild-type controls, and the level in DAO-lacking animals is reduced to half in DAO-lacking SR knockout mice, indicating that SR synthesizes D-serine but is actively degraded by DAO in the regions. In short, the dominant regulator of D-serine is SR in the cerebral cortex and is DAO in the cerebellum and spinal cord in mice.

9.1.2.2 Accumulation of D-Serine in ALS

D-Serine accumulation in the ventral spinal cord was first reported using immunohistochemistry with a polyclonal antibody to D-serine in an animal model of familial ALS, G93A-SOD1 transgenic mice (mSOD1 mice), and in three patients with sporadic ALS and one patient with familial ALS with a mutation of A4T in *SOD1* (Sasabe et al. 2007). Immunoreactivity to D-serine is accumulated in motoneurons in the presymptomatic stage of the mSOD1 mice and found prominently in activated astrocytes and microglia in the advanced symptomatic stage. In postmortem tissue of the patients with ALS, increased immunoreactivity to D-serine is observed in motoneurons as well as surrounding glial cells, whereas a paper using another antibody shows that D-serine immunoreactivity is rather decreased in ALS patients (Paul et al. 2014). The accumulation of D-serine in mSOD1 mice was verified using an enzymatic assay with DAO (Sasabe et al. 2007) and also in a later study using a two-dimensional high-performance liquid chromatography (2D-HPLC) (Sasabe et al. 2012), which enables supersensitive determination of D-serine with chiral specificity (Morikawa et al. 2003; Miyoshi et al. 2009). In accordance with the immunohistochemical findings, quantification with 2D-HPLC shows that the D-serine level in the spinal cord of mSOD1 mice is higher than that in non-transgenic wild-type animals even in presymptomatic stage and is progressively elevated up to threefold of wild type with progression of motoneuronal degeneration (Fig. 9.1). The D-serine increase is accompanied by mild increase of L-serine (Sasabe et al. 2012), but not by alteration of glycine (unpublished data). On the other hand, D-alanine and D-aspartate, which also bind to NMDA receptors, show no alterations in their levels in spinal cord of the mSOD1 mice (Sasabe et al. 2012). Such D-serine-specific increase can be exclusively found in the spinal cord of mSOD1, but not in their cortex nor in the spinal cords of another model of ALS, overexpressed with A315T-TDP-43, or models for Parkinson's/Alzheimer's diseases (Sasabe et al. 2012). Thus, D-serine accumulation is assumed to be specific to a certain physiology of motoneuronal degeneration.

Compared to the well-characterized mouse model of ALS, D-serine levels in patients with ALS are little understood except for immunohistological findings. Our preliminary results obtained from limited number of non-fixed frozen postmortem samples using 2D-HPLC suggest that ratio of D- to L-serine levels in the spinal cords of sporadic ALS patients is about three times higher than that of non-ALS donors (unpublished observation). On the other hand, the D-serine level or D-/L-serine ratio in the cerebrospinal fluid showed no detectable difference between sporadic ALS patients and non-ALS donors (unpublished observation), implying that D-serine level in cerebrospinal fluid may not reflect the spinal lesion.

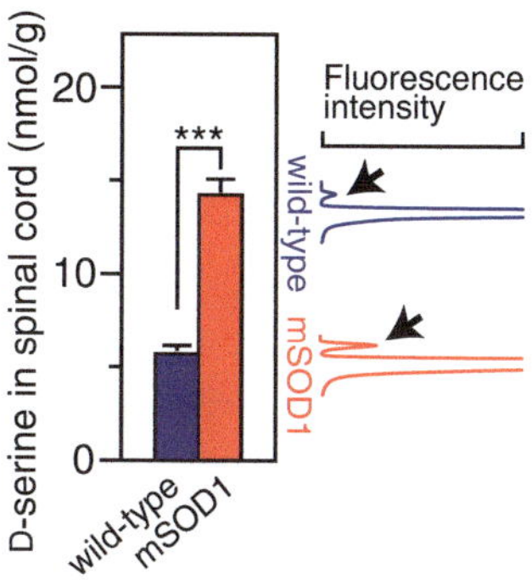

Fig. 9.1 Increased D-serine level in the spinal cord of an ALS model animal. D-Serine in the spinal cords of mSOD1 mice and wild-type controls was quantified using a 2D-HPLC system at 5 months of age (corresponding to the end stage in mSOD1 mice) (*t*-test, $***P < 0.001$). The *right lines* are representative chromatogram in the second dimension of the HPLC. The *arrows* mean the peaks specific to D-serine (Modified from Sasabe et al. 2012)

9.1.2.3 Genetic Relevance of D-Serine Metabolism in ALS

Among ALS cases, 10 % of them show familial transmission. Mutations in *SOD1*, *TDP-43*, and *FUS* occur in 20–30 % of familial form of the disease (Leblond et al. 2014; Renton et al. 2014). A massive intronic hexanucleotide expansion in *C9orf72* accounts for up to 50 % of familial ALS (Leblond et al. 2014). The rest of the cases with familial ALS comprise several genes, including vesicle-associated membrane protein-associated protein B (*VAPB*), optineurin (*OPTN*), valosin-containing protein (*VCP*), *DAO*, and etc. (Leblond et al. 2014; Renton et al. 2014). Mitchell et al. reported in their seminal study a unique mutation "R199W" in *DAO* associated with classical adult-onset ALS in a three-generational kindred of familial ALS (Mitchell et al. 2010). The familial ALS is transmitted as a dominant trait (Mitchell et al. 2010). The locus on chromosome 12 was identified through linkage analysis by a whole genome screen with microsatellite markers in families with confirmed ALS and lacking known familial ALS mutations (Mitchell et al. 2010). The R199W mutation in exon 7 of *DAO* was found in three affected individuals, one confirmed obligate carrier, and three "at-risk" individuals in the kindred, but not in 1002 unrelated individuals (780 non-ALS cases, 23 sporadic ALS cases, and 199 familial ALS cases lacking SOD1 mutations) (Mitchell et al. 2010). The kindred with R199W-DAO exhibits both upper and lower motoneuronal symptoms with bulbar sings. The disease onset was early 40s and the affected cases showed a severe phenotype with an average age at death of 44 years (Mitchell et al. 2010).

Arg199 is highly conserved between mammalian and lower organisms and lies close to the FAD binding site and between residues Tyr228 and His307. The R199W-mutation nullifies enzymatic activity of DAO and the mutant suppresses the activity of wild-type enzyme (Mitchell et al. 2010). DAO activity in the spinal cord of a patient with both R199W mutation and wild-type alleles is 1/25 of that in control spinal cords with wild-type alleles only (Mitchell et al. 2010), suggesting that R199W is a dominant negative mutant. Although D-serine levels in the familial ALS cases have not been determined yet, significant disturbance of the degradation by the mutation is considered to increase D-serine levels in the spinal cord.

9.1.3 Mechanism Underlying D-Serine Accumulation

9.1.3.1 Metabolism of D-Serine in the Motor Pathways

The cell type-specific regulation of D-serine remains uncertain. A serine-shuttle model introduced by Wolosker explains how D-serine is synthesized and degraded in the rodent forebrain (Wolosker 2011), where SR, but not DAO, works as a principal regulator of D-serine. In the model, D-serine is synthesized in neurons from L-serine, which is transported from astrocytes where L-serine is biosynthesized from a glycolytic intermediate. The model explains that neuronal SR also catalyzes D-serine degradation by α,β-elimination activity and maintains the levels of D-serine in the SR-rich forebrain regions. On the other hand, the cellular regulation of D-serine in DAO-rich regions, such as the motor pathways in the brainstem and spinal cord, is not fully understood. In the DAO-rich regions, the expression level of SR is relatively low compared to the forebrain (Miya et al. 2008). Considering DAO is a major regulator of D-serine and is expressed in the astrocytes in the DAO-rich regions (Miyoshi et al. 2012; Sasabe et al. 2014), D-serine released from neurons could be uptaken by astrocytes, degraded into hydroxypyruvate, and utilized in energy metabolism.

Motor pathways consist of pyramidal and extrapyramidal systems. ALS primarily involves pathology of the pyramidal system, also called as the corticospinal pathway, which directly connects between the cerebral cortex and spinal cord. The pyramidal system is well-developed in humans for fine movement of fingers compared to rodents, in which reticulospinal tract dominantly controls motor function, mostly consisting of gross movement (Lemon 2008; Alstermark et al. 2004; Yang and Lemon 2003). Intriguingly, histological analysis based on enzymatic activity shows that DAO activity is prominent in astrocytes of corticospinal tracts in humans, whereas it is principally found in reticulospinal tracts in mice (Sasabe et al. 2014). Although the distribution of SR is not well defined, D-serine in the motor pathways should be retained at a low level due to relatively high astrocytic DAO activity in physiological conditions.

9.1.3.2 Aberrant Metabolism of D-Serine in ALS

D-Serine accumulation results from both overproduction and suppressed degradation. Mechanisms underlying the accumulation have been characterized mostly in the mSOD1 mice. Since D-serine accumulation is observed in neurons and glial cells both in mSOD1 and in patients with ALS (Sasabe et al. 2007), the accumulation processes are supposed to involve these types of cells. Immunohistochemistry shows that microglial population with SR expression grows with progression of motoneuronal degeneration in mSOD1 mice, implying that microglia overproduce D-serine (Sasabe et al. 2007). In accordance with the report that SR expression is induced in microglia by proinflammatory stimuli through a c-jun N-terminal kinase and activator protein 1 (JNK-AP1) signal transduction pathway (Wu and Barger 2004), phosphorylation of JNK is observed in activated microglia in the motor pathway of mSOD1 mice (Sasabe et al. 2007). Interestingly, besides inflammatory stimuli, the forced expression of mSOD1 also increases the expression of SR in microglia (Sasabe et al. 2007), suggesting that mSOD1 triggers the proinflammatory signal of microglia or phosphorylates JNK through unknown mechanisms. On the other hand, degradation of D-serine is also suppressed in the mSOD1 mice. The histological analysis based on enzymatic activity shows that DAO activity is strikingly suppressed in activated astrocytes in the reticulospinal motor pathway of mSOD1 mice (Fig. 9.2) (Sasabe et al. 2012). The mRNA expression of DAO is progressively reduced in the spinal cord of mSOD1 (Sasabe et al. 2012), suggesting that the DAO inactivation results from loss of its expression. Such reduction of DAO mRNA can be attenuated by inhibition of extracellular signal-regulated kinase (ERK) (Sasabe et al. 2012). Indeed, phospho-ERK-positive astrocytes lack DAO activity in mSOD1 mice (Sasabe et al. 2012), indicating that proliferating signals in the activated astrocytes may downregulate DAO. The contribution of DAO in the accumulation of D-serine was evaluated by creating DAO-lacking mSOD1 mice (Sasabe et al. 2012). While D-serine level in mSOD1 mice is threefold higher than that in the wild-type animals, the D-serine level in DAO-lacking mSOD1 mice shows only 10 % increase than that in DAO-lacking mice. Therefore, overproduction by SR comprises only a small portion, and reduction of DAO expression is the major contributor of D-serine accumulation in

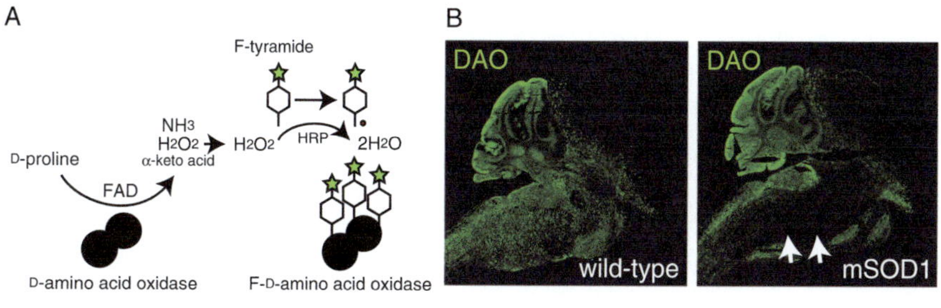

Fig. 9.2 Reduction of DAO activity in an ALS model animal. (**a**) *Cartoon* shows how DAO can be labeled with fluorescent dye based on its activity. Degradation of D-proline (externally added) produces hydrogen peroxide. Hydrogen peroxide in the presence of peroxidase (HRP) activates fluorescence-conjugated tyramide (F-tyramide), which triggers covalent binding of F-tyramide to DAO. (**b**) Shown are images of DAO visualized by the activity-based technique in the sagittal section of mouse brain from mSOD1 mice and wild-type animals. *Arrows* indicate reticular formation including reticulospinal pathway in the brain stem (Modified from Sasabe et al. 2012)

mSOD1 mice. In patients with ALS, D-serine accumulation and its mechanisms remain to be fully characterized. Considering that activations of microglia and astrocytes are a common feature of sporadic and familial ALS (Philips and Rothstein 2014), the derangement of D-serine in mSOD1 may represent ALS pathophysiology. Given the R199W-DAO functions dominant negatively, reduction of DAO activity followed by D-serine accumulation may be causative of the familial ALS.

9.1.4 Toxicity of D-Serine Against Motoneurons

Classically, motoneurons are known to be vulnerable to glutamate excitotoxicity, which has long been implicated in sporadic and familial ALS (Vucic et al. 2014; Sasabe and Aiso 2010; Bogaert et al. 2010). Glutamate exerts toxicity through excessive activation of ionotropic glutamate receptors, including NMDA and α-amino-3-hydroxy-5-methyl-4-isoxazoleproprionic acid (AMPA) receptors, which leads to motoneuronal degeneration through activation of Ca^{2+}-dependent enzymatic pathways. Furthermore, such excessive Ca^{2+} disables mitochondrial calcium-buffering system, resulting in being vulnerable to mitochondria-mediated apoptosis, oxidative stress, and electron-chain dysfunction (Clapham 2007; Emerit et al. 2004). The crucial role of D-serine in NMDA-evoked excitotoxicity against neurons has been reported. Complete removal of D-serine abolished NMDA-elicited excitotoxicity in organotypic hippocampal slices (Shleper et al. 2005), and cell death caused by in vivo injection of NMDA is significantly attenuated in SR knockout mice (Inoue et al. 2008). NMDA is more toxic to ChAT-positive neurons in primary cultured cells from mSOD1 spinal cords than those from wild-type spinal cords (Sasabe et al. 2007). The additive toxicity in the tissue from mSOD1 is attenuated by co-incubation with a specific inhibitor for SR (Sasabe et al. 2007), suggesting that overproduction of D-serine accelerates NMDA-mediated excitotoxicity. The view of D-serine toxicity in mSOD1 mice is partly supported by in vivo study using SR knockout mSOD1 mice. Thompson et al. have evaluated motor function and survival of SR knockout mSOD1 in comparison to mSOD1 animals (Thompson et al. 2012). SR knockout significantly prolongs the survival of mSOD1 mice, whereas it expedites the onset of the motor symptom. Contrary to the SR knockout phenotype in mSOD1 mice, D-serine challenge test in the drinking water also results in acceleration of the onset and elongated survival. The authors explain the paradoxical findings that oral intake of D-serine rather reduces D-serine accumulation in the spinal cord of mSOD1 mice through induction of a D-serine transporter, alanine-serine-cysteine transporter 1 (Asc1). Together with our unpublished findings that every-other-day intraperitoneal injections of D-serine shorten the survival of mSOD1 mice with dramatic accumulation of D-serine in the spinal cord, intestinal absorption of D-serine may trigger unknown neuroprotective signals on the central nervous system.

On the other hand, recent findings provide new aspect of mechanisms underlying D-serine-mediated neurotoxicity. Mice lacking DAO activity due to G181R mutation with C57BL6 background exhibit mild motoneuronal degeneration with reduction in number and size of motoneurons, accompanied by ubiquitin-positive aggregates (Sasabe et al. 2012). Since D-serine level in the spinal cord of DAO-lacking animals is ≈ten-fold higher than that of wild-type animals (Sasabe et al. 2012; Miyoshi et al. 2012), D-serine increase potentially causes neurodegeneration. More recent study by Paul et al. shows that human DAO with the D199W mutation expressed in neurons or glia initiates apoptosis (Paul et al. 2014). Furthermore, when expressed in glia, the mutant protein is also able to initiate apoptosis in the neighboring neurons that lack the mutant protein. The D199W-DAO is suggested to trigger motoneuronal apoptosis through promoting autophagy, which can be nullified by an inhibitor of D-serine-binding site in NMDA receptors. These results support the idea that D-serine triggers the accumulation of ubiquitinated protein aggregates, activates autophagy, and leads to apoptotic cell death through NMDA receptors, whose mechanisms could be distinguished from excitotoxicity.

9.2 Conclusion

In this paper, we reviewed the role of D-serine in the etiology of ALS with recent findings (Fig. 9.3). Currently, the evidence that link D-serine to ALS pathology are as follows: (1) D-serine is accumulated in the spinal cord of an animal model of ALS (G93A-SOD1 transgenic mice) mainly due to downregulation of the D-serine-degrading enzyme, DAO, in activated glial cells; (2) the dominant negative mutation of D199W in DAO gene causes familial ALS that exhibits adult-onset ALS phenotype with rapid progression; (3) D-serine is toxic to motoneurons potentially through enhancement of glutamate excitotoxicity, NMDA receptor-related acceleration of autophagy, or both; and (4) lack of DAO activity triggers mild motoneuronal degeneration in mice. Since such evidence involve mechanisms that mediate mSOD1, glutamate excitotoxicity, nonneuronal cells, and/or autophagy, D-serine-related pathology would have nice correlation to currently proposed pathophysiological hypotheses in ALS. To the best of our knowledge, the role of D-serine has been established in the pathology mostly in familial ALS with mutant SOD1 and DAO, and therefore, whether D-serine-related pathology represents ALS or whether modulation of D-serine has therapeutic benefits in general for ALS patients warrants future examinations.

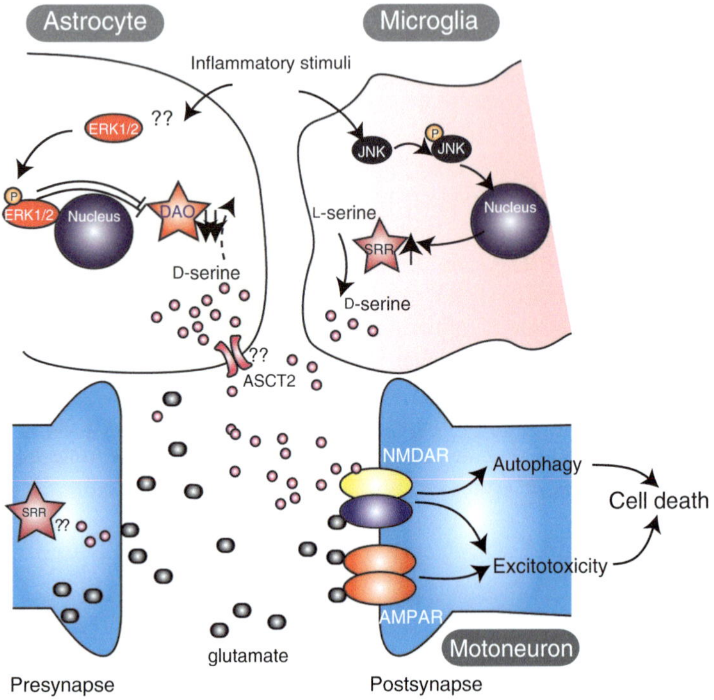

Fig. 9.3 Schematic depicting D-serine accumulation and its toxic mechanism in ALS. Degradation of D-serine in activated astrocytes is degreased due to downregulation of DAO mediated by phosphorylation of ERK, whereas production of D-serine in microglia is increased by upregulation of SR through phosphorylation of JNK. Increased D-serine binds to NMDA receptors and may trigger excitotoxicity, enhanced autophagy, or both, leading to motoneuronal cell death (Modified from Sasabe et al. 2012)

References

Alstermark B, Ogawa J, Isa T (2004) Lack of monosynaptic corticomotoneuronal EPSPs in rats: disynaptic EPSPs mediated via reticulospinal neurons and polysynaptic EPSPs via segmental interneurons. J Neurophysiol 91(4):1832–1839. doi:10.1152/jn.00820.2003

Basu AC, Tsai GE, Ma CL, Ehmsen JT, Mustafa AK, Han L, Jiang ZI, Benneyworth MA, Froimowitz MP, Lange N, Snyder SH, Bergeron R, Coyle JT (2009) Targeted disruption of serine racemase affects glutamatergic neurotransmission and behavior. Mol Psychiatry 14 (7):719–727. doi:10.1038/mp.2008.130

Bensimon G, Lacomblez L, Meininger V (1994) A controlled trial of riluzole in amyotrophic lateral sclerosis. ALS/Riluzole Study Group. N Engl J Med 330(9):585–591. doi:10.1056/NEJM199403033300901

Bogaert E, d'Ydewalle C, Van Den Bosch L (2010) Amyotrophic lateral sclerosis and excitotoxicity: from pathological mechanism to therapeutic target. CNS Neurol Disord Drug Targets 9(3):297–304

Chio A, Logroscino G, Traynor BJ, Collins J, Simeone JC, Goldstein LA, White LA (2013) Global epidemiology of amyotrophic lateral sclerosis: a systematic review of the published literature. Neuroepidemiology 41(2):118–130. doi:10.1159/000351153

Clapham DE (2007) Calcium signaling. Cell 131(6):1047–1058. doi:10.1016/j.cell.2007.11.028

DeJesus-Hernandez M, Mackenzie IR, Boeve BF, Boxer AL, Baker M, Rutherford NJ, Nicholson AM, Finch NA, Flynn H, Adamson J, Kouri N, Wojtas A, Sengdy P, Hsiung GY, Karydas A, Seeley WW, Josephs KA, Coppola G, Geschwind DH, Wszolek ZK, Feldman H, Knopman DS, Petersen RC, Miller BL, Dickson DW, Boylan KB, Graff-Radford NR, Rademakers R (2011) Expanded GGGGCC hexanucleotide repeat in noncoding region of C9ORF72 causes chromosome 9p-linked FTD and ALS. Neuron 72(2):245–256. doi:10.1016/j.neuron.2011.09.011

Emerit J, Edeas M, Bricaire F (2004) Neurodegenerative diseases and oxidative stress. Biomed Pharmacother 58(1):39–46

Furukawa H, Gouaux E (2003) Mechanisms of activation, inhibition and specificity: crystal structures of the NMDA receptor NR1 ligand-binding core. EMBO J 22(12):2873–2885. doi:10.1093/emboj/cdg303

Hashimoto A, Nishikawa T, Konno R, Niwa A, Yasumura Y, Oka T, Takahashi K (1993) Free D-serine, D-aspartate and D-alanine in central nervous system and serum in mutant mice lacking D-amino acid oxidase. Neurosci Lett 152(1–2):33–36

Inoue R, Hashimoto K, Harai T, Mori H (2008) NMDA- and beta-amyloid1-42-induced neurotoxicity is attenuated in serine racemase knock-out mice. J Neurosci 28(53):14486–14491. doi:10.1523/JNEUROSCI.5034-08.2008

Johnson JW, Ascher P (1987) Glycine potentiates the NMDA response in cultured mouse brain neurons. Nature 325(6104):529–531. doi:10.1038/325529a0

Kabashi E, Valdmanis PN, Dion P, Spiegelman D, McConkey BJ, Vande Velde C, Bouchard JP, Lacomblez L, Pochigaeva K, Salachas F, Pradat PF, Camu W, Meininger V, Dupre N, Rouleau GA (2008) TARDBP mutations in individuals with sporadic and familial amyotrophic lateral sclerosis. Nat Genet 40(5):572–574. doi:10.1038/ng.132

Kiernan MC, Vucic S, Cheah BC, Turner MR, Eisen A, Hardiman O, Burrell JR, Zoing MC (2011) Amyotrophic lateral sclerosis. Lancet 377(9769):942–955. doi:10.1016/S0140-6736(10)61156-7

Kwiatkowski TJ Jr, Bosco DA, Leclerc AL, Tamrazian E, Vanderburg CR, Russ C, Davis A, Gilchrist J, Kasarskis EJ, Munsat T, Valdmanis P, Rouleau GA, Hosler BA, Cortelli P, de Jong PJ, Yoshinaga Y, Haines JL, Pericak-Vance MA, Yan J, Ticozzi N, Siddique T, McKenna-Yasek D, Sapp PC, Horvitz HR, Landers JE, Brown RH Jr (2009) Mutations in the FUS/TLS gene on chromosome 16 cause familial amyotrophic lateral sclerosis. Science 323(5918):1205–1208. doi:10.1126/science.1166066

Lacomblez L, Bensimon G, Leigh PN, Guillet P, Meininger V (1996) Dose-ranging study of riluzole in amyotrophic lateral sclerosis. Amyotrophic Lateral Sclerosis/Riluzole Study Group II. Lancet 347(9013):1425–1431

Leblond CS, Kaneb HM, Dion PA, Rouleau GA (2014) Dissection of genetic factors associated with amyotrophic lateral sclerosis. Exp Neurol 262:91–101. doi:10.1016/j.expneurol.2014.04.013

Lemon RN (2008) Descending pathways in motor control. Annu Rev Neurosci 31:195–218. doi:10.1146/annurev.neuro.31.060407.125547

Lomen-Hoerth C, Anderson T, Miller B (2002) The overlap of amyotrophic lateral sclerosis and frontotemporal dementia. Neurology 59(7):1077–1079

Lomen-Hoerth C, Murphy J, Langmore S, Kramer JH, Olney RK, Miller B (2003) Are amyotrophic lateral sclerosis patients cognitively normal? Neurology 60(7):1094–1097

Matsui T, Sekiguchi M, Hashimoto A, Tomita U, Nishikawa T, Wada K (1995) Functional comparison of D-serine and glycine in rodents: the effect on cloned NMDA receptors and the extracellular concentration. J Neurochem 65(1):454–458

McBain CJ, Kleckner NW, Wyrick S, Dingledine R (1989) Structural requirements for activation of the glycine coagonist site of N-methyl-D-aspartate receptors expressed in Xenopus oocytes. Mol Pharmacol 36(4):556–565

Mitchell J, Paul P, Chen HJ, Morris A, Payling M, Falchi M, Habgood J, Panoutsou S, Winkler S, Tisato V, Hajitou A, Smith B, Vance C, Shaw C, Mazarakis ND, de Belleroche J (2010) Familial amyotrophic lateral sclerosis is associated with a mutation in D-amino acid oxidase. Proc Natl Acad Sci U S A 107(16):7556–7561. doi:10.1073/pnas.0914128107

Miya K, Inoue R, Takata Y, Abe M, Natsume R, Sakimura K, Hongou K, Miyawaki T, Mori H (2008) Serine racemase is predominantly localized in neurons in mouse brain. J Comp Neurol 510(6):641–654. doi:10.1002/cne.21822

Miyoshi Y, Hamase K, Tojo Y, Mita M, Konno R, Zaitsu K (2009) Determination of D-serine and D-alanine in the tissues and physiological fluids of mice with various D-amino-acid oxidase activities using two-dimensional high-performance liquid chromatography with fluorescence detection. J Chromatogr B Anal Technol Biomed Life Sci 877(24):2506–2512. doi:10.1016/j.jchromb.2009.06.028

Miyoshi Y, Konno R, Sasabe J, Ueno K, Tojo Y, Mita M, Aiso S, Hamase K (2012) Alteration of intrinsic amounts of D-serine in the mice lacking serine racemase and D-amino acid oxidase. Amino Acids 43(5):1919–1931. doi:10.1007/s00726-012-1398-4

Morikawa A, Hamase K, Inoue T, Konno R, Niwa A, Zaitsu K (2001) Determination of free D-aspartic acid, D-serine and D-alanine in the brain of mutant mice lacking D-amino acid oxidase activity. J Chromatogr B Biomed Sci Appl 757(1):119–125

Morikawa A, Hamase K, Zaitsu K (2003) Determination of D-alanine in the rat central nervous system and periphery using column-switching high-performance liquid chromatography. Anal Biochem 312(1):66–72

Mothet JP, Parent AT, Wolosker H, Brady RO Jr, Linden DJ, Ferris CD, Rogawski MA, Snyder SH (2000) D-serine is an endogenous ligand for the glycine site of the N-methyl-D-aspartate receptor. Proc Natl Acad Sci U S A 97(9):4926–4931

Nagata Y, Horiike K, Maeda T (1994) Distribution of free D-serine in vertebrate brains. Brain Res 634(2):291–295

Panatier A, Theodosis DT, Mothet JP, Touquet B, Pollegioni L, Poulain DA, Oliet SH (2006) Glia-derived D-serine controls NMDA receptor activity and synaptic memory. Cell 125(4):775–784. doi:10.1016/j.cell.2006.02.051

Papouin T, Ladepeche L, Ruel J, Sacchi S, Labasque M, Hanini M, Groc L, Pollegioni L, Mothet JP, Oliet SH (2012) Synaptic and extrasynaptic NMDA receptors are gated by different endogenous coagonists. Cell 150(3):633–646. doi:10.1016/j.cell.2012.06.029

Paul P, Murphy T, Oseni Z, Sivalokanathan S, de Belleroche JS (2014) Pathogenic effects of amyotrophic lateral sclerosis-linked mutation in D-amino acid oxidase are mediated by D-serine. Neurobiol Aging 35(4):876–885. doi:10.1016/j.neurobiolaging.2013.09.005

Philips T, Rothstein JD (2014) Glial cells in amyotrophic lateral sclerosis. Exp Neurol. doi:10.1016/j.expneurol.2014.05.015

Phukan J, Elamin M, Bede P, Jordan N, Gallagher L, Byrne S, Lynch C, Pender N, Hardiman O (2012) The syndrome of cognitive impairment in amyotrophic lateral sclerosis: a population-based study. J Neurol Neurosurg Psychiatry 83(1):102–108. doi:10.1136/jnnp-2011-300188

Renton AE, Majounie E, Waite A, Simon-Sanchez J, Rollinson S, Gibbs JR, Schymick JC, Laaksovirta H, van Swieten JC, Myllykangas L, Kalimo H, Paetau A, Abramzon Y, Remes AM, Kaganovich A, Scholz SW, Duckworth J, Ding J, Harmer DW, Hernandez DG, Johnson JO, Mok K, Ryten M, Trabzuni D, Guerreiro RJ, Orrell RW, Neal J, Murray A, Pearson J, Jansen IE, Sondervan D, Seelaar H, Blake D, Young K, Halliwell N, Callister JB, Toulson G, Richardson A, Gerhard A, Snowden J, Mann D, Neary D, Nalls MA, Peuralinna T, Jansson L, Isoviita VM, Kaivorinne AL, Holtta-Vuori M, Ikonen E, Sulkava R, Benatar M, Wuu J, Chio A, Restagno G, Borghero G, Sabatelli M, Consortium I, Heckerman D, Rogaeva E, Zinman L, Rothstein JD, Sendtner M, Drepper C, Eichler EE, Alkan C, Abdullaev Z, Pack SD, Dutra A, Pak E, Hardy J, Singleton A, Williams NM, Heutink P, Pickering-Brown S, Morris HR, Tienari PJ, Traynor BJ (2011) A hexanucleotide repeat expansion in C9ORF72 is the cause of chromosome 9p21-linked ALS-FTD. Neuron 72(2):257–268. doi:10.1016/j.neuron.2011.09.010

Renton AE, Chio A, Traynor BJ (2014) State of play in amyotrophic lateral sclerosis genetics. Nat Neurosci 17(1):17–23. doi:10.1038/nn.3584

Rosen DR, Siddique T, Patterson D, Figlewicz DA, Sapp P, Hentati A, Donaldson D, Goto J, O'Regan JP, Deng HX et al (1993) Mutations in Cu/Zn superoxide dismutase gene are associated with familial amyotrophic lateral sclerosis. Nature 362(6415):59–62. doi:10.1038/362059a0

Sasabe J, Aiso S (2010) Aberrant control of motoneuronal excitability in amyotrophic lateral sclerosis: excitatory glutamate/D-serine vs. inhibitory glycine/gamma-aminobutanoic acid (GABA). Chem Biodivers 7(6):1479–1490. doi:10.1002/cbdv.200900306

Sasabe J, Chiba T, Yamada M, Okamoto K, Nishimoto I, Matsuoka M, Aiso S (2007) D-serine is a key determinant of glutamate toxicity in amyotrophic lateral sclerosis. EMBO J 26 (18):4149–4159. doi:10.1038/sj.emboj.7601840

Sasabe J, Miyoshi Y, Suzuki M, Mita M, Konno R, Matsuoka M, Hamase K, Aiso S (2012) D-amino acid oxidase controls motoneuron degeneration through D-serine. Proc Natl Acad Sci U S A 109(2):627–632. doi:10.1073/pnas.1114639109

Sasabe J, Suzuki M, Imanishi N, Aiso S (2014) Activity of D-amino acid oxidase is widespread in the human central nervous system. Front Synaptic Neurosci 6:14. doi:10.3389/fnsyn.2014.00014

Shleper M, Kartvelishvily E, Wolosker H (2005) D-serine is the dominant endogenous coagonist for NMDA receptor neurotoxicity in organotypic hippocampal slices. J Neurosci 25 (41):9413–9417. doi:10.1523/JNEUROSCI.3190-05.2005

Sreedharan J, Brown RH Jr (2013) Amyotrophic lateral sclerosis: problems and prospects. Ann Neurol 74(3):309–316. doi:10.1002/ana.24012

Sreedharan J, Blair IP, Tripathi VB, Hu X, Vance C, Rogelj B, Ackerley S, Durnall JC, Williams KL, Buratti E, Baralle F, de Belleroche J, Mitchell JD, Leigh PN, Al-Chalabi A, Miller CC, Nicholson G, Shaw CE (2008) TDP-43 mutations in familial and sporadic amyotrophic lateral sclerosis. Science 319(5870):1668–1672. doi:10.1126/science.1154584

Thompson M, Marecki JC, Marinesco S, Labrie V, Roder JC, Barger SW, Crow JP (2012) Paradoxical roles of serine racemase and D-serine in the G93A mSOD1 mouse model of amyotrophic lateral sclerosis. J Neurochem 120(4):598–610. doi:10.1111/j.1471-4159.2011.07601.x

Vance C, Rogelj B, Hortobagyi T, De Vos KJ, Nishimura AL, Sreedharan J, Hu X, Smith B, Ruddy D, Wright P, Ganesalingam J, Williams KL, Tripathi V, Al-Saraj S, Al-Chalabi A, Leigh PN, Blair IP, Nicholson G, de Belleroche J, Gallo JM, Miller CC, Shaw CE (2009) Mutations in FUS, an RNA processing protein, cause familial amyotrophic lateral sclerosis type 6. Science 323(5918):1208–1211. doi:10.1126/science.1165942

Vucic S, Rothstein JD, Kiernan MC (2014) Advances in treating amyotrophic lateral sclerosis: insights from pathophysiological studies. Trends Neurosci 37(8):433–442. doi:10.1016/j.tins.2014.05.006

Wolosker H (2011) Serine racemase and the serine shuttle between neurons and astrocytes. Biochim Biophys Acta 1814(11):1558–1566. doi:10.1016/j.bbapap.2011.01.001

Wu S, Barger SW (2004) Induction of serine racemase by inflammatory stimuli is dependent on AP-1. Ann N Y Acad Sci 1035:133–146. doi:10.1196/annals.1332.009

Yang HW, Lemon RN (2003) An electron microscopic examination of the corticospinal projection to the cervical spinal cord in the rat: lack of evidence for cortico-motoneuronal synapses. Exp Brain Res 149(4):458–469. doi:10.1007/s00221-003-1393-9

Part III
Function of D-Aspartate

Chapter 10
Overview

Hiroshi Homma

Abstract As an introduction to Part III, the flow of D-aspartate research is described. The primary works and review articles concerning the emergence of D-aspartate and its action in the nonmammalian and mammalian animals are summarized.

Keywords D-Aspartate • Flow of research

As an introduction to this chapter, I look back over the course of D-aspartate (D-Asp) research and briefly summarize the primary reports and subsequent review articles for readers who hope to start D-Asp research.

D-Asp was first discovered by D'Aniello's group in the brain of cephalopods (octopus) (D'Aniello and Giuditta 1977). Thereafter, it was found in the nervous system, endocrine tissues, and reproductive organs of a variety of invertebrates and nonmammalian vertebrates, as reviewed in detail by D'Aniello (D'Aniello 2007). In mammals, D-Asp was first detected in rat brain and peripheral tissues (such as the liver) as well as in human blood (Dunlop et al. 1986). Dunlop et al. recognized that the D-Asp contents of tissues exhibit characteristic alterations during growth. Then, the group of Nishikawa carried out extensive research on the developmental changes in the D-Asp contents of various tissues in rat and human (Hashimoto et al. 1993, 1995). Thereafter, an anti-D-Asp antibody was prepared, and the localization of D-Asp was analyzed in rat tissues such as the brain and pineal gland (Schell et al. 1997; Lee et al 1997). Subsequent studies on the localization of D-Asp in a variety of tissues indicated that it is localized in a distinct type of cells at a particular stage of the developmental process (Homma 2002; Furuchi and Homma 2005).

Among nonmammalian vertebrates, D-Asp actions are related to testosterone levels in the ovary of frog (*Rana esculenta*) (Di Fiore et al. 1998) and with the testosterone and estradiol contents and ovarian aromatase activity in lizard (*Podarcis s. sicula*) (Assisi et al. 2001), suggesting a physiological role of D-Asp

H. Homma (✉)
Laboratory of Biomolecular Science, Graduate School of Pharmaceutical Life Sciences, Kitasato University, 5-9-1 Shirokane, Minato-ku, Tokyo 108-8641, Japan
e-mail: hommah@pharm.kitasato-u.ac.jp

T. Yoshimura et al. (eds.), *D-Amino Acids*, DOI 10.1007/978-4-431-56077-7_10

in reproductive organs. Thereafter, its action was reported in other tissues such as the testis and an exocrine organ (Harderian gland), as reviewed by Di Fiore's group (Raucci and Di Fiore 2011; Di Fiore et al. 2014). In this chapter, a review article is described by Di Fiore et al., entitled "Endocrine activity of D-aspartate in nonmammalian animals," in which the role of D-Asp in endocrine control of reproduction in amphibians, reptiles, and birds is summarized and reviewed.

D'Aniello et al. suggested that D-Asp acts as a neurotransmitter by using *Aplysia* as an experimental material (D'Aniello et al. 2011). Sweedler et al. detected D-amino acids at subcellular levels in a giant axon of *Aplysia* (Miao et al. 2005) and similarly studied and reviewed the role of D-Asp as a neurotransmitter (Ota et al. 2012). In addition, they cloned the gene encoding aspartate racemase of *Aplysia* (Wang et al. 2011). In this chapter, Sweedler and his associates summarize the analytical techniques for D-amino acids and review the biochemical and molecular biological aspects of D-Asp. They also provide comprehensive explanations for the role of D-Asp in nonmammalian animals.

The physiological activities of D-Asp were also reported in several endocrine tissues and reproductive glands of mammals, such as the pineal gland (Takigawa et al. 1998; Ishio et al. 1998), pituitary gland (Wang et al. 2000; Long et al. 2000; D'Aniello et al. 1996, 2000), and testis (D'Aniello et al. 1996; Nagata et al. 1999a, b), suggesting that D-Asp is involved in the regulation of hormonal secretion. Furthermore, clinical reports indicated the involvement of D-Asp in human reproduction and related diseases (D'Aniello et al. 2005, 2007). Several review articles have been published concerning the physiological function of D-Asp in mammals (Homma 2007; D'Aniello 2007; Katane and Homma 2011; Ota et al. 2012; Di Fiore et al. 2014). In this chapter, Homma and Katane describe a review article entitled "Homeostasis of free D-aspartate in mammalian cells".

Recently, there has been a major advance in research of the physiological action of D-Asp in the mammalian central nervous system. Knockout (KO) mice of D-Asp oxidase, a major degradative enzyme of acidic D-amino acids, were established (Huang et al. 2006; Errico et al. 2006), and the effects of its deletion were analyzed. It has been suggested that D-Asp is an agonist for the N-methyl-D-aspartate (NMDA) receptor, a subtype of the glutamate receptor in the mammalian brain. In KO mice, tissue levels of D-Asp (especially the brain level) are markedly increased, and the effects on brain function and behavioral changes were investigated. In addition to gene deletion, Usiello et al. studied the pharmacological effects of chronic administration of D-Asp, and detailed reviews of the physiological function of D-Asp were published (Errico et al. 2009, 2012). In this chapter, a review article is described entitled "Neuromodulatory activity of D-aspartate in mammals," concerning the correlation among D-Asp, NMDA receptor, and schizophrenia.

Although studies of the enzymes involved in D-Asp metabolism are introduced and summarized in the other chapters of this book, some review articles of enzymatic research of D-Asp biosynthesis and degradation were published (Ota et al. 2012; Katane and Homma 2010).

D-Asp was discovered earlier than D-serine. However, research of D-Asp lags far behind that of D-serine. Further studies of D-Asp and other acidic D-amino acids are expected to identify the protein(s) that binds directly to D-Asp and controls its biological actions and to characterize the enzymes related to D-Asp metabolism, such as its biosynthesis, and the molecular apparatuses for D-Asp transport and flow. Based on these findings and investigations, related diseases will be clarified, and improved drugs and relevant functional foods will be developed.

References

Assisi L, Botte V, D'Aniello A, Di Fiore MM (2001) Enhancement of aromatase activity by D-aspartic acid in the ovary of the lizard *Podarcis s. sicula*. Reproduction 121:803–808

D'Aniello A (2007) D-Aspartic acid: an endogenous amino acid with an important endocrine role. Brain Res Rev 53:215–234

D'Aniello A, Giuditta A (1977) Identification of D-aspartic acid in the brain of *Octopus vulgaris* lam. J Neurochem 29:1053–1057

D'Aniello S, Somorjai I, Garcia-Fernàndez J, Top E, D'Aniello A (2011) D-Aspartic acid is a novel endogenous neurotransmitter. FASEB J 25:1014–1027

D'Aniello A, Di Cosmo A, Di Cristo C, Annunziato L, Petrucelli L, Fisher G (1996) Involvement of D-aspartic acid in the synthesis of testosterone in rat testes. Life Sci 59:97–104

D'Aniello G, Tolino A, D'Aniello A, Errico F, Fisher GH, Di Fiore MM (2000) The role of D-aspartic acid and N-methyl-D-aspartic acid in the regulation of prolactin release. Endocrinology 141:3862–3870

D'Aniello G, Ronsini S, Guida F, Spinelli P, D'Aniello A (2005) Occurrence of D-aspartic acid in human seminal plasma and spermatozoa: possible role in reproduction. Fertil Steril 84:1444–1449

D'Aniello G, Grieco N, Di Filippo MA, Cappiello F, Topo E, D'Aniello E, Ronsini S (2007) Reproductive implication of D-aspartic acid in human pre-ovulatory follicular fluid. Hum Reprod 22:3178–3183

Di Fiore MM, Assisi L, Botte V, D'Aniello A (1998) D-Aspartic acid is implicated in the control of testosterone production by the vertebrate gonad. Studies on the female green frog, *Rana esculenta*. J Endocrinol 157:199–207

Di Fiore MM, Santillo A, Chieffi Baccari G (2014) Current knowledge of D-aspartate in glandular tissues. Amino Acids 46:1805–1818

Dunlop DS, Neidle A, McHale D, Dunlop DM (1986) The presence of free D-aspartic acid in rodent and man. Biochem Biophys Res Commun 141:27–35

Errico F, Pirro MT, Affuso A, Spinelli P, De Felice M, D'Aniello A, Di Lauro R (2006) A physiological mechanism to regulate D-aspartic acid and NMDA levels in mammals revealed by D-aspartate oxidase deficient mice. Gene 374:50–57

Errico F, Napolitano F, Nisticò R, Centonze D, Usiello A (2009) D-aspartate: an atypical amino acid with neuromodulatory activity in mammals. Rev Neurosci 20:429–440

Errico F, Napolitano F, Nisticò R, Usiello A (2012) A new insights on the role of free D-aspartate in the mammalian brain. Amino Acids 43:1861–1871

Furuchi T, Homma H (2005) Free D-aspartate in mammals. Biol Pharm Bull 28:1566–1570

Hashimoto A, Kumashiro S, Nishikawa T, Oka T, Takahashi K, Mito T, Takashima S, Doi N, Mizutani Y, Yamazaki T, Kaneko T, Ootomo E (1993) Embryonic development and postnatal changes in free D-aspartate and D-serine in the human prefrontal cortex. J Neuochem 61:348–351

Hashimoto A, Oka T, Nishikawa T (1995) Anatomical distribution and postnatal changes in endogenous free D-aspartate and D-serine in rat brain and periphery. Eur J Neurosci 7:1657–1663

Homma H (2002) D-Aspartate in the mammalian body. Viva Origino 30:204–215. Available online at http://www.origin-life.gr.jp/3004/3004204/3004204.pdf

Homma H (2007) Biochemistry of D-aspartate in mammalian cells. Amino Acids 32:3–11

Huang AS, Beigneux A, Weil ZM, Kim PM, Molliver ME, Blackshaw S, Nelson RJ, Young SG, Snyder SH (2006) D-Aspartate regulates melanocortin formation and function: behavioral alterations in D-aspartate oxidase-deficient mice. J Neurosci 26:2814–2819

Ishio S, Yamada H, Hayashi M, Yatsushiro S, Noumi T, Yamaguchi A, Moriyama Y (1998) D-Aspartate modulates melatonin synthesis in rat pinealocytes. Neurosci Lett 249:143–146

Katane M, Homma H (2010) D-Aspartate oxidase: the sole catabolic enzyme acting on free D-aspartate in mammals. Chem Biodivers 7:1435–1449

Katane M, Homma H (2011) D-Aspartate—an important bioactive substance in mammals: a review from an analytical and biological point of view. J Chromatogr B Analyt Technol Biomed Life Sci 879:3108–3121

Lee JA, Homma H, Sakai K, Fukushima T, Santa T, Tashiro K, Iwatsubo T, Yoshikawa M, Imai K (1997) Immunohistochemical localization of D-aspartate in the rat pineal gland. Biochem Biophys Res Commun 231:505–508

Long Z, Lee JA, Okamoto T, Nimura N, Imai K, Homma H (2000) D-Aspartate in a prolactin-secreting clonal strain of rat pituitary tumor cells (GH_3). Biochem Biophys Res Commun 276:1143–1147

Miao H, Rubakhin SS, Sweedler JV (2005) Subcellular analysis of D-aspartate. Anal Chem 77:7190–7194

Nagata Y, Homma H, Matsumoto M, Imai K (1999a) Stimulation of steroidogenic acute regulatory protein (StAR) gene expression by D-aspartate in rat Leydig cells. FEBS Lett 454:317–320

Nagata Y, Homma H, Lee JA, Imai K (1999b) D-Aspartate stimulation of testosterone synthesis in rat Leydig cells. FEBS Lett 444:160–164

Ota N, Shi T, Sweedler JV (2012) D-aspartate acts as a signaling molecule in nervous and neuroendocrine systems. Amino Acids 43:1873–1886

Raucci F, Di Fiore MM (2011) D-Asp: a new player in reproductive endocrinology of the amphibian *Rana esculenta*. J Chromatogr B Analyt Technol Biomed Life Sci 879:3268–3276

Schell MJ, Cooper OB, Snyder SH (1997) D-aspartate localizations imply neuronal and neuroendocrine roles. Proc Natl Acad Sci U S A 92:3948–3952

Takigawa Y, Homma H, Lee JA, Fukushima T, Santa T, Iwatsubo T, Imai K (1998) D-Aspartate uptake into cultured rat pinealocytes and the concomitant effect on L-aspartate levels and melatonin secretion. Biochem Biophys Res Commun 248:641–647

Wang H, Wolosker H, Pevsner J, Snyder SH, Selkoe DJ (2000) Regulation of rat magnocellular neurosecretory system by D-aspartate: evidence for biological role(s) of a naturally occurring free D-amino acid in mammals. J Endocrinol 167:247–252

Wang L, Ota N, Romanova EV, Sweedler JV (2011) A novel pyridoxal 5′-phosphate-dependent amino acid racemase in the *Aplysia californica* central nervous system. J Biol Chem 286:13765–13774

Chapter 11
Endocrine Activity of D-Aspartate in Nonmammalian Animals

Maria Maddalena Di Fiore, Lavinia Burrone, Alessandra Santillo, and Gabriella Chieffi Baccari

Abstract D-Aspartate is a natural D-amino acid endogenously present in animal tissues and able to bind to NMDA receptors. The biological actions of D-aspartate have been demonstrated in the central nervous system and in peripheral tissue where the amino acid is under the tight control of two enzymes responsible for its synthesis and degradation, D-aspartate racemase and D-aspartate oxidase, respectively. Furthermore, D-aspartate intervenes on aromatase enzyme functioning.

In the last 20 years, many reports have pointed out this D-amino acid as mediators of critical signals involved together with sex hormones in various aspects of animal and human reproduction. Although mammalian models have been mainly used in the investigation of the role of D-aspartate in reproduction, this amino acid has also been identified in the gonads of nonmammalian animals, in particular in amphibians, reptiles, and birds.

The identification of its occurrence in reproductive cells and organs of vertebrates highlights the key role played by these endogenous compounds along the evolutionary axis. In this review we have summarized the current knowledge of the main actions of D-aspartate on female and male reproductive processes along the evolutionary axis of nonmammalian animals, focusing on the endocrine activity and discussing the interplay between D-aspartate functions and steroid hormones in regulating the reproduction.

Keywords D-Aspartate • Reproduction • Sex hormones • Testis • Ovary

11.1 Background in Mammals

Although it is beyond the scope of this review, a very brief accent is being here.

In the last two decades, growing evidence has been accumulated to show the occurrence of D-aspartate (D-Asp) in endocrine tissues and its central role in

---

M.M. Di Fiore (✉) • L. Burrone • A. Santillo • G. Chieffi Baccari
Dipartimento di Scienze e Tecnologie Ambientali, Biologiche e Farmaceutiche, Seconda Università di Napoli (SUN), via Vivaldi, 43, 81100 Caserta, Italy
e-mail: MariaM.DiFiore@unina2.it

T. Yoshimura et al. (eds.), *D-Amino Acids*, DOI 10.1007/978-4-431-56077-7_11

controlling endocrine functions (for review, see Homma 2007; D'Aniello 2007; Ota et al. 2012; Di Fiore et al. 2014). The investigations have been mostly dedicated to mammals, demonstrating that D-Asp is involved in several endocrine steps, which follow one another to guarantee the synthesis and/or the release of several hormones. In mammals, an increase in local concentrations of D-Asp facilitates the endocrine secretion of anterior pituitary hormones, i.e., LH and GH (D'Aniello et al. 2000a) and PRL (D'Aniello et al. 2000b), whereas it inhibits the secretion of POMC/α-MSH from the intermediate pituitary (Huang et al. 2006) and of oxytocin from the posterior pituitary (Wang et al. 2000). D-Asp also acts as a negative regulator for melatonin synthesis in the pineal gland (Takigawa et al. 1998; Ishio et al. 1998).

D-Asp has been demonstrated to play a peculiar role on the control of reproductive functions in mammals. Recently, in rat testis, it has been demonstrated that it induces testicular NMDA receptor–extracellular signal-regulated kinase pathway and upregulates androgen receptor expression (Santillo et al. 2014).

Several reports have showed that D-Asp exerts potent positive effects upon both experimental animal and human reproduction, affecting the gonadotropin synthesis and release, the gonadic steroid production, the spermatogenesis, and the ovulation (D'Aniello et al. 1998, 2000a, b, 2005, 2007, 2012; Lamanna et al. 2006, 2007a, b; Falvo et al. 2015; Santillo et al. 2016). In mammals D-aspartate modulates reproductive functions by acting at multiple levels: (1) it can affect the secretion of gonadotropin-releasing hormone (GnRH) by acting at the hypothalamic level, generating an indirect cascade role on the hypothalamic-pituitary axis; (2) a direct action on pituitary is possible, since D-aspartate regulates pituitary hormone secretions; and (3) it has a local role on gonads, inducing sex steroid hormone secretion (progesterone, testosterone, 17β-estradiol) and therefore modulating the reproductive functions through steroidogenesis.

11.2 Role of D-Asp in *Ciona intestinalis*

Only one study has been carried out on endocrine activity exerted by D-Asp in *Ciona intestinalis*. It is an emerging animal model, situated in a key phylogenetic position, and considered a descent of a chordate ancestor.

C. intestinalis is a marine protochordate belonging to the subphylum Urochordata (Tunicata), class Ascidiacea. This species is considered to be phylogenetically about 500 million years distant from the highest vertebrate group, and it is postulated that the ascidians share common ancestry with vertebrates. From an endocrine point of view, *C. intestinalis* possesses steroid hormones such as progesterone, testosterone, and 17β-estradiol (Delrio et al. 1971; Cangialosi et al. 2010), a humanlike thyroid-stimulating hormone (Di Fiore et al. 1997), and two different forms of GnRH with amino acid compositions similar to that occurring in birds and mammals (Di Fiore et al. 2000). Therefore, because of a dynamic developmental pattern and a simplified anatomical structure, *C. intestinalis* represents an excellent experimental model for developmental and phylogenetic studies.

D'Aniello and co-workers (2003) have demonstrated the occurrence of endogenous D-Asp and of its methylated form, N-methyl-D-aspartic acid (NMDA), in the neural complex and in the gonads of *C. intestinalis*, indicating their involvement in hormonal activity. D-Asp is synthesized from L-Asp by an aspartate racemase; it is transferred to the neural gland where it gives rise to NMDA by an NMDA synthase. NMDA, in turn, passes from the neural gland into the gonads where it induces the synthesis and release of GnRH. The GnRH in turn modulates the release and synthesis of testosterone and progesterone in the gonads (D'Aniello et al. 2003).

Given the simplicity of *C. intestinalis* endocrine system with respect to mammals, the discovery of D-Asp in this animal has been important to gain insights into the role of endogenous D-Asp in the endocrine system of metazoans and particularly in the chordate lineage.

11.3 Role of D-Aspartate in the Endocrine Regulation of Reproduction in Nonmammalian Animals

11.3.1 Nonmammalian Animals as Models of Study

It has been generally recognized that a comparative approach gains a deep insight on adaptative phenomena leading to evolutionarily track as well as provides an insight into the understanding of physiological mechanisms in building general models. Besides evolutionary speculations, the major advantage of nonmammalian animals in the study of reproductive function is that most of them are seasonal breeders with a physiological switch on/off of gonadic activity. Environmental factors such as temperature and photoperiod deeply control reproductive functions; both can easily be modified in the laboratory, leading to a controlled suppression and/or induction of reproductive functions.

Lastly, gonad architecture is simpler than mammals, and this certainly facilitates morpho-functional studies. Indeed, nonmammalian vertebrates have been recognized to possess morphological features to better study relationships between different neuroendocrine-paracrine systems. In this respect, the use of experimental model other than mammals has been useful to elucidate some functional aspects of D-aspartate. Seasonality of reproduction in nonmammalian animals has been the subject of numerous studies, including investigations at the brain level, well known to be the production site of neurosteroids. In the present review, we will mainly describe the current knowledge about the role of D-aspartate in endocrine control of reproduction in amphibians, reptiles, and birds.

11.3.1.1 *Pelophylax esculentus*

In the amphibian green frog *Pelophylax esculentus* (new nomenclature of *Rana esculenta*), mating usually occurs in early spring. Occurrence and role of D-Asp have been investigated during the three main phases of their reproductive cycle that are prereproductive (October), reproductive (February–March), and post-reproductive (July), in both sexes.

It has been demonstrated that D-Asp undergoes regular changes in its concentration throughout the reproductive cycle in both ovary (Di Fiore et al. 1998) and testis (Raucci et al. 2004).

It is to be taken into account that the ovaries of *P. esculentus* undergo an annual cycle of growth and developmental maturation depending on the reproductive season. At pre-reproduction, the majority of follicles appear to be previtellogenic; no estimable accumulation of yolk plates is observed in the oocyte cytoplasm (Fig. 11.1a). At reproduction, the follicles appeared fully engaged in vitellogenic and the zona pellucida is fully distinguishable (Fig. 11.1a). In post-reproductive frogs, the follicles were smaller and no indication of vitellogenesis was observed (Fig. 11.1a). D-Asp concentrations undergo significant variations during the main phases of the sexual cycle. In spawning females (March), its concentration was low, and during the post-reproductive period (June), it increased and reached its peak level in prereproductive phase (October) (Fig. 11.1b). The concentrations of D-Asp in the ovary and of testosterone in both ovary and plasma were inversely correlated during the reproductive cycle: when endogenous D-Asp was low (reproductive period), testosterone was high in the ovary (Fig. 11.1b) and plasma (23.1 ± 2.76 ng/ml), and, in contrast, when the D-Asp concentration was high (prereproductive period), the testosterone concentration was low in the ovary (Fig. 11.1b) and plasma (5.0 ± 1.3 ng/ml). In vivo experiments, consisting of injection of D-Asp (2 μmol/g body weight) into the dorsal lymphatic sac, demonstrated that this amino acid accumulated significantly in the ovary. After 3 h, D-Asp treatment caused a decrease in plasma testosterone levels of about 80 %. This inhibition was reversible: within 18 h after the amino acid injection, as the D-Asp concentration in the ovary decreased, the testosterone titer was restored. In vitro experiments, performed in isolated ovarian follicles, confirmed this phenomenon. Other amino acids (L-Asp, D-Glu, L-Glu, D-Ala, and L-Ala) used instead of D-Asp were ineffective (Di Fiore et al. 1998).

Furthermore, it has been demonstrated that D-Asp enhances the maturation of frog oocytes, determining a higher accumulation of carbohydrate yolk plates in the ooplasm (Raucci and Di Fiore 2011).

The study, carried out on the male of *P. esculentus*, investigated the involvement of D-Asp in testicular steroidogenesis and its action on both morphology and glandular activity of thumb pad, a typical testosterone-dependent secondary sexual character in this amphibian species (Raucci et al. 2004).

Figure 11.2a reports morphological modification of germinal epithelium of the testis through the sexual cycle. In the prereproductive period, seminiferous

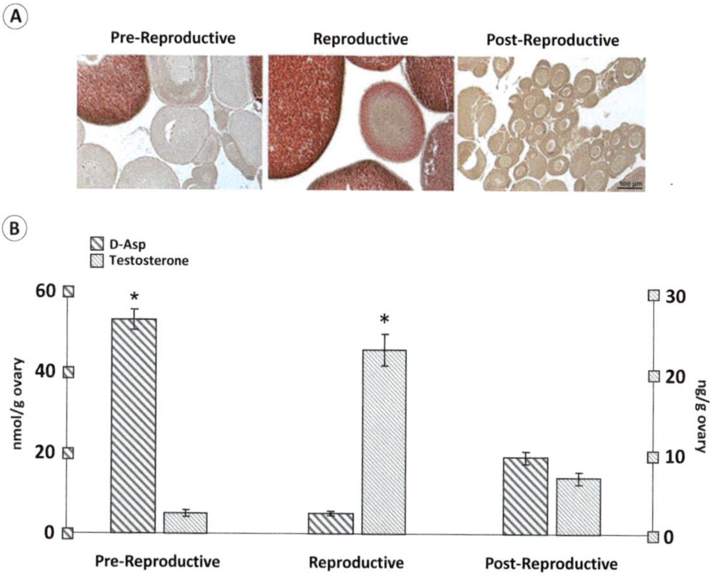

Fig. 11.1 (**a**) Oocyte morphology. Hematoxylin/eosin-stained sections of follicles of *Pelophylax esculentus* during the sexual cycle. (**b**) D-Asp and testosterone levels in the green frog ovary during the prereproductive, reproductive, and post-reproductive periods. Each point represents the mean value ± S.D. from five individual female frogs. *P < 0.01 prereproductive versus reproductive and post-reproductive; *P < 0.01 reproductive versus prereproductive and post-reproductive

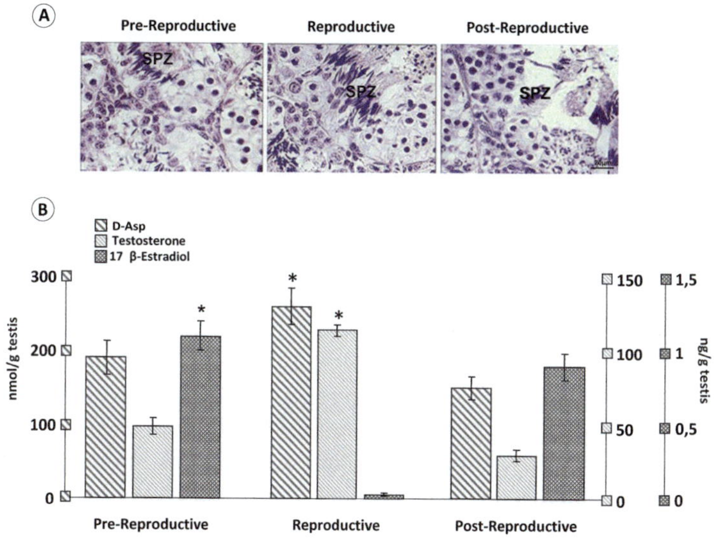

Fig. 11.2 (**a**) Testis morphology. Hematoxylin/eosin-stained sections of *Pelophylax esculentus* testis during the sexual cycle (SPZ, spermatozoa). (**b**) D-Asp, testosterone, and 17β-estradiol levels in the green frog testis during prereproductive, reproductive, and post-reproductive periods. Each point represents the mean value ± S.D. from five individual male frogs. *$P < 0.01$ prereproductive versus reproductive; *$P < 0.01$ reproductive versus post-reproductive (D-Asp). *$P < 0.01$ reproductive versus prereproductive and post-reproductive (testosterone)

ampoules of the testis contained predominantly primary and secondary spermatogonia (Fig. 11.2a). In the reproductive period, testicular epithelium contained the maximal number of primary spermatogonia, several secondary spermatogonia, and primary and secondary spermatocytes, and the lumen was full of spermatozoa (SPZ) (Fig. 11.2a). In the post-reproductive period, high numbers of cysts still contained primary and secondary spermatocytes and a very few SPZ were present (Fig. 11.2a). In the testis, in all three phases of the reproductive cycle, substantial amounts of free D-Asp were found in testicular extracts, and its concentration varied significantly during the reproductive cycle: it was lower in pre- and post-reproductive periods, but when animals are engaged in sexual activity (reproductive period), D-Asp reached the peak level (Fig. 11.2b). Moreover, testis D-Asp concentrations through the sexual cycle positively matched with testis testosterone levels (Fig. 11.2b).

In vivo short-term experiments consisting of a single injection of D-Asp (2 μmol/g body weight) demonstrated that this amino acid accumulated significantly in the testis, and after 3 h its uptake is coupled with a testosterone increase in both testis and plasma. Then, within 18 h of amino acid administration, testis D-Asp concentration decreased to pre-stimulation levels as well as testis and serum testosterone levels.

In in vivo long-term experiments, D-Asp was chronically administered to frogs caught during the three phases of the reproductive cycle. D-Asp induced testosterone increase and 17β-estradiol decrease in the testis during the pre- and post-reproductive period and vice versa during the reproductive period. The stimulatory effect of D-Asp on testosterone production by the testis is consistent with the stimulation of spermatogenesis and the maturation of thumb pads occurring in D-Asp-treated frogs. In these animals, there was an increase of seminiferous ampoule area and a higher number of spermatids and sperm. Moreover, in spermatogonia I and II and in spermatocytes, proliferating cell nuclear antigen (PCNA) intense immunopositivity was observed. In addition, the thumb pads of D-Asp-treated frogs compared with controls showed a significantly thicker epithelial lining, a wider area of their glands with taller secretory cells (Raucci et al. 2004).

In conclusion D-Asp is endowed with the control of the synthesis and/or the release of testosterone by the gonad as function of the sex. In the ovary a reverse relation is founded between D-Asp and testosterone levels; conversely, in the testis this link seems to be direct (Raucci and Di Fiore 2011).

To examine the molecular involvement of D-Asp on steroidogenesis, we have carried out a study on the cyclic gene expression of key enzymes involved in the steroidogenic pathway regulation in the testis of *P. esculentus* (Burrone et al. 2012a). Particularly, we analyzed the expression of StAR (steroidogenic acute regulatory protein, required for shuttling cholesterol across the mitochondrial membrane), P450 aro (P450 aromatase, the enzyme responsible of testosterone conversion into 17β-estradiol), and 5αRed (5α-reductase, an enzyme which converts testosterone into 5α-dihydrotestosterone) mRNAs in *P. esculentus* testis, either in relation to the reproductive cycle or to D-Asp treatment (Fig. 11.3). Basal StAR mRNA levels were higher in reproductive (Fig. 11.3a) than in post-

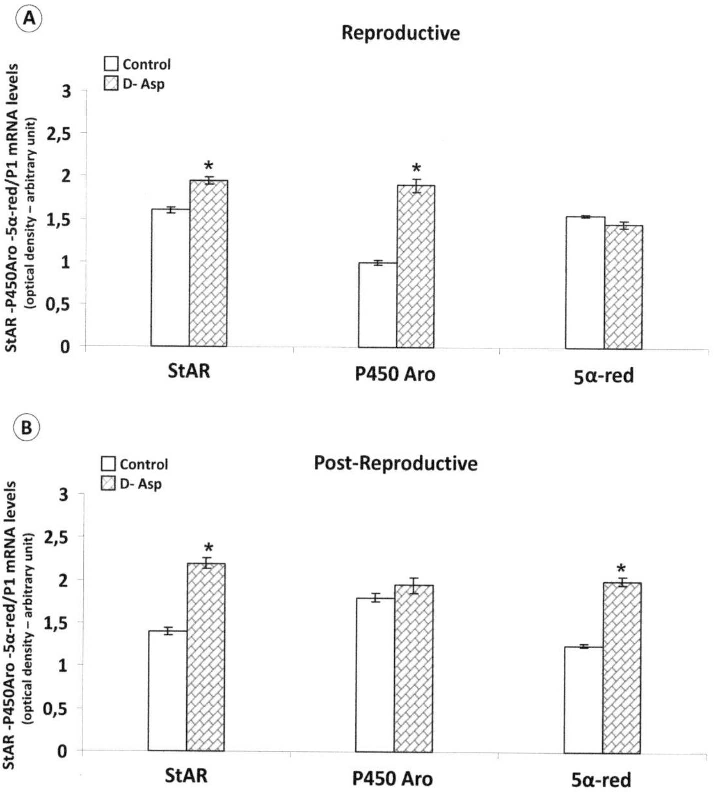

Fig. 11.3 RT-PCR-based measurements of StAR, P450 aro, and 5α-red mRNA in the testis from control and D-Asp-treated frogs during the reproductive (**a**) and post-reproductive (**b**) periods. Ribosomal protein P1 mRNA levels were measured as the internal standard. Each point represents the mean value ± SEM of five measurements. $^*P < 0.05$ versus controls

reproductive (Fig. 11.3b) periods. D-Asp treatment increased StAR mRNA expression in both the reproductive (Fig. 11.3a) and post-reproductive (Fig. 11.3b) periods. Although basal P450 aromatase mRNA levels were higher in the post-reproductive period (Fig. 11.3b), following D-Asp administration, they increased only in the reproductive period. The level of 5αRed mRNA was higher in reproductive (Fig. 11.3a) than in post-reproductive (Fig. 11.3b) frogs, and it increased after D-Asp treatment only in the post-reproductive phase (Fig. 11.3b). The results of this study demonstrated that D-Asp could modulate frog male sex steroid levels through StAR, P450 aromatase, and 5αRed gene expressions, depending on the endocrine status of the animals. In fact, in the reproductive period, because the testosterone concentrations and 5αRed mRNA levels are already elevated, the D-Asp administration did not augment 5αRed mRNA expression, whereas it activated the aromatase to prevent a testosterone surplus. In the post-reproductive period, where testosterone concentrations and 5αRed mRNA levels were lower, the D-Asp administration increased their levels, mimicking the physiology of the reproductive period (Burrone et al. 2012a).

11.3.1.2 *Podarcis sicula sicula*

In the ovary of the lizard *Podarcis sicula sicula*, endogenous D-Asp content varied with the sexual cycle phase (Assisi et al. 2001). In reproductive females, it was at the highest values, significantly lower in post-reproductive females, and intermediate in females at the prereproductive phase (Fig. 11.4). Testosterone level was high in both pre- and post-reproductive females and low in the reproductive ones in both ovary (Fig. 11.4) and plasma (prereproductive 5.54 ± 0.29 ng/ml, reproductive 1.54 ± 0.13 ng/ml, post-reproductive 7.16 ± 0.53 ng/ml). Contrariwise, 17-β-estradiol concentrations were low in pre- and post-reproductive females and reached the highest value in reproductive females in ovary (Fig. 11.4) and plasma (prereproductive 1.50 ± 0.15 ng/ml, reproductive 2.37 ± 0.13 ng/ml, post-reproductive 0.75 ± 0.06 ng/ml). Therefore, the D-Asp trend was inversely related to testosterone and directly parallel to 17β-estradiol patterns (Assisi et al. 2001).

Another study carried out on lizard ovary demonstrated that D-Asp is endogenously present in the oocytes, and its distribution varies during the reproductive cycle and following intraperitoneal administration (Raucci and Di Fiore 2010). At previtellogenesis, it is observed in the cytoplasm and nucleus of pyriform cells, in intermediate cells, in some small cells of the granulosa, in the ooplasm, and in some thecal elements. At vitellogenesis, D-Asp is localized in the proximity of the zona pellucida, in the theca, and in the ooplasm. Injected D-Asp is mainly captured by pyriform cells and ooplasm of previtellogenic oocytes, but a moderate accumulation is evident in the cytoplasm of some small granulosa cells and in the theca (Raucci and Di Fiore 2010). Therefore, taken together, the results of studies in lizard ovary show that D-Asp may be related to the synchrony of reproduction,

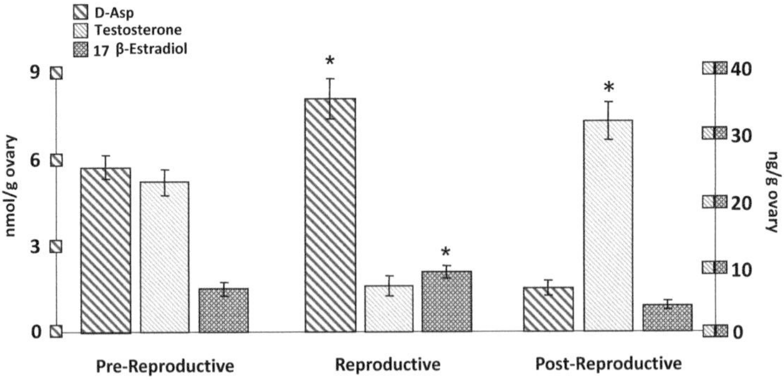

Fig. 11.4 D-Asp, testosterone, and 17β-estradiol levels in the lizard (*Podarcis s. sicula*) ovary during the prereproductive, reproductive, and post-reproductive periods. Each point represents the mean value ± S.D. from five individual female lizards. $*P < 0.01$ reproductive versus post-reproductive; $*P < 0.01$ post-reproductive versus reproductive

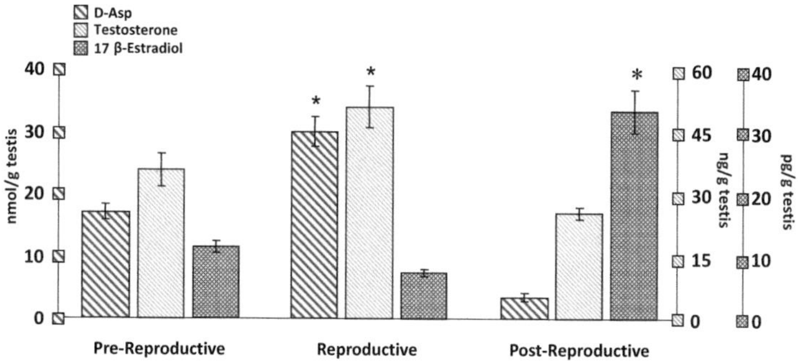

Fig. 11.5 D-Asp, testosterone, and 17β-estradiol levels in the lizard (*Podarcis s. sicula*) testis during the prereproductive, reproductive, and post-reproductive periods. Each point represents the mean value ± S.D. from five individual male lizards. *$P < 0.01$ reproductive versus prereproductive and post-reproductive; *$P < 0.01$ post-reproductive versus prereproductive and reproductive

either enhancing the growth and maturation of follicular epithelium or influencing its endocrine functions.

In the testis of *P. s. sicula*, endogenous D-Asp underwent significant variations depending on the reproductive cycle phase (Fig. 11.5) (Raucci et al. 2005). The D-Asp level was significantly higher during the reproductive period than in pre- and post-reproductive periods, and they are directly correlated with testosterone and inversely correlated with 17β-estradiol levels (Fig. 11.5).

In adult males of *P. s. sicula*, the i.p. injection of D-Asp (2 μmol/g body weight) was followed by the amino acid significant, although temporary, uptake by the testis in all sexual cycle stages. Particularly, in the prereproductive phase, D-Asp was rapidly taken up by the testis so that 3 h after injection its concentration was about seven times greater than the value observed in animals injected with the saline alone (from 17.0 ± 1.2 nmol/g tissue to 120.1 ± 9.9 nmol/g tissue). D-Asp levels were still high at 6 and 15 h after injection and then reached near baseline value within 24 h (23.0 ± 1.7 nmol/g tissue). In the reproductive period, D-Asp was accumulated in the testis, peaking at same set time observed in the prereproductive period (3 h), although its uptake was only twice as much; basal values were reached within 15–24 h. D-Asp was also rapidly taken up by the testis in the post-reproductive period, and 3 h after injection its levels were about 30 times greater than the endogenous content. At 6 h after injection its levels were still high; they decreased at 15 and 24 h. D-Asp administration affected the levels of sex hormones. In the prereproductive period, 3 h after D-Asp injection, a significant increase of testicular testosterone was observed (from 35.0 ± 4.0 to 112.7 ± 10.9 ng/g tissue). This effect also appeared in circulation (plasma) when the testosterone peaked at 6 h after injection (from 8.0 ± 0.9 to 45.3 ± 4.7 ng/ml plasma). Plasma testosterone concentration reached baseline within 24 h. The changes in 17β-estradiol after injection of D-Asp were different. Estradiol concentrations in testes decreased at 3 h (from 11.5 ± 0.9 pg/g to 3.0 ± 0.7 pg/g tissue) and in the plasma at 6 h (from 0.7 ± 0.2 to

0.2 ± 0.1 ng/ml plasma). However, this effect appeared to be reversible because plasma and testicular levels of 17β-estradiol returned to baseline values within 24 h of treatment. In the reproductive and post-reproductive period, both testosterone and 17β-estradiol increased.

In order to characterize the germinal cells involved in the capture and accumulation of D-Asp, immunocytochemistry against D-Asp was performed. D-Asp immunoreactivity was localized in both cytoplasm and nucleus of Leydig cells, in peritubular cells, and in spermatogonium cytoplasm (Raucci and Di Fiore 2009). After D-Asp treatment, immunostaining revealed exogenous D-Asp highly accumulated by the testis at 3 h after injection. At this time the increase of immunostaining intensity was specifically observed in the Leydig cells, in peritubular elements, and in some spermatogonia and Sertoli cells.

11.3.1.3 *Anas platyrhynchos*

There is only one paper reporting the endocrine activity of D-Asp in a bird, the mallard duck *Anas platyrhynchos* (Di Fiore et al. 2008). Free D-Asp was endogenously contained in *Anas* testis and varied between the two sexual reproductive periods supporting its correlation with reproductive activity. D-Asp concentration was higher during the reproductive period than during the post-reproductive (Fig. 11.6).

Testosterone level in testicular extracts well correlated with D-Asp levels (Fig. 11.6). In fact, the high testicular testosterone content occurring during the reproductive period was coupled to a high D-Asp level in the gonad. By contrast, in sexually inactive *Anas* (post-reproductive), the low testosterone content in the testis was coupled to a low D-Asp content.

The immunohistochemical analyses of D-Asp in the testes of reproductive and post-reproductive mallards revealed the presence of the amino acid mostly in Leydig cells. A feeble positive reaction was also noticed in Sertoli cells, in peritubular cells surrounding the seminiferous tubule, in spermatogonia, and in

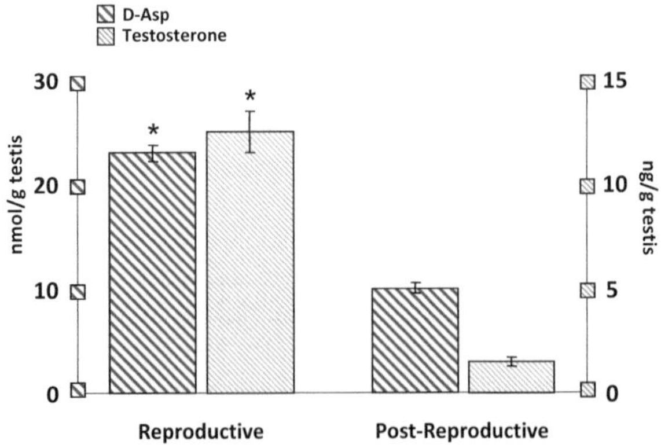

Fig. 11.6 D-Asp and testosterone levels in the duck (*A. platyrhynchos*) testis during the reproductive and post-reproductive periods. Each point represents the mean value ± S.D. from six individual male ducks. *$P < 0.01$ reproductive versus post-reproductive

spermatids. Although no difference in the number of Leydig cells was found between the two phases of the reproductive cycle considered, a significant increase in D-Asp immunopositive Leydig cells was detected during the reproductive but not during the post-reproductive period (Di Fiore et al. 2008).

To verify the exogenous effects of D-Asp on testicular testosterone release, slices of duck testes were in vitro incubated for 60 and 120 min in a medium containing D-Asp, where it induced a significant increase in testosterone synthesis with respect to controls (Di Fiore et al. 2008).

The results of this study suggested the involvement of D-Asp in testosterone production in the adult captive wild-strain mallard drake. Further, the localization of this molecule in the Leydig cells in different periods of the reproductive cycle strongly supports the direct involvement in biosynthesis and/or release of testosterone.

11.4 Enzymes Involved in Biosynthesis, Degradation, and Functioning of D-Aspartate

11.4.1 D-Aspartate Racemase

The endogenous presence of D-Asp in gonads of above reported nonmammalian animals suggests a local biosynthesis of this amino acid, being generated from L-Asp through the activity of the D-aspartate racemase, an enzyme which converts D-Asp to L-Asp.

In a study carried out on *R. esculenta* (Raucci et al. 2004), D-Asp was incubated with a homogenate of testis, and the formation of D-Asp was determined. Racemase activity occurred in the testis of *R. esculenta* with the maximum at pH 7.0, while at pH 6.0 and 8.0 the activity was about 80 % of pH 7.0. In addition, a variation of racemase concentration was observed in different periods of the year. High activity was observed during the reproductive cycle, where the values were found to be 25.3 ± 3.9 U/g, as opposed to 9.1 ± 2.0 U/g in the prereproductive period and 5.5 ± 1.2 U/g in the post-reproductive period.

We have reported that D-Asp synthesis also occurred in the testis of the lizard *Podarcis s. sicula* (Raucci et al. 2005). Testis tissue homogenates were incubated with L-Asp under different pH values. The conversion rate (L-Asp/D-Asp) was the highest when the in vitro incubation was carried out at pH 6.0. Testicular tissue converted L-Asp into D-Asp in all phases of the cycle reaching its maximal level in the reproductive period (239.0 ± 21.4 nmol/g tissue) when the endogenous content of both free D-Asp and testosterone was maximum and spermatogenesis was active.

11.4.2 D-Aspartate Oxidase

D-Aspartate oxidase (D-AspO, Ddo; EC 1.4.3.1) is a flavin-dependent enzyme that catalyzes the oxidative deamination of D-Asp to produce oxaloacetate, ammonia, and hydrogen peroxide. Localizations of D-AspO are reciprocal to D-Asp, suggesting that the enzyme depletes endogenous stores of the amino acid. Therefore, D-AspO-enriched tissues, low in D-Asp, may represent areas of high turnover where D-Asp may be physiologically important. Frog testis contained D-AspO and specifically expressed D-AspO activity in response to D-Asp (Burrone et al. 2010; Di Giovanni et al. 2010). The enzyme activity increased sixfold after D-Asp administration (2 μmol/g body weight).

11.4.3 Aromatase

Aromatase (P450 aromatase, estrogen synthase), the enzyme that catalyzes the transformation of testosterone into estradiol (for a review, see Simpson et al. 1994), is an enzyme involved in the reproductive physiology and is expressed both in gonads and the brain.

The results from the study in the oviparous lizard *Podarcis s. sicula* support the hypothesis that D-Asp plays a direct local role in the enhancement of ovarian aromatase activity (Assisi et al. 2001). This hypothesis is supported by the relationships among the concentrations of D-Asp, testosterone, and estradiol that were observed (1) during the reproductive cycle of the lizard, (2) under experimental treatments, and (3) in experiments performed in vitro. Lizard ovarian follicles converted testosterone to estradiol. The conversion rate was significantly increased by the addition of D-Asp into the incubation wells containing fresh ovarian tissue. An increase in the rate of conversion was also observed when D-Asp was added to incubation wells containing acetone powder extracts from the ovarian follicles. This conversion rate, evaluated on the basis of substrate concentration, was four times greater in the presence of D-Asp than in the controls.

Using RT-PCR, P450 aromatase mRNA expression levels were measured in testes from control and D-Asp-treated frogs (*P. esculentus*) during both the reproductive and post-reproductive periods, respectively. Basal mRNA levels of aromatase were significantly higher in post-reproductive than in reproductive testes (Burrone et al. 2012a). At reproduction, P450 aromatase mRNA levels increased of about twofold in D-Asp-injected frogs versus controls, whereas, in the post-reproductive period, no differences were observed following D-Asp treatment. However, in the reproductive period, when the testosterone production is already at its maximum, additional D-Asp administration increased testosterone levels, causing the stimulation of P450 aromatase gene expression, which converts the testosterone into 17β-estradiol (Burrone et al. 2012a).

Brain aromatase is essential for sex hormone-mediated regulation of physiological and behavioral processes, such as sexual differentiation of the brain, activation of male sexual behavior, and regulation of gonadotropic hormone secretion, as well as neurogenesis.

Given the role of aromatase in reproductive physiology, our research group has evaluated the possible existence of a crosstalk between D-Asp and brain aromatase, by analyzing either D-Asp content or the distribution in the brain sections or the expression of aromatase in the frog *P. esculentus* following D-aspartate treatment (Burrone et al. 2012b). We found that D-aspartate enhanced brain aromatase expression (mRNA and protein) through the CREB pathway. Then, P450 aromatase induced 17β-estradiol production from testosterone, with a consequent increase of its receptor. Therefore, the regulation of D-aspartate-mediated P450 aromatase expression could be an important step in the control of neuroendocrine regulation of the reproductive axis (Burrone et al. 2012b).

In another paper (Santillo et al. 2013), we have reported that there is a direct correlation among brain levels of D-Asp, P450 aromatase, 17β-estradiol, and estradiol receptor-α (ERα) in the male frogs during the reproductive as well as the post-reproductive phases of the breeding cycle, with the highest levels being observed in the post-reproductive period.

D-Asp i.p. administration to frogs ready for reproduction induced an increase of brain P450 aromatase protein expression with concomitant enhancement of both 17β-estradiol levels and ERα expression; at the same time, brain testosterone levels and androgen receptor expression decreased. In contrast, in the post-reproductive frogs, D-Asp treatment did not modify any of these parameters. Therefore, brain aromatase expression and activity fluctuate seasonally and with the reproductive state. Furthermore, the neurosteroid production (testosterone and 17β-estradiol) and the expression of androgen and estrogen receptors in relation to D-Asp treatment in different phases of reproductive cycle have also been analyzed (Santillo et al. 2013).

Taken together, these results imply that the regulation of P450 aro expression by D-Asp could be an important step in the control of 17β-estradiol levels in the frog brain.

Although our results are restricted to one species of amphibian, we can assert that D-aspartate, favoring aromatase regulation, is likely to be the key molecule enabling brain sexualization throughout the reproductive lifespan of amphibians.

11.5 Concluding Remarks

The data reported in nonmammalian animals strongly support the modulatory role of D-aspartate on several endocrine activities correlated to reproductive functions.

Although the general physiological role of D-aspartate is far from being understood in nonmammalian animals, the investigations in organisms different from mammals actually provide new insights into the biology of reproduction.

Basic information derived from comparative investigations in nonmammalian animals, besides bearing value for evolutionary and wildlife biological studies, could contribute to a better understanding of the role and of the mechanisms of action of D-aspartate-related molecules and stimulate the development of new studies in this field.

We hope that this review, in addition to having summarized the understanding of D-aspartate role in the complex mechanisms involved in the endocrine activity of nonmammalian animals, may also give a contribution to the already enlightening data on breeding biology.

References

Assisi L, Botte V, D'Aniello A, Di Fiore MM (2001) Enhancement of aromatase activity by D-aspartic acid in the ovary of the lizard *Podarcis s. sicula*. Reproduction 121(5):803–808

Burrone L, Di Giovanni M, Di Fiore MM, Baccari GC, Santillo A (2010) Effects of D-aspartate treatment on D-aspartate oxidase, superoxide dismutase, and caspase 3 activities in frog (*Rana esculenta*) tissues. Chem Biodivers 7(6):1459–1466

Burrone L, Raucci F, Di Fiore MM (2012a) Steroidogenic gene expression following D-aspartate treatment in frog testis. Gen Comp Endocrinol 175(1):109–117

Burrone L, Santillo A, Pinelli C, Baccari GC, Di Fiore MM (2012b) Induced synthesis of P450 aromatase and 17β-estradiol by D-aspartate in frog brain. J Exp Biol 215(Pt 20):3559–3565

Cangialosi MV, Puccia E, Mazzola A, Mansueto V, Arukwe A (2010) Screening of ovarian steroidogenic pathway in *Ciona intestinalis* and its modulation after tributyltin exposure. Toxicol Appl Pharmacol 245(1):124–133

D'Aniello A (2007) D-Aspartic acid: an endogenous amino acid with an important neuroendocrine role. Brain Res Rev 53(2):215–234

D'Aniello A, Di Fiore MM, Fisher GH, Milone A, Seleni A, D'Aniello S, Perna AF, Ingrosso D (2000a) Occurrence of D-aspartic acid and N-methyl-D-aspartic acid in rat neuroendocrine tissues and their role in the modulation of luteinizing hormone and growth hormone release. FASEB J 14:699–714

D'Aniello G, Tolino A, D'Aniello A, Errico F, Fisher GH, Di Fiore MM (2000b) The role of D-aspartic acid and N-Methyl-D-Aspartic acid in the regulation of prolactin release. Endocrinology 141:3862–3870

D'Aniello G, Ronsini S, Guida F, Spinelli P, D'Aniello A (2005) Occurrence of D-aspartic acid in human seminal plasma and spermatozoa: possible role in reproduction. Fertil Steril 84 (5):1444–1449

D'Aniello G, Grieco N, Di Filippo MA, Cappiello F, Topo E, D'Aniello E, Ronsini S (2007) Reproductive implication of D-aspartic acid in human pre-ovulatory follicular fluid. Hum Reprod 22(12):3178–3183

D'Aniello G, Ronsini S, Notari T, Grieco N, Infante V, D'Angelo N, Mascia F, Di Fiore MM, Fisher G, D'Aniello A (2012) D-Aspartate, a key element for the improvement of sperm quality. Adv Sex Med 2:47–53

D'Aniello A, Di Fiore MM, D'Aniello G, Colin FE, Lewis G, Setchell BP (1998) Secretion of D-aspartic acid by the rat testis and its role in endocrinology of the testis and spermatogenesis. FEBS Lett 436(1):23–27

D'Aniello A, Spinelli P, De Simone A, D'Aniello S, Branno M, Aniello F, Fisher GH, Di Fiore MM, Rastogi RK (2003) Occurrence and neuroendocrine role of D-aspartic acid and N-methyl-D-aspartic acid in Ciona intestinalis. FEBS Lett 552(2–3):193–198

Delrio G, d'Istria M, Milone M, Chieffi G (1971) Identification and biosynthesis of steroid hormones in the gonads of *Ciona intestinalis*. Experientia 27(11):1348–1350

Di Fiore MM, Perrone L, D'Aniello A (1997) Presence of a human-like thyroid stimulating hormone (TSH) in *Ciona intestinalis*. Life Sci 61(6):623–629

Di Fiore MM, Assisi L, Botte V, D'Aniello A (1998) D-aspartic acid is implicated in the control of testosterone production by the vertebrate gonad. Studies on the female green frog, Rana esculenta. J Endocrinol 157(2):199–207

Di Fiore MM, Rastogi RK, Ceciliani F, Messi E, Botte V, Botte L, Pinelli C, D'Aniello B, D'Aniello A (2000) Mammalian and chicken I forms of gonadotropin-releasing hormone in the gonads of a protochordate, *Ciona intestinalis*. Proc Natl Acad Sci U S A 97(5):2343–2348

Di Fiore MM, Lamanna C, Assisi L, Botte V (2008) Opposing effects of D-aspartic acid and nitric oxide on tuning of testosterone production in mallard testis during the reproductive cycle. Reprod Biol Endocrinol 6:28

Di Fiore MM, Santillo A, Chieffi G (2014) Current knowledge of D-aspartate in glandular tissues. Amino Acids 46(8):1805–1818. doi:10.1007/s00726-014-1759-2

Di Giovanni M, Burrone L, Chieffi Baccari G, Topo E, Santillo A (2010) Distribution of free D-aspartic acid and D-aspartate oxidase in frog *Rana esculenta* tissues. J Exp Zool A Ecol Genet Physiol 313(3):137–143

Falvo S, Di Fiore MM, Burrone L, Baccari GC, Longobardi S, Santillo A (2015) Androgen and oestrogen modulation by D-aspartate in rat epididymis. Reprod Fertil Dev. doi:10.1071/RD15092

Homma H (2007) Biochemistry of D-aspartate in mammalian cells. Amino Acids 32(1):3–11

Huang AS, Beigneux A, Weil ZM, Kim PM, Molliver ME, Blackshaw S, Nelson RJ, Young SG, Snyder SH (2006) D-aspartate regulates melanocortin formation and function: behavioral alterations in D-aspartate oxidase-deficient mice. J Neurosci 26(10):2814–2819

Ishio S, Yamada H, Hayashi M, Yatsushiro S, Noumi T, Yamaguchi A, Moriyama Y (1998) D-aspartate modulates melatonin synthesis in rat pinealocytes. Neurosci Lett 249 (2–3):143–146

Lamanna C, Assisi L, Botte V, Di Fiore MM (2006) Endogenous testicular D-aspartic acid regulates gonadal aromatase activity in boar. J Endocrinol Invest 29(2):141–146

Lamanna C, Assisi L, Botte V, Di Fiore MM (2007a) Involvement of D-Asp in P450 aromatase activity and estrogen receptors in boar testis. Amino Acids 32(1):45–51

Lamanna C, Assisi L, Vittoria A, Botte V, Di Fiore MM (2007b) D-Aspartic acid and nitric oxide as regulators of androgen production in boar testis. Theriogenology 67(2):249–254

Ota N, Shi T, Sweedler JV (2012) D-Aspartate acts as a signaling molecule in nervous and neuroendocrine systems. Amino Acids 43(5):1873–1886

Raucci F, Di Fiore MM (2009) The reproductive activity in the testis of *Podarcis s. sicula* involves D-aspartic acid: a study on c-kit receptor protein, tyrosine kinase activity and PCNA protein during annual sexual cycle. Gen Comp Endocrinol 161(3):373–383

Raucci F, Di Fiore MM (2010) The maturation of oocyte follicular epithelium of *Podarcis s. sicula* is promoted by D-aspartic acid. J Histochem Cytochem 58(2):157–171

Raucci F, Di Fiore MM (2011) D-Asp: a new player in reproductive endocrinology of the amphibian *Rana esculenta*. J Chromatogr B Analyt Technol Biomed Life Sci 879 (29):3268–3276

Raucci F, Assisi L, D'Aniello S, Spinelli P, Botte V, Di Fiore MM (2004) Testicular endocrine activity is upregulated by D-aspartic acid in the green frog, *Rana esculenta*. J Endocrinol 182:365–376

Raucci F, D'Aniello S, Di Fiore MM (2005) Endocrine roles of D-Aspartic acid in the testis of lizard *Podarcis s. sicula*. J Endocrinol 187(3):347–359

Santillo A, Pinelli C, Burrone L, Chieffi Baccari G, Di Fiore MM (2013) D-Aspartic acid implication in the modulation of frog brain sex steroid levels. Gen Comp Endocrinol 181:72–76

Santillo A, Falvo S, Chieffi P, Burrone L, Chieffi Baccari G, Longobardi S, Di Fiore MM (2014) D-aspartate affects NMDA receptor-extracellular signal-regulated kinase pathway and upregulates androgen receptor expression in the rat testis. Theriogenology 81(5):744–751

Santillo A, Falvo S, Chieffi P, Di Fiore MM, Senese R, Chieffi Baccari G (2016) D-Aspartate induces proliferative pathways in spermatogonial GC-1 cells. J Cell Physiol 231(2):490–495

Simpson ER, Mahendroo MS, Means GD, Kilgore MW, Hinshelwood MM, Graham-Lorence S, Amarneh B, Ito Y, Fisher CR, Michael MD et al (1994) Aromatase cytochrome P450, the enzyme responsible for estrogen biosynthesis. Endocr Rev 15:342–355

Takigawa Y, Homma H, Lee J-A, Fukushima T, Santa T, Iwatsubo T, Imai K (1998) D-Aspartate uptake into cultured rat pinealocytes and the concomitant effect on L-aspartate levels and melatonin secretion. Biochem Biophys Res Commun 248:641–647

Wang H, Wolosker H, Pevsner J, Snyder SH, Selkoe DJ (2000) Regulation of rat magnocellular neurosecretory system by D-aspartate: evidence for biological role(s) of a naturally occurring free D-amino acid in mammals. J Endocrinol 167(2):247–252

Chapter 12
Free D-Aspartate in Nonmammalian Animals: Detection, Localization, Metabolism, and Function

Amit V. Patel, Takayuki Kawai, Stanislav S. Rubakhin, and Jonathan V. Sweedler

Abstract Many functions of amino acids, including protein synthesis, require that they be in the L-form. As a result, in most biological systems, the levels of free D-amino acids (DAAs) are enzymatically suppressed. However, the site-specific synthesis, accumulation, and release of DAAs do occur. In fact, the accumulation of DAAs in the nervous, exocrine, and endocrine systems suggests that they perform specific functions. The focus here is on the well-studied DAA, D-aspartate; we review the advancements in the analytical approaches used for its detection and characterization and discuss the role it plays in the structural and functional organization of numerous biological systems of nonmammalian animals. The view that D-Asp has specific functions is supported by a large body of experimental data showing its endogenous synthesis, accumulation, release, stimulation of follower cells, uptake, and enzymatic catabolism. A variety of biological models, each having distinct anatomies, morphologies, biochemistries, and behaviors, have been used to investigate the fundamental mechanisms of D-Asp involvement in the normal and pathological functioning of cells and organisms. Many physiological and behavioral effects induced by D-Asp have been documented, demonstrating it has neurotransmitter, hormonal, and neuromodulator roles. Similar to many classical neurotransmitters, D-Asp has physiological roles that are conserved throughout the evolutionary tree, with nearly all studied animals shown to possess and use D-Asp.

A.V. Patel • S.S. Rubakhin • J.V. Sweedler (✉)
Department of Chemistry and Beckman Institute, University of Illinois at Urbana-Champaign, 600 South Mathews Ave. 63-5, Urbana, IL 61801, USA
e-mail: jsweedle@illinois.edu

T. Kawai
Department of Chemistry and Beckman Institute, University of Illinois at Urbana-Champaign, 600 South Mathews Ave. 63-5, Urbana, IL 61801, USA

Integrated Biodevice Unit, Quantitative Biology Center, RIKEN, D-E-206, 2-2-3 Minatojima-Minamimachi, Chuo-ku, Kobe, Hyogo 650-0047, Japan

T. Yoshimura et al. (eds.), *D-Amino Acids*, DOI 10.1007/978-4-431-56077-7_12

Keywords D-aspartate • D-amino acids • Bioanalytical methods • Neuromodulation • Neurotransmission • Signaling molecule • Central nervous system

12.1 Introduction

Aspartic acid (Asp), or 2-aminobutanedioic acid, is one of the nonessential proteinogenic (L-stereoisomer) endogenous α-amino acids. Of the common amino acids, Asp exhibits the lowest isoelectric point of 2.77. There are two conjugate bases for Asp, aspartate(1-) and aspartate(2-). As is true for most of the other proteinogenic amino acids, Asp has two enantiomers, an L-form and a D-form. The physical properties of both are identical, with the exception of the direction the substance rotates plane-polarized light. Despite reports on the parameters of D-aspartate (D-Asp) as early as 1852 (Pasteur 1852; Dalton and Schmidt 1933), and the finding of the degradative enzyme D-aspartate oxidase (D-AspO) in rabbit kidney and liver in 1949 (Still et al. 1949), for more than 100 years, only L-aspartate (L-Asp) was considered as an endogenous form of Asp in animals. However, the detection of significant amounts of free D-Asp in the mollusk *Octopus vulgaris* in 1977 (D'Aniello and Giuditta 1977) challenged this view.

Nonmammalian animal models are often used to investigate D-Asp involvement in the structural organization and functional activities of biological systems. This diverse group of organisms unites a variety of animal species—insects, mollusks, amphibians, reptiles, and birds—offering countless options for investigating the fundamental aspects of organismal function and evolution on different levels of organization, including cellular, tissue, organ, organismal, and biota. For example, the mollusk *Aplysia californica* possesses a well-characterized neuronal network, allowing for the experimental determination of functional links between the activity of specific neurons and behavior. The insect *Drosophila melanogaster* and roundworm *Caenorhabditis elegans* have genomes and transcriptomes that are amenable to extensive molecular manipulation, as well as short life spans, making them ideal for high-throughput investigations. These and other extensively characterized model systems make possible experiments that target diverse fundamental questions under controlled conditions. Not surprisingly, nonmammalian animal models have been widely used in D-Asp-related research to determine its localization, metabolism, and function, as well as its role in cell-to-cell signaling (D'Aniello and Giuditta 1977, 1978; D'Aniello et al. 1993b; Shibata et al. 2003; Miao et al. 2006a; Spinelli et al. 2006; Di Fiore et al. 1998; Raucci and Di Fiore 2010, 2011; Assisi et al. 2001).

D-Asp is one of several endogenous D-amino acids (DAAs) found throughout the metazoan. D-Alanine (D-Ala), D-asparagine (D-Asn), D-serine, D-glutamate (D-Glu), and D-glutamine (D-Gln) are also detected in different nonmammalian vertebrates and invertebrates, including cephalopods (D'Aniello and Giuditta 1977, 1978; D'Aniello et al. 2010), gastropods (D'Aniello et al. 1993b; Shibata et al. 2003;

Miao et al. 2006a; Spinelli et al. 2006), agnathans (Villar-Cerviño et al. 2010), amphibians (Di Fiore et al. 1998; Raucci and Di Fiore 2011), and reptiles (Assisi et al. 2001; Raucci and Di Fiore 2010).

D-Asp has been detected in representatives from many phyla (for a review, see D'Aniello 2007), including the roundworm *C. elegans* (Saitoh et al. 2012), the common cuttlefish *Sepia officinalis* (D'Aniello and Giuditta 1978), the Cape rock lobster *Jasus lalandii* (Okuma and Abe 1994), the amphioxus *Branchiostoma lanceolatum* (D'Aniello and Garcia-Fernandez 2007), the sea squirt *Ciona intestinalis* (D'Aniello et al. 1992, 2003), the edible frog *Rana esculenta* (Di Fiore et al. 1998; Di Giovanni et al. 2010; Raucci et al. 2004, 2005); Raucci and Di Fiore 2011, the Italian wall lizard *Podarcis s. sicula* (Santillo et al. 2006), the European hake *Merluccius merluccius*, the common sole *Solea solea* (D'Aniello et al. 1995), and the chicken *Gallus gallus domesticus* (Kera et al. 1996). D-Asp has also been found in a variety of mammalian species. There are several recent reviews describing different aspects of D-Asp physiology, biochemistry, and pathology in mammals, an area of D-Asp research where investigations by Homma and his group have played a crucial role (Errico et al. 2012; Homma 2007; Ota et al. 2012; Di Fiore et al. 2014; Katane and Homma 2011; 2010; Errico et al. 2013; Furuchi and Homma 2005; Errico et al. 2009; Tverdislov et al. 2011).

Detectable amounts of D-Asp are predominantly localized in the neuronal, endocrine, and exocrine (Baccari et al. 2003) tissues and associated with cells producing significant amounts of secreted biochemicals, including cell-to-cell signaling molecules. For example, D-Asp has been found in the retina of octopus (D'Aniello et al. 2005), as well as in the liver, kidney, and brain of squid (D'Aniello et al. 2005, 2010) and frog (Di Fiore et al. 1998; Raucci et al. 2005; Burrone et al. 2010; Di Giovanni et al. 2010; Raucci and Di Fiore 2011; Santillo et al. 2013) and neuroendocrine cells of the sea hare (Miao et al. 2005, 2006a; Fieber et al. 2010; Scanlan et al. 2010; Wang et al. 2011), and lizard (Raucci 2005; Santillo et al. 2006; Raucci and Di Fiore 2010; Raucci and Fiore 2009). The tissue levels of free D-Asp typically increase in the earlier stages of an individual's development and decrease with age (Neidle and Dunlop 1990).

The success of D-Asp research is partially due to progress in the development and application of a variety of analytical tools. One of the main challenges in the investigation of amino acid enantiomers is the similarity of most of their physicochemical properties. However, the difference in the shape of these molecules allows their chiral separation via affinity recognition or enzymatic transformation, enabling the detection and characterization of individual enantiomers. Utilization of complex approaches based on chiral high-performance liquid chromatography (LC)-laser-induced fluorescence (LIF), capillary electrophoresis (CE)-LIF, and immunohistochemical staining, in conjunction with clever experimental designs such as radionuclide pulse-chase experiments and D-AspO sample treatment, has allowed the studies of D-Asp synthesis, uptake, localization, and degradation with greater sensitivity, selectivity, and confidence. Using these and other technologies, the roles of D-Asp in neurotransmission, neuromodulation, and hormonal regulation have been investigated (for a review, see Ota et al. 2012), and D-Asp

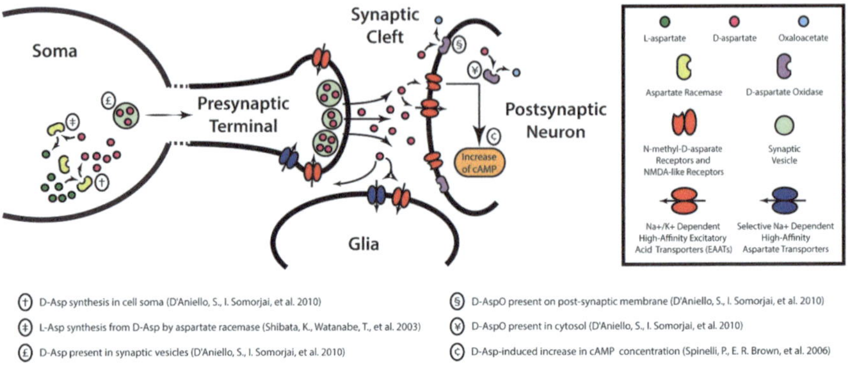

Fig. 12.1 Schematic of potential D-Asp involvement in synaptic signaling. Biosynthesis of D-Asp from L-Asp in cell soma by the action of aspartate racemase, release from presynaptic terminals, and interactions with receptors and transporters are depicted. Additionally, D-Asp involvement in signal transduction through triggering of an increase in cyclic adenosine monophosphate (cAMP) concentration in a postsynaptic neuron and signal termination via the D-aspartate oxidase-mediated degradation of D-Asp are also shown

neurotransmitter function has been asserted (D'Aniello et al. 2010). The schematic in Fig. 12.1 depicts experimentally suggested mechanisms reported to be involved in D-Asp neurotransmission, including D-Asp endogenous synthesis and known and putative transporters and receptors, as well as enzymes that are required for signal termination.

We begin with a discussion of the development and implementation of analytical techniques used to investigate D-Asp, followed by information on D-Asp localization and metabolism, as well as its physiological functions in nonmammalian animals.

12.2 Analytical Techniques for D-Asp Detection and Characterization in Biological Systems

The physical and chemical properties of L- and D-Asp are virtually identical, aside from the direction in which they rotate plane-polarized light. Therefore, D-Asp research requires methods that are capable of measuring enantiomeric species. A variety of suitable approaches based on enzymatic assays, affinity recognition, and chiral separations, including CE and LC, have been introduced (Konno 2007). Here, the emphasis will be on the more commonly employed methods for analysis of D-Asp in biological systems.

12.2.1 Enzymatic Assays

Due to their ability to alter the structure of D-Asp with high selectivity, enzymatic assays, often using the enzymes D-AspO and D-amino acid transaminase, are an effective approach for D-Asp analysis (D'Aniello and Giuditta 1977, 1978; Jones et al. 1994). For example, using a D-AspO-based colorimetric assay in which D-Asp was catabolized by D-AspO in the presence of flavin adenine dinucleotide, D'Aniello and Giuditta observed D-Asp in the nervous tissues of octopus and squid (D'Aniello and Giuditta 1977, 1978). This resulted in the formation of oxaloacetate, which was subsequently reacted with 2,4-dinitrophenylhydrazine to form a colored hydrazone compound, which was then quantified spectrophotometrically (D'Aniello et al. 1993c; Nagata et al. 1985). Enzymatic approaches are typically not specific for a single DAA, for example, D-AspO is known to also catabolize other DAAs and derivatives of DAAs, such as N-methyl-D-aspartate (NMDA), D-Glu, D-Asn, and D-Gln. Typically D-AspO exhibits different specificities for each substrate. Therefore, it is important to separate the DAAs prior to enzymatic analysis to prevent signal overlap, thus enabling the characterization and quantification of individual species.

12.2.2 Gas Chromatography (GC)

Due to the speed and resolving power of GC analysis, this rather mature separation technique has been widely used to separate amino acid enantiomers or their derivatives in complex mixtures. Chiral separations with GC are made possible by using columns where the common stationary phases are derivatized with compounds such as cyclodextrins (e.g., alkylated beta-cyclodextrins). Chiral columns in which DAAs elute before their L-forms, and vice versa, are commercially available and can be used with optically inactive derivatization reagents to facilitate detection (Brückner and Hausch 1993; Frank et al. 1977; Ali et al. 2010; Schieber et al. 1997). Additionally, traditional achiral columns can separate diastereomers formed by derivatization of D- and L-Asp with optically active reagents (Hoopes et al. 1978; Payan et al. 1985; Bertrand et al. 2008). However, it should be noted that derivatization with such reagents sometimes racemizes the amino acids (Payan et al. 1985).

Although there are a large number of detectors that can be utilized with GC, the flame ionization detector (FID) is one of the most widely used (Brückner and Hausch 1993; Frank et al. 1977; Hoopes et al. 1978; Payan et al. 1985; Skoog et al. 2007). More recently, the use of a mass spectrometer as a detector for GC, known as GC-mass spectrometry (MS), is an important approach. Compared to the FID, MS adds confidence to the analyte identification (Kaspar et al. 2008; Waldhier et al. 2010; Zampolli et al. 2007; Ali et al. 2010; Bertrand et al. 2008; Schieber et al. 1997; Waldhier et al. 2011). While providing an increased confidence in identification, it offers a poorer detectability compared to the FID; 300 nM is one example of GC-MS limits of detection (LODs) for D-Asp analysis (Kaspar et al. 2008).

12.2.3 Liquid Chromatography (LC)

Currently, LC is the most commonly used technique for DAA analyses due to its versatility, robustness, flexibility, and wide availability. As with GC, both chiral and achiral columns can be used. Chiral columns with stationary phases possessing optically active chemicals such as amino acids (Welch 1994), crown ethers (Shinbo et al. 1987), cyclodextrins (Armstrong et al. 1987), and proteins (Allenmark and Andersson 1991) are used in the separation of intact enantiomers. For separations with achiral columns, D- and L-Asp must first be derivatized with chemical modifiers such as o-phthalaldehyde (OPA) and/or chiral thiols like N-acyl-L-cysteine to form diastereomers (Aswad 1984; Brückner and Westhauser 2003; Buck and Krummen 1987; Nimura and Kinoshita 1986; Tsesarskaia et al. 2009; Saitoh et al. 2012). Two-dimensional high-performance LC separation using both achiral and chiral columns increases separation efficiency and simplifies analyte identification (Hamase et al. 2009; Miyoshi et al. 2012; Han et al. 2011).

Analyses of D-Asp by LC typically utilize fluorescence or MS for detection. Fluorescence detection allows for quantification with high sensitivity, for example, an LOD of 150 nM was achieved for OPA-derivatized compounds (Morikawa et al. 2001) and an LOD of 1 nM for 4-fluoro-7-nitrobenzofurazan-modified molecules (Han et al. 2011). Confidence in the identification of D-Asp in experiments relying on fluorescence-based detection can be enhanced by treating the sample with D-AspO, resulting in a specific decrease in the D-Asp peak area (Spinelli et al. 2006). Importantly, the use of MS detection adds information on the mass-to-charge ratios of the analytes. This information, along with retention time, aids analyte identification. The use of tandem MS (MS/MS) further enhances analyte identification capabilities (Song et al. 2008). The reported LOD for MS/MS detection of D-Asp is 1.20 μM (Song et al. 2007), which is poorer than obtained using LC with fluorescence-based detection or with GC-MS.

12.2.4 Capillary Electrophoresis (CE)

CE is yet another separation technique that has been used to analyze D-Asp in biological systems. This approach has a number of notable advantages, such as high separation efficiency, relatively short separation times, and low sample volumes. As with LC and GC, there are chiral and achiral separation modalities available (reviewed in (Wan and Blomberg 2000)). Chiral CE analyte separation, using micellar electrokinetic chromatography, for example, utilizes the interaction between enantiomers and chiral pseudostationary phases, with cyclodextrin and its various derivatives (Kitagawa and Otsuka 2011; Wan and Blomberg 2000) most commonly employed as the chiral selectors. Chiral surfactants such as deoxycholate (Terabe et al. 1989) and *N*-dodecoxycarbonylvaline (Swartz et al. 1995) are also capable of providing chiral separations in CE. The achiral separation is applicable in the analysis of diastereomeric derivatives of amino acids, which are produced by derivatization of analytes with optically pure reagents or chiral reagents like o-phthaldialdehyde with N-acetyl-cysteine (Kang and Buck 1992) or chiral thiols (Tivesten et al. 1997).

LIF has been the most commonly utilized detection modality in CE analyses of D-Asp. Ultraviolet absorption (Yang et al. 2010) and MS (Simo et al. 2005; Moini et al. 2003) are also used for D-Asp detection following CE separation, although with significantly lower sensitivity. When LIF detection is employed, the sample is initially derivatized with fluorescent reagents such as fluorescein isothiocyanate (Cheng and Dovichi 1988) or naphthalene-2,3-dicarboxaldehyde (Ueda et al. 1991). The typical LODs in CE-LIF analyses of D-Asp are on the pM level, which are sufficient for even subcellular analyses (Miao et al. 2005). In addition to comparing migration times using standards and enzymatic (D-AspO) treatments, immunoprecipitation of D-Asp increases the confidence in the D-Asp measurements (Miao et al. 2006b).

12.2.5 Immunohistochemistry

The previously described enzymatic assays, GC, LC, and CE, are useful for the characterization and quantification of endogenous free D-Asp; however, they do not allow for the determination of D-Asp's distribution in a given system with high spatial resolution. In cases where details on analyte distribution are needed with high specificity, sensitivity, and spatial resolution, immunohistochemical staining is the method of choice. With proper tissue sample acquisition and fixation, and treatment with a selective antibody, information on the localization of D-Asp can be acquired. Accordingly, to enable the direct immunohistochemical analysis of D-Asp, selective anti-D-Asp antibodies have been developed using the glutaraldehyde conjugate of D-Asp as an antigen (Masuda et al. 2003; Lee et al. 1997; Schell et al. 1997). The obtained antibodies exhibit specificity only to the glutaraldehyde conjugate of D-Asp, and not to that of L-Asp, thus enabling specific immunohistochemical staining after tissue treatment with glutaraldehyde. The successful application and efficacy of the developed antibodies for the purpose of D-Asp visualization is shown in Fig. 12.2, which depicts immunostained neurons from *Aplysia limacina* showing differences in D-Asp localization from cell to cell. Specifically, the intensity of D-Asp immunostaining is distinct in the nucleus and the cytoplasm of different neurons, suggesting a unique function for D-Asp in different neuronal populations.

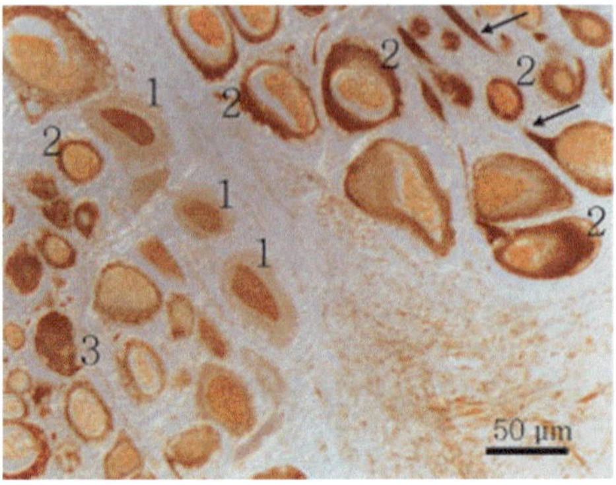

Fig. 12.2 Paraffin section of a cerebral ganglion from *Aplysia limacina* immunostained with anti-D-Asp antibody. Immunohistochemical staining visualized the differential localization of D-Asp; some cells are stained in the (*1*) nucleus, others in the (*2*) cytoplasm, and others still in both the (*3*) cytoplasm and nucleus, and positive staining is also observed in the (*arrows*) neuropil (Modified with permission from Spinelli et al. 2006, copyright 2006, John Wiley & Sons)

12.2.6 Indirect Detection Approaches

The presence of D-Asp in a biological system can be inferred based on the presence of metabolic enzymes that are related to D-Asp, such as D-AspO and D-Asp racemase. These proteins can be detected with the use of affinity probes, which are commonly used in immunohistochemistry and related approaches such as immunoelectron microscopy, and the use of Western blotting and a range of molecular reporters such as protein coexpression with green fluorescent protein (Zaar 1996; Zaar et al. 2002; Yamamoto et al. 2011; Kim et al. 2010; Wang et al. 2011). These approaches are used to localize D-Asp-related proteins and hence, indirectly, the presence of D-Asp itself.

12.3 D-Asp Localization, Metabolism, and Fluxes

The application of the approaches described above has facilitated significant progress in characterizing and understanding the metabolism, concentration fluxes, and functions of D-Asp in neuronal, endocrine, and exocrine tissues. Information on the localization, fluxes, and levels of a given analyte in cells, tissues, and organs provides important clues on its putative function as well as its relation to various pathological states. The multiple mechanisms of D-Asp synthesis, uptake, spatial containment, release, and catabolism, as well as the different modes of regulation of extracellular and intracellular D-Asp levels, indicate that this molecule is an important player in the homeostatic balance of multicellular organisms (Homma 2007). These mechanisms also suggest the capability of D-Asp to mechanistically interfere with important biological processes such as protein synthesis.

12.3.1 Localization of D-Asp

Investigations of D-Asp localization have been carried out on vertebrate (Santillo et al. 2006; Di Giovanni et al. 2010) and invertebrate (D'Aniello et al. 2005; Scanlan et al. 2010) models as well as on cell cultures (Takigawa et al. 1998; Gadea et al. 2004; Adachi et al. 2004). These studies often involved immunohistochemical staining with anti-D-Asp antibodies in order to visualize D-Asp localization in cells and tissues and used a variety of analytical techniques to detect the presence of endogenous and radiolabeled D-Asp. Using these approaches, D-Asp has been detected in a large number of nervous, exocrine, and endocrine tissues of nonmammalian animals, including the gonads and Harderian glands of tiger prawn *Penaeus esculentus* (Di Fiore et al. 1998; Raucci et al. 2004; Di Giovanni et al. 2010) and the lizard *P. s. sicula* (Raucci 2005; Raucci and Fiore 2009), as well as the central nervous system of sea hare (Scanlan et al. 2010) and green frog (Santillo et al. 2013). Di Fiore et al. (2014) reviewed and summarized reports of D-Asp concentrations in endocrine and exocrine glands of several different species. Endogenous D-Asp concentrations in the exocrine Harderian glands of frog and lizard are 130 and 19 nmol/g tissue, respectively. Levels of D-Asp in endocrine glands are often similar to concentrations of the compound in exocrine glands. As examples (all in nmol/g tissue): the adrenal glands of chicken (23–30), pancreas of pigeon (15–18), chicken (6–10), frog (3–58), duck (11–23), and lizard (2–8 in ovaries and 3–30 in testes). Frog testes exhibit the highest levels of D-Asp at 140–236 nmol/g. D-Asp levels in the abovementioned glands are in the same concentration range as in the exocrine and endocrine glands of mammals, except for the rat pineal gland where its concentration may reach 3524 nmol/g tissue. Importantly, D-Asp concentrations in these glands may vary with age and functional state, as well as with the levels of D-Asp in the extracellular environment.

Free D-Asp formation and accumulation generally occur in a site-specific manner. For example, D-Asp was detected in the optic lobe of the cephalopod (D'Aniello et al. 2005), suggesting it may play a role in cephalopod vision. The detected D-Asp most likely had an endogenous origin. In other studies, D-Asp was found to form and accumulate in specific *A. californica* ganglia (Scanlan et al. 2010) and in the somata of selected neurons (D'Aniello et al. 2010).

12.3.2 Biosynthesis of D-Asp

There are several sources of D-Asp in organisms. Exogenous sources include the environment, e.g., seawater (Azua et al. 2014), food (Man and Bada 1987; Friedman and Levin 2012), and gut microbiota (Friedman 1999), whereas the endogenous sources are due to enzymatic conversion of L-Asp to D-Asp and the freeing of D-Asp from DAA-containing proteins. DAA residues of proteins are formed via enzymatic (Shikata et al. 1995) and nonenzymatic (Takahashi 2014; Stephenson and Clarke 1989) processes, which are often age dependent. The exogenous origin of some D-Asp is supported by results of experiments carried out in both mammalian and nonmammalian animals, which have demonstrated the absorption of this small molecule by the intestine and its transport to various solid tissues, including the brain, muscle, kidney, and liver (D'Aniello et al. 1993a). The enzymatic conversion of L- to D-Asp has been shown in a number of models, including pheochromocytoma cells (Long et al. 1998). The use of radiolabeled L-Asp in pharmacological, biological, and analytical studies has demonstrated that D-Asp can be formed from L-Asp via enzymatic activity in the nervous system, as shown in Fig. 12.3 (Scanlan et al. 2010). This observation corroborates findings of other studies performed on invertebrates such as European squid *Loligo vulgaris* (D'Aniello et al. 2005, 2010) and sea hare *A. californica* (Miao et al. 2006a) as well as vertebrates, e.g., the Italian wall lizard *P. s. sicula* (Raucci 2005; Raucci and Fiore 2009) and the edible frog *Pelophylax esculentus* (Raucci et al. 2004, 2005).

In light of the findings from the aforementioned studies, it has been hypothesized that in nonmammalian animals, D-Asp is synthesized by enzymes with characteristics similar to known racemases and by DAA aminotransferases. One such D-Asp racemase has been cloned from the foot muscle of the clam *Anadara broughtonii* and subsequently purified and characterized (Shibata et al. 2003; Abe 2006). More recently, two pyridoxal 5′-phosphate-dependent D-Asp racemases were reported and characterized, one from the mouse brain (Kim et al. 2010). The other was from the head ganglia of *A. californica* (Wang et al. 2011); interestingly, this D-Asp racemase was found to also catalyze serine racemization.

In addition to D-Asp racemases, free D-Asp can be produced during the enzymatic degradation of DAA-containing peptides and proteins, perhaps where D-aspartyl endopeptidase (EC 3.4.99.B2) plays a role. DAA-containing peptides are typically formed by age-dependent posttranslational deamidation, isomerization, and racemization of L-aspartyl residues (Katane and Homma 2011; Geiger and Clarke 1987; Reissner and Aswad 2003). As a result, D-aspartyl and D-isoaspartyl residues are produced. While this mechanism may be responsible for only a small percentage of free D-Asp, it can impact D-Asp levels in some tissues and cells.

12.3.3 Catabolism of D-Asp

In addition to its formation, the ability to remove or degrade D-Asp is also important. For example, the catabolism of intercellular signals allows for signal termination and preparation for a new signaling event. D-Asp clearance combines diffusion away from the site of release, uptake by neurons or glia, and/or enzymatic modification, including breakdown. D-Asp catabolism is predominantly mediated by the enzyme D-AspO, a flavoenzyme which selectively catalyzes the oxidation of D-Asp, D-Glu, and NMDA to their corresponding α-keto acids (D'Aniello et al. 1993a). In contrast, the oxidative transformation of neutral and basic DAAs into α-keto acids is carried out by a different flavoenzyme, D-amino acid oxidase

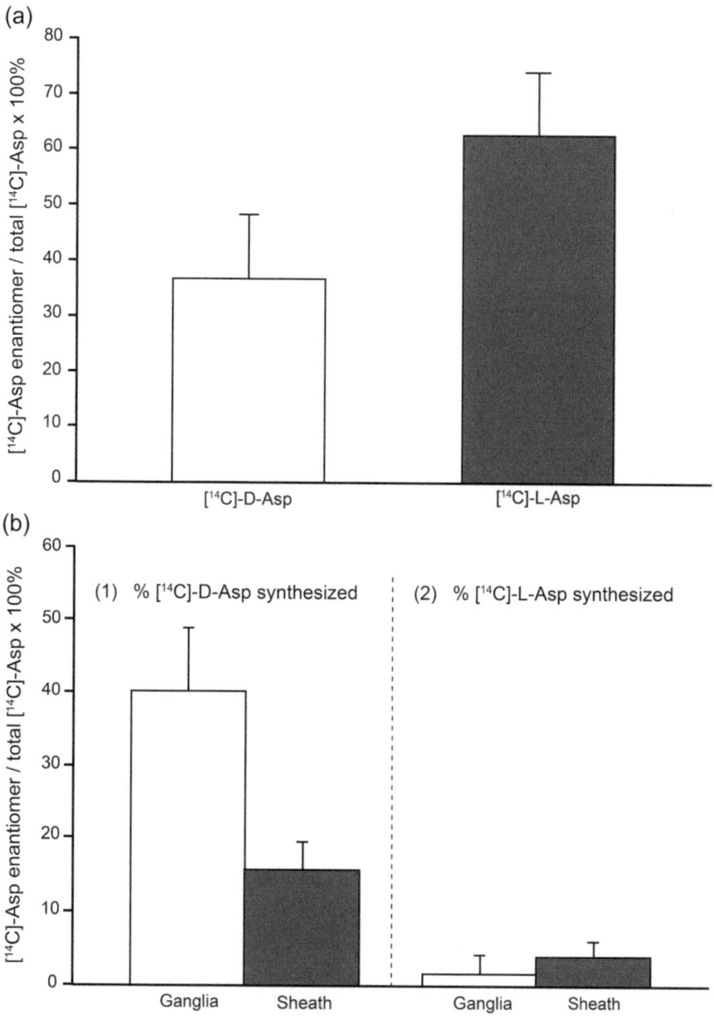

Fig. 12.3 Biosynthesis of D-Asp in a cerebral ganglion of *Aplysia californica*. (**a**) Final percentages of radiolabeled Asp enantiomers after incubation of intact cerebral ganglion in [^{14}C]-L-Asp, showing biosynthesis of D-Asp from L-Asp. (**b**) Final percentages of radiolabeled Asp enantiomers after incubation of desheathed cerebral hemiganglia and related sheath tissue in (*1*) [^{14}C]-L-Asp and (*2*) [^{14}C]-L-Asp, showing significantly larger conversion of [^{14}C]-L-Asp to [^{14}C]-D-Asp in the hemiganglion over the sheath tissue (Reprinted with permission from Scanlan et al. 2010, copyright 2010, Wiley-Blackwell)

(DAAO). Both D-AspO and DAAO are reported in *C. elegans* (Katane et al. 2010; Saitoh et al. 2012) and *L. vulgaris* (D'Aniello et al. 2010) and are likely ubiquitous throughout the metazoan. Interesting distributions of these enzymes occur; for example, D-AspO is present in higher concentrations in *L. vulgaris* brain cytosol fractions.

D-AspOs from different species exhibit a variety of catalytic efficiencies. For example, multiple genes encoding functional D-AspOs have been found in *C. elegans*, each with unique specificities, kinetic parameters, and expected subcellular localizations (Katane et al. 2010). Although the physiological functions of the multiple D-AspOs may be redundant, the aforementioned differences in compartmentalization, concentration, and catalytic properties suggest the possibility of distinct roles for each enzyme. It is apparent that the existence of multiple D-AspOs with functional differences can allow for the modulation of local concentrations of D-Asp and thereby play integral roles in signaling, possibly by attenuating or terminating D-Asp signaling via its degradation.

Hypothetically, two other enzymes could influence the levels of free D-Asp. One is the protein-L-isoaspartate (D-aspartate) O-methyltransferase (EC 2.1.1.77), which is involved in the repair of age-related damage of protein aspartate and asparagine residues (Gomez et al. 2008). Another enzyme, D-amino acid transaminase (EC 2.6.1.21), catalyzes reactions in which D-Ala and D-Glu are formed from D-Asp. However, so far, the presence of this enzyme has only been reported in microorganisms and plants, e.g., *Arabidopsis thaliana* (Funakoshi et al. 2008).

The important role of D-AspO in the oxidative degradation of D-Asp as a consequence of its potential involvement in signaling pathways has led to investigations of the relationship between D-Asp levels and D-AspO activity in nonmammalian species. For example, intraperitoneal injection of D-Asp into green frog led to an increase of D-AspO activity in a variety of organs: brain, kidney, liver, pancreas, Harderian gland, heart, spleen, and testes (Di Giovanni et al. 2010; Burrone et al. 2010). In contrast, no increase in DAAO or L-amino acid oxidase activities were observed following D-Asp administration (Di Giovanni et al. 2010).

These data demonstrate that D-AspO may have multiple physiological purposes. It is possible that D-AspO is responsible for the general metabolism of endogenous and exogenous D-Asp to prevent the accumulation of high concentrations of D-Asp, which may have a negative impact on normal functioning of biological systems. Additionally, D-AspO may also be involved in the controlled degradation of D-Asp for the purposes of chemical signaling modulation, including signal termination.

12.3.4 D-Asp Flux

Understanding the movement of D-Asp into and out of specific locations, its flux, is important. Amino acid transporters such as solute carrier family 1 members 1–5 (also reported as excitatory amino acid transporters 1–5 (EAATs 1–5)) likely are involved in its import, mediated by the Na+/K+ electrochemical gradient (for reviews, see EMBL-EBI 2014; Boudko 2010; Stevens 2010; Krehbiel and Matthews 2003). For example, dEAAT1 and dEAAT2 are present in the nervous system of *D. melanogaster* (Umesh et al. 2003; Besson et al. 2000) (see Fig. 12.4). *Drosophila* sodium-dependent high-affinity aspartate transporter dEAAT2 has a

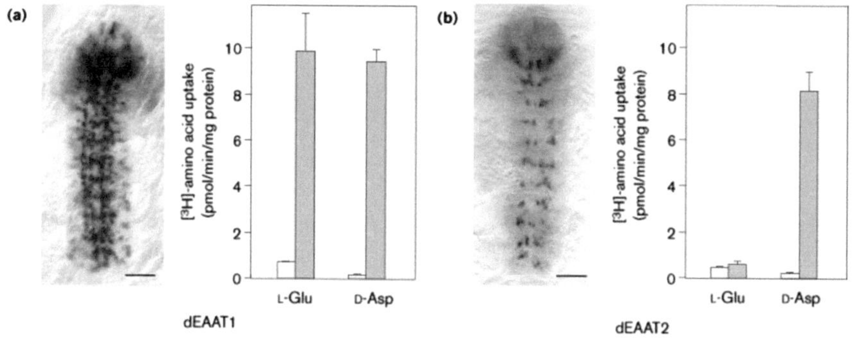

Fig. 12.4 Central nervous system expression patterns and transport properties of *Drosophila melanogaster* excitatory amino acid transporters (EAATs). *Left panels*: ventral views of *Drosophila* embryos stained with in situ hybridization with antisense (**a**) dEAAT1 or (**b**) dEAAT2 RNA probe. A differential expression pattern for dEAAT1 and dEAAT2 is apparent, with substantially higher expression levels for dEAAT1 in the embryonic nerve. *Right panels*: uptake of radiolabeled L-Glu and D-Asp into *Drosophila* S2 cells transfected with (**a**) dEAAT1 or (**b**) dEAAT2 cDNAs. Cells were transfected with (*white bars*, control) mock plasmid or (*gray bars*) *Drosophila* EAAT expression vectors. Selective transport of D-Asp relative to L-Glu is apparent for dEAAT2, whereas dEAAT1 shows comparable transport for the two excitatory amino acids (Reprinted with permission from Besson et al. 2000, copyright 2000, Elsevier)

much lower affinity to glutamate than D- or L-Asp. In another example, a single EAAT was expressed in the neuropil regions of the thoracic ganglia of the mosquito *Aedes aegypti* (Umesh et al. 2003). The transporter mediates high-affinity uptake of D-Asp as well as L-Asp and L-Glu, where both of the L-amino acids compete with D-Asp uptake. Therefore, the resulting excitatory amino acid import will depend on the extent and profile of transporter expression as well as levels of these compounds in the extracellular space, where D-Asp is typically present at lower concentrations. Not surprisingly, a large body of information on D-Asp transmembrane transport is available for mammalian models (for reviews, see Broer 2008; Saunders et al. 2013). This knowledge can be used in genomic and transcriptomic searches to aid the identification of molecular homologs in nonmammalian species.

D-Asp can be transported along nerves to release sites (Miao et al. 2006a; D'Aniello et al. 2010; Scanlan et al. 2010) and subsequently released upon stimulation. The vesicular nature of D-Asp release is also supported by the measurement of the subcellular localization of D-Asp in synaptosomes and synaptic vesicles of the nervous systems of *A. limacina* (Spinelli et al. 2006) and *L. vulgaris* (D'Aniello et al. 2010). Along with D-Asp, these synaptic vesicles possess relatively high levels of L-Asp and L-Glu, creating competitive conditions for neurotransmitter import by EAATs.

12.4 Physiological Functions of Free D-Asp

A number of roles for free D-Asp in central physiological processes on the cellular, tissue, organ, and organismal level have been suggested and in some cases experimentally confirmed. A large number of reports point to D-Asp operating as an environmental signal, neuromodulator, hormone, and neurotransmitter and being involved in mechanisms of the biosynthesis of hormones, endocrine and exocrine secretion (Monteforte et al. 2009), neurogenesis, memory (Topo et al. 2010), reproduction, vision, individual development, and stress (Erwan et al. 2012).

12.4.1 Neuronal Function

In neurons, D-Asp may generate an immediate cellular electrophysiological response through activation of the N-methyl-D-aspartate receptor (NMDAR), a type of ionotropic glutamate receptor (Foster and Fagg 1987), NMDAR-like receptors (Carlson and Fieber 2011), or sodium-dependent EAATs. The presence of other or unknown receptors and/or transporters that are specific to D-Asp is indicated by results of experiments where 25% of studied *A. californica* neurons exhibited D-Asp-induced currents but were not sensitive to L-Glu application (Carlson and Fieber 2011). Similarly, noticeable specificity to D-Asp is characteristic of *Drosophila* sodium-dependent dEAAT2 (Besson et al. 2000) (Fig. 12.4). Indeed, partial involvement of EAATs in the formation of D-Asp-induced L-Glu-independent cation currents in cultured *A. californica* neurons was demonstrated using a transporter blocker (Carlson et al. 2012).

The involvement of NMDA receptors in D-Asp-induced physiological and behavioral effects is supported by the results of experiments involving systemic in vivo intracerebroventricular administration of D-Asp to neonatal chicks. Specifically, when subjected to stressful conditions, D-Asp dose-dependent calming of the animals was observed, including a decrease in vocalization and an increase in the duration of time spent in resting positions with eyes open (Erwan et al. 2012). NMDA injection was also found to yield sedative effects (Yamane et al. 2009). The effects of L-Asp mimicked those of D-Asp; however, the periods of sleeplike state were observed to be longer, indicating possible mechanistic differences for the two enantiomers.

12.4.2 Environmental Signals

Decapod crustaceans such as the red rock crab (*Cancer productus*) and the Pacific rock crab (*Romaleon antennarium*) possess chemoreceptors located at their claw's moveable fingers, or dactyls. Application of D-Asp induced an almost five times higher electrophysiological response in the corresponding nerve than application of L-Asp (Case 1964). Importantly, the response to D-Asp is pH dependent and not visibly modulated by the presence of L-Asp. It is plausible that D-Asp is present in seawater at nM concentrations; for example, the top layer of the ocean near Hawaii has been found to contain 8 nM of total dissolved D-Asp (Benner and Herndl 2011). This makes it apparent that D-Asp, along with other amino acids, may work as an environmental signal for marine animals. In the ocean environment, bacterioplankton are considered the main biotic source of D-Asp, and some other DAAs, and are found in the cell walls of these microorganisms (Perez et al. 2003; Azua et al. 2014).

12.4.3 Organism Development

D-Asp has been shown to play a role in development. In an early study, a transient increase in D-Asp levels during the embryonic stage in chicken was detected, with maximal concentration in the retina and brain between the 10th and 15th day of incubation (Neidle and Dunlop 1990). D-Asp in the chicken retina was observed to reach 20% of the total Asp levels at day 13. No similar changes in D-Asp content were found to occur in the heart, muscle, and liver during this stage of organism development. Additionally, *C. elegans* demonstrate a transient increase in D-Asp content, with a maximum level at the L3 larva stage (Saitoh et al. 2012). Interestingly, there is strong evidence to suggest that expression of the excitatory amino acid transporters dEAAT1 and dEAAT2 leads to formation of glia in the embryonic *Drosophila* central nervous system and may be involved in the generation of the functional diversity of these glial cells (Soustelle et al. 2002). The temporal alignment of such events may be significant for proper development of nervous system cellular circuitry.

12.4.4 Reproduction

The high concentration of D-Asp observed in the testes and ovaries of various nonmammalian animals suggests an involvement in reproduction (Di Fiore et al. 1998; Raucci et al. 2004; Raucci and Di Fiore 2011). Quantitative measurements of free D-Asp in the seasonal breeding green frog and Italian wall lizard uncovered a correlation between D-Asp levels and the phases of the reproductive cycle (Raucci and Di Fiore 2011) (Fig. 12.5). Moreover, D-Asp levels were found to correlate with androgen production in a sex- and species-dependent manner. D-Asp application in vivo enhanced expression of P450 aromatase mRNA and protein levels in male frog brain, which led to an increase in 17β-estradiol synthesis as well as a corresponding increase in expression of its receptor (Burrone et al. 2012). The modulatory effects of P450 aromatase expression were found to be a result of D-Asp

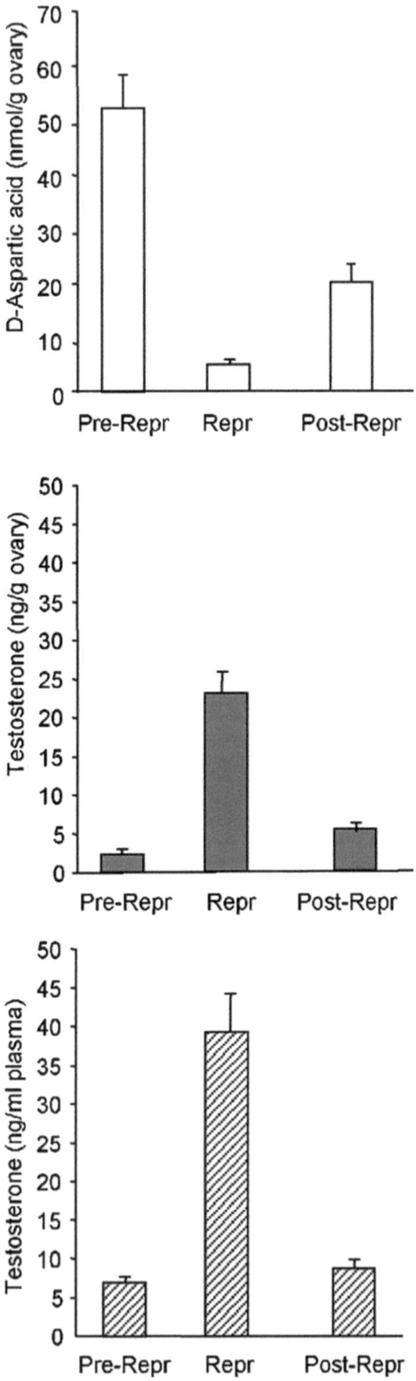

Fig. 12.5 Free D-Asp is present in the ovary extract of *Rana esculenta* in each phase of the sexual cycle. The variations in D-Asp concentration depend on the phase of the reproductive cycle, with the highest D-Asp concentrations observed during the prereproductive period, at which time testosterone is at its minimum level. During the reproduction phase, D-Asp concentration decreases substantially, while the testosterone levels in the ovary and plasma increase greatly. During the post-reproductive period, the concentration of D-Asp moderately increases relative to the reproductive period, and the ovarian and plasma testosterone levels reduce to moderately low concentrations (Modified with permission from Raucci and Di Fiore 2011, copyright 2011 from Elsevier)

participation in the CREB pathway. In a separate experiment, a D-Asp intraperitoneal injection was observed to modulate the levels of testosterone and 17β-estradiol in Italian wall lizard (Raucci 2005; Schell et al. 1997). The data made evident that D-Asp regulates steroidogenesis and spermatogenesis. The observed involvement of D-Asp in reproduction in a number of different species suggests that this particular functional role of D-Asp is likely evolutionarily conserved. For example, *C. elegans* has three functional D-AspOs: D-AspO-1, D-AspO-2, and D-AspO-3 (Katane et al. 2010). Modulation of D-AspO-1 expression alters D-Asp levels in animals, which correlates with changes in egg-laying capacity as well as hatching rate (Saitoh et al. 2012).

12.4.5 Vision

As mentioned earlier, another physiological function in which D-Asp may play a role relates to vision. This potential role for D-Asp is supported by its abundance in the retina of multiple animals, including the chicken (Neidle and Dunlop 1990) and common cuttlefish (D'Aniello et al. 2005). D-Asp was also found to be present in the optic lobe of *S. officinalis* (D'Aniello et al. 2005). There are several origins of D-Asp in the retina of this animal, including transport from the optical lobe as well as endogenous synthesis by the action of D-Asp racemase. The levels of D-Asp, L-Asp, L-Glu, and D-Asp racemase all decrease in both the retina and optic lobe after sustained light deprivation. However, other amino acids exhibit no change with regard to concentration. These observations strongly point toward an important role for D-Asp in vision.

12.5 Summary

D-Asp is an enigmatic molecule that has been observed throughout the metazoan, with a number of studies helping to define its formation, localization, and degradation. Uncovering the different roles of D-Asp has been made possible by the development and application of a number of measurement approaches that provide information on the specific chiral form of Asp in an animal tissue. The measurement approaches have included immunohistochemical staining, LC-LIF, and CE-LIF.

Now that such approaches exist, studies have turned to more functional questions. The use of nonmammalian animal models has enabled the characterization of D-Asp in organisms with well-understood nervous systems, such as *A. californica*, and in organisms with short life spans and well-characterized genomes, such as *C. elegans*. The result of the D-Asp research has been the thorough investigation of the biosynthesis, localization, metabolism, controlled degradation, and concentration fluxes of this DAA. Together, these findings provide a more complete understanding of free D-Asp and its involvement in important physiological processes, from the cellular to the organismal level. For example, the cell-to-cell signaling roles of free D-Asp have been demonstrated in a number of nonmammalian animals. In addition, several physiological and behavioral effects have been linked to D-Asp. Specifically, relationships to cell-to-cell and environment-to-organism signaling, as well as neurogenesis, reproduction, and vision, have been revealed.

Acknowledgments This work was supported by the National Science Foundation (NSF), Division of Chemistry, under grant CHE-11-11705 (with co-funding from the Division of Biological Infrastructure), by Award No. RO1 NS031609 from the National Institute of Neurological Disorders and Stroke (NINDS) and Award Number P30 DA018310 from the National Institute on Drug Abuse (NIDA). The content is solely the responsibility of the authors and does not necessarily represent the official views of the NSF, NINDS, NIDA, or the National Institutes of Health.

References

Abe K (2006) Cloning and expression of the pyridoxal 5′-phosphate-dependent aspartate racemase gene from the bivalve mollusk *Scapharca broughtonii* and characterization of the recombinant enzyme. J Biochem 139(2):235–244. doi:10.1093/jb/mvj028

Adachi M, Koyama H, Long Z, Sekine M, Furuchi T, Imai K, Nimura N, Shimamoto K, Nakajima T, Homma H (2004) L-Glutamate in the extracellular space regulates endogenous D-aspartate homeostasis in rat pheochromocytoma MPT1 cells. Arch Biochem Biophys 424 (1):89–96. doi:10.1016/j.abb.2004.01.016

Ali H, Pätzold R, Brückner H (2010) Gas chromatographic determination of amino acid enantiomers in bottled and aged wines. Amino Acids 38(3):951–958. doi:10.1007/s00726-009-0304-1

Allenmark S, Andersson S (1991) Chiral amino acid microanalysis by direct optical resolution of fluorescent derivatives on BSA-based (resolvosil) columns. Chromatographia 31 (9-10):429–433. doi:10.1007/BF02262384

Armstrong DW, Yang X, Han SM, Menges RA (1987) Direct liquid chromatographic separation of racemates with an alpha-cyclodextrin bonded phase. Anal Chem 59(21):2594–2596. doi:10.1021/ac00148a014

Assisi L, Botte V, D'Aniello A, Di Fiore MM (2001) Enhancement of aromatase activity by D-aspartic acid in the ovary of the lizard Podarcis s. Sicula. Reproduction 121(5):803–808. doi:10.1530/rep.0.1210803

Aswad DW (1984) Determination of D- and L-aspartate in amino acid mixtures by high-performance liquid chromatography after derivatization with a chiral adduct of o-phthaldialdehyde. Anal Biochem 137(2):405–409. http://dx.doi.org/10.1016/0003-2697(84)90106-4

Azua I, Goiriena I, Bana Z, Iriberri J, Unanue M (2014) Release and consumption of D-amino acids during growth of marine prokaryotes. Microb Ecol 67(1):1–12. doi:10.1007/s00248-013-0294-0

Baccari G, Di Fiore M, Monteforte R, Raucci F, D'Aniello A (2003) D-aspartic acid induces merocrine secretion in the frog harderian gland. Rend Fis Acc Lincei 14(3):205–215. doi:10.1007/bf02904524

Benner R, Herndl GJ (2011) Bacterially derived dissolved organic matter in the microbial carbon pump. In: Jiao N, Azam F, Sanders S (eds) Microbial carbon pump in the ocean. Science/AAAS, Washington, DC, pp 46–48. doi:10.1126/science.opms.sb0001

Bertrand M, Chabin A, Brack A, Westall F (2008) Separation of amino acid enantiomers VIA chiral derivatization and non-chiral gas chromatography. J Chromatogr 1180(1–2):131–137. http://dx.doi.org/10.1016/j.chroma.2007.12.004

Besson MT, Soustelle L, Birman S (2000) Selective high-affinity transport of aspartate by a Drosophila homologue of the excitatory amino-acid transporters. Curr Biol 10(4):207–210

Boudko DY (2010) Molecular ontology of amino acid transport. In: Gerencser GA (ed) Epithelial transport physiology. Humana Press, Totowa, NJ, pp 379–472. doi:10.1007/978-1-60327-229-2_16

Broer S (2008) Amino acid transport across mammalian intestinal and renal epithelia. Physiol Rev 88(1):249–286. doi:10.1152/physrev.00018.2006

Brückner H, Hausch M (1993) Gas chromatographic characterization of free D-amino acids in the blood serum of patients with renal disorders and of healthy volunteers. J Chromatogr 614 (1):7–17. http://dx.doi.org/10.1016/0378-4347(93)80218-S

Brückner H, Westhauser T (2003) Chromatographic determination of L- and D-amino acids in plants. Amino Acids 24(1–2):43–55. doi:10.1007/s00726-002-0322-8

Buck RH, Krummen K (1987) High-performance liquid chromatographic determination of enantiomeric amino acids and amino alcohols after derivatization with o-phthaldialdehyde and various chiral mercaptans: application to peptide hydrolysates. J Chromatogr 387(0):255–265. http://dx.doi.org/10.1016/S0021-9673(01)94529-7

Burrone L, Di Giovanni M, Di Fiore MM, Chieffi Baccari G, Santillo A (2010) Effects of D-aspartate treatment on D-aspartate oxidase, superoxide dismutase, and caspase 3 activities in frog (*Rana esculenta*) tissues. Chem Biodivers 7:1459–1466. doi:10.1002/cbdv.200900331

Burrone L, Santillo A, Pinelli C, Baccari GC, Di Fiore MM (2012) Induced synthesis of P450 aromatase and 17β-estradiol by D-aspartate in frog brain. J Exp Biol 215(20):3559–3565. doi:10.1242/jeb.073296

Carlson SL, Fieber LA (2011) Physiological evidence that D-aspartate activates a current distinct from ionotropic glutamate receptor currents in Aplysia californica neurons. J Neurophysiol 106 (4):1629–1636. doi:10.1152/jn.00403.2011

Carlson SL, Kempsell AT, Fieber LA (2012) Pharmacological evidence that D-aspartate activates a current distinct from ionotropic glutamate receptor currents in Aplysia californica. Brain Behav 2(4):391–401. doi:10.1002/brb3.60

Case J (1964) Properties of the dactyl chemoreceptors of Cancer antennarius stimpson and C productus randall. Biol Bull 127(3):428–446. doi:10.2307/1539246

Cheng Y, Dovichi N (1988) Subattomole amino acid analysis by capillary zone electrophoresis and laser-induced fluorescence. Science 242(4878):562–564. doi:10.1126/science.3140381

D'Aniello A (2007) D-aspartic acid: an endogenous amino acid with an important neuroendocrine role. Brain Res Rev 53(2):215–234. doi:10.1016/j.brainresrev.2006.08.005

D'Aniello S, Garcia-Fernandez J (2007) D-aspartic acid and L-amino acids in the neural system of the amphioxus Branchiostoma lanceolatum. Amino Acids 32(1):21–26

D'Aniello A, Giuditta A (1977) Identification of D-aspartic acid in the brain of Octopus vulgaris Lam. J Neurochem 29(6):1053–1057. doi:10.1111/j.1471-4159.1977.tb06508.x

D'Aniello A, Giuditta A (1978) Presence of D-aspartate in squid axoplasm and in other regions of the cephalopod nervous system. J Neurochem 31(4):1107–1108. doi:10.1111/j.1471-4159.1978.tb00155.x

D'Aniello A, Vetere A, Padula L (1992) Occurrence of free D-amino acids in the gametes, embryos, larvae and adults of the sea-squirt Ciona intestinalis. Comp Biochem Phys B 102 (4):795–797. doi:10.1016/0305-0491(92)90082-3

D'Aniello A, D'Onofrio G, Pischetola M, D'Aniello G, Vetere A, Petrucelli L, Fisher G (1993a) Biological role of D-amino acid oxidase and D-aspartate oxidase. Effects of D-amino acids. J Biol Chem 268:26941–26949

D'Aniello A, Nardi G, Vetere A, Ferguson GP (1993b) Occurrence of free D-aspartic acid in the circumsoesophageal ganglia of Aplysia fasciata. Life Sci 52(8):733–736. doi:10.1016/0024-3205(93)90235-u

D'Aniello A, Vetere A, Petrucelli L (1993c) Further study on the specificity of D-amino acid oxidase and of D-aspartate oxidase and time course for complete oxidation of D-amino acids. Comp Biochem Phys B 105(3–4):731–734. http://dx.doi.org/10.1016/0305-0491(93)90113-J

D'Aniello A, Nardi G, DeSantis A, Vetere A, diCosmo A, Marchelli R, Dossena A, Fisher G (1995) Free L-amino acids and D-aspartate content in the nervous system of Cephalopoda. A comparative study. Comp Biochem Phys B 112(4):661–666. doi:10.1016/0305-0491(95)00227-8

D'Aniello A, Spinelli P, De Simone A, D'Aniello S, Branno M, Aniello F, Fisher GH, Di Fiore MM, Rastogi RK (2003) Occurrence and neuroendocrine role of D-aspartic acid and N-methyl-D-aspartic acid in Ciona intestinalis. FEBS Lett 552(2–3):193–198. doi:10.1016/s0014-5793(03)00921-9

D'Aniello S, Spinelli P, Ferrandino G, Peterson K, Tsesarskia M, Fisher G, D'Aniello A (2005) Cephalopod vision involves dicarboxylic amino acids: D-aspartate, L-aspartate and L-glutamate. Biochem J 386:331–340. doi:10.1042/BJ20041070

D'Aniello S, Somorjai I, Garcia-Fernandez J, Topo E, D'Aniello A (2010) D-aspartic acid is a novel endogenous neurotransmitter. FASEB J 25(3):1014–1027. doi:10.1096/fj.10-168492

Dalton JB, Schmidt CLA (1933) The solubilities of certain amino acids in water, the densities of their solutions at twenty-five degrees, and the calculated heats of solution and partial molal volumes. J Biol Chem 103(2):549–578

Di Fiore MM, Assisi L, Botte V, D'Aniello A (1998) D-aspartic acid is implicated in the control of testosterone production by the vertebrate gonad. Studies on the female green frog, *Rana esculenta*. J Endocrinol 157(2):199–207. doi:10.1677/joe.0.1570199

Di Fiore MM, Santillo A, Chieffi Baccari G (2014) Current knowledge of D-aspartate in glandular tissues. Amino Acids 46(8):1805–1818. doi:10.1007/s00726-014-1759-2

Di Giovanni M, Burrone L, Chieffi Baccari G, Topo E, Santillo A (2010) Distribution of free D-aspartic acid and D-aspartate oxidase in frog Rana esculenta tissues. J Exp Zool A Ecol Genet Physiol 303:137–143. doi:10.1002/jez.585

EMBL-EBI (2014) GO:0070779 D-aspartate import. http://www.ebi.ac.uk/QuickGO/GTerm?id=GO:0070779#info=2&term=info. Accessed 30 Nov 2014

Errico F, Napolitano F, Nistico R, Centonze D, Usiello A (2009) D-aspartate: an atypical amino acid with neuromodulatory activity in mammals. Rev Neurosci 20(5–6):429–440

Errico F, Napolitano F, Nistico R, Usiello A (2012) New insights on the role of free D-aspartate in the mammalian brain. Amino Acids 43(5):1861–1871. doi:10.1007/s00726-012-1356-1

Errico F, Di Maio A, Marsili V, Squillace M, Vitucci D, Napolitano F, Usiello A (2013) Bimodal effect of D-aspartate on brain aging processes: insights from animal models. J Biol Regul Homeost Agents 27(2):49–59

Erwan E, Tomonaga S, Yoshida J, Nagasawa M, Ogino Y, Denbow DM, Furuse M (2012) Central administration of L- and D-aspartate attenuates stress behaviors by social isolation and CRF in neonatal chicks. Amino Acids 43(5):1969–1976. doi:10.1007/s00726-012-1272-4

Fieber LA, Carlson SL, Capo TR, Schmale MC (2010) Changes in D-aspartate ion currents in the Aplysia nervous system with aging. Brain Res 1343:28–36. doi:10.1016/j.brainres.2010.05.001

Foster AC, Fagg GE (1987) Comparison of L-[^{3}H]glutamate, D-[^{3}H]aspartate, DL-[^{3}H]AP5 and [^{3}H]NMDA as ligands for NMDA receptors in crude postsynaptic densities from rat brain. Eur J Pharmacol 133(3):291–300

Frank H, Nicholson GJ, Bayer E (1977) Rapid gas chromatographic separation of amino acid enantiomers with a novel chiral stationary phase. J Chromatogr Sci 15(5):174–176. doi:10.1093/chromsci/15.5.174

Friedman M (1999) Chemistry, nutrition, and microbiology of D-amino acids. J Agric Food Chem 47(9):3457–3479

Friedman M, Levin CE (2012) Nutritional and medicinal aspects of D-amino acids. Amino Acids 42(5):1553–1582. doi:10.1007/s00726-011-0915-1

Funakoshi M, Sekine M, Katane M, Furuchi T, Yohda M, Yoshikawa T, Homma H (2008) Cloning and functional characterization of Arabidopsis thaliana D-amino acid aminotransferase - D-aspartate behavior during germination. FEBS J 275(6):1188–1200. doi:10.1111/j.1742-4658.2008.06279.x

Furuchi T, Homma H (2005) Free D-aspartate in mammals. Biol Pharm Bull 28(9):1566–1570

Gadea A, López E, López-Colomé A (2004) Glutamate-induced inhibition of D-aspartate uptake in Müller glia from the retina. Neurochem Res 29(1):295–304. doi:10.1023/B:NERE.0000010458.45085.e8

Geiger T, Clarke S (1987) Deamidation, isomerization, and racemization at asparaginyl and aspartyl residues in peptides. Succinimide-linked reactions that contribute to protein degradation. J Biol Chem 262(2):785–794

Gomez TA, Banfield KL, Clarke SG (2008) The protein L-isoaspartyl-*O*-methyltransferase functions in the Caenorhabditis elegans stress response. Mech Ageing Dev 129(12):752–758. doi:10.1016/j.mad.2008.09.019

Hamase K, Morikawa A, Etoh S, Tojo Y, Miyoshi Y, Zaitsu K (2009) Analysis of small amounts of D-amino acids and the study of their physiological functions in mammals. Anal Sci 25 (8):961–968. doi:10.2116/analsci.25.961

Han H, Miyoshi Y, Ueno K, Okamura C, Tojo Y, Mita M, Lindner W, Zaitsu K, Hamase K (2011) Simultaneous determination of D-aspartic acid and D-glutamic acid in rat tissues and physiological fluids using a multi-loop two-dimensional HPLC procedure. J Chromatogr B Analyt Technol Biomed Life Sci 879(29):3196–3202. http://dx.doi.org/10.1016/j.jchromb.2011.01.023

Homma H (2007) Biochemistry of D-aspartate in mammalian cells. Amino Acids 32(1):3–11. doi:10.1007/s00726-006-0354-6

Hoopes EA, Peltzer ET, Bada JL (1978) Determination of amino acid enantiomeric ratios by gas liquid chromatography of the N-trifluoroacetyl-L-prolyl-peptide methyl esters. J Chromatogr Sci 16(11):556–560. doi:10.1093/chromsci/16.11.556

Jones WM, Ringe D, Soda K, Manning JM (1994) Determination of free D-amino acids with a bacterial transaminase: their depletion leads to inhibition of bacterial growth. Anal Biochem 218(1):204–209. http://dx.doi.org/10.1006/abio.1994.1161

Kang L, Buck RH (1992) Separation and enantiomer determination of OPA-derivatised amino acids by using capillary zone electrophoresis. Amino Acids 2(1–2):103–109. doi:10.1007/BF00806080

Kaspar H, Dettmer K, Gronwald W, Oefner PJ (2008) Automated GC–MS analysis of free amino acids in biological fluids. J Chromatogr B Analyt Technol Biomed Life Sci 870(2):222–232. http://dx.doi.org/10.1016/j.jchromb.2008.06.018

Katane M, Homma H (2010) D-aspartate oxidase: the sole catabolic enzyme acting on free D-aspartate in mammals. Chem Biodivers 7(6):1435–1449. doi:10.1002/cbdv.200900250

Katane M, Homma H (2011) D-aspartate – an important bioactive substance in mammals: a review from an analytical and biological point of view. J Chromatogr B Analyt Technol Biomed Life Sci 879(29):3108–3121. doi:10.1016/j.jchromb.2011.03.062

Katane M, Saitoh Y, Seida Y, Sekine M, Furuchi T, Homma H (2010) Comparative characterization of three D-aspartate oxidases and one D-amino acid oxidase from Caenorhabditis elegans. Chem Biodivers 7(6):1424–1434. doi:10.1002/cbdv.200900294

Kera Y, Aoyama H, Watanabe N, Yamada RH (1996) Distribution of D-aspartate oxidase and free D-glutamate and D-aspartate in chicken and pigeon tissues. Comp Biochem Physiol B: Biochem Mol Biol 115(1):121–126

Kim PM, Duan X, Huang AS, Liu CY, Ming GL, Song HJ, Snyder SH (2010) Aspartate racemase, generating neuronal D-aspartate, regulates adult neurogenesis. Proc Natl Acad Sci U S A 107 (7):3175–3179. doi:10.1073/pnas.0914706107

Kitagawa F, Otsuka K (2011) Recent progress in capillary electrophoretic analysis of amino acid enantiomers. J Chromatogr B Analyt Technol Biomed Life Sci 879(29):3078–3095, http://dx.doi.org/10.1016/j.jchromb.2011.03.016

Konno R (2007) D-amino acids: a new frontier in amino acids and protein research: practical methods and protocols. Nova Biomedical Books, New York

Krehbiel CR, Matthews JC (2003) Absorption of amino acids and peptides. In: D'Mello JPF (ed) Amino acids in animal nutrition, 2nd edn. CABI Publishers, Wallingford/Cambridge, MA, pp 41–70

Lee J-A, Homma H, Sakai K, Fukushima T, Santa T, Tashiro K, Iwatsubo T, Yoshikawa M, Imai K (1997) Immunohistochemical localization of D-aspartate in the rat pineal gland. Biochem Biophys Res Commun 231(2):505–508, http://dx.doi.org/10.1006/bbrc.1996.5902

Long Z, Homma H, Lee J, Fukushima T, Santa T, Iwatsubo T, Yamada R, Imai K (1998) Biosynthesis of D-aspartate in mammalian cells. FEBS Lett 434(3):231–235

Man EH, Bada JL (1987) Dietary D-amino acids. Annu Rev Nutr 7:209–225. doi:10.1146/annurev.nu.07.070187.001233

Masuda W, Nouso C, Kitamura C, Terashita M, Noguchi T (2003) Free D-aspartic acid in rat salivary glands. Arch Biochem Biophys 420(1):46–54, http://dx.doi.org/10.1016/j.abb.2003.09.032

Miao H, Rubakhin SS, Sweedler JV (2005) Subcellular analysis of D-aspartate. Anal Chem 77 (22):7190–7194. doi:10.1021/ac0511694

Miao H, Rubakhin SS, Scanlan CR, Wang LP, Sweedler JV (2006a) D-aspartate as a putative cell-cell signaling molecule in the Aplysia californica central nervous system. J Neurochem 97 (2):595–606. doi:10.1111/j.1471-4159.2006.03891.x

Miao H, Rubakhin SS, Sweedler JV (2006b) Confirmation of peak assignments in capillary electrophoresis using immunoprecipitation. Application to D-aspartate measurements in neurons. J Chromatogr A 1106(1–2):56–60. doi:10.1016/j.chroma.2005.09.037

Miyoshi Y, Koga R, Oyama T, Han H, Ueno K, Masuyama K, Itoh Y, Hamase K (2012) HPLC analysis of naturally occurring free D-amino acids in mammals. J Pharm Biomed Anal 69:42–49, http://dx.doi.org/10.1016/j.jpba.2012.01.041

Moini M, Schultz CL, Mahmood H (2003) CE/electrospray ionization-MS analysis of underivatized D/L-amino acids and several small neurotransmitters at attomole levels through the use of 18-crown-6-tetracarboxylic acid as a complexation reagent/background electrolyte. Anal Chem 75(22):6282–6287. doi:10.1021/ac034708i

Monteforte R, Santillo A, Di Giovanni M, D'Aniello A, Di Maro A, Chieffi Baccari G (2009) D-aspartate affects secretory activity in rat Harderian gland: molecular mechanism and functional significance. Amino Acids 37(4):653–664. doi:10.1007/s00726-008-0185-8

Morikawa A, Hamase K, Inoue T, Konno R, Niwa A, Zaitsu K (2001) Determination of free D-aspartic acid, D-serine and d-alanine in the brain of mutant mice lacking D-amino-acid oxidase activity. J Chromatogr B Biomed Sci Appl 757(1):119–125, http://dx.doi.org/10.1016/S0378-4347(01)00131-1

Nagata Y, Akino T, Ohno K (1985) Microdetermination of serum D-amino acids. Anal Biochem 150(1):238–242, http://dx.doi.org/10.1016/0003-2697(85)90465-8

Neidle A, Dunlop DS (1990) Developmental changes in free D-aspartic acid in the chicken embryo and in the neonatal rat. Life Sci 46(21):1517–1522. doi:10.1016/0024-3205(90)90424-P

Nimura N, Kinoshita T (1986) O-Phthalaldehyde—N-acetyl-L-cysteine as a chiral derivatization reagent for liquid chromatographic optical resolution of amino acid ernantiomers and its application to conventional amino acid analysis. J Chromatogr 352:169–177, http://dx.doi.org/10.1016/S0021-9673(01)83377-X

Okuma E, Abe H (1994) Simultaneous determination of D- and L-amino acids in the nervous tissues of crustaceans using precolumn derivatization with (+)-1-(9-fluorenyl)ethyl chloroformate and reversed-phase ion-pair high-performance liquid chromatography. J Chromatogr B Biomed Appl 660(2):243–250. doi:10.1016/0378-4347(94)00304-1

Ota N, Shi T, Sweedler J (2012) D-aspartate acts as a signaling molecule in nervous and neuroendocrine systems. Amino Acids 43(5):1873–1886. doi:10.1007/s00726-012-1364-1

Pasteur L (1852) Untersuchungen über Asparaginsäure und Aepfelsäure. Justus Liebigs Ann Chem 82(3):324–335. doi:10.1002/jlac.18520820306

Payan IL, Cadilla-Perezrios R, Fisher GH, Man EH (1985) Analysis of problems encountered in the determination of amino acid enantiomeric ratios by gas chromatography. Anal Biochem 149(2):484–491, http://dx.doi.org/10.1016/0003-2697(85)90603-7

Perez MT, Pausz C, Herndl GJ (2003) Major shift in bacterioplankton utilization of enantiomeric amino acids between surface waters and the ocean's interior. Limnol Oceanogr 48(2):755–763

Raucci F (2005) Endocrine roles of D-aspartic acid in the testis of lizard Podarcis s. sicula. J Endocrinol 187(3):347–359. doi:10.1677/joe.1.06115

Raucci F, Di Fiore MM (2010) The maturation of oocyte follicular epithelium of Podarcis s. sicula is promoted by D-aspartic acid. J Histochem Cytochem 58(2):157–171. doi:10.1369/jhc.2009.954636

Raucci F, Di Fiore MM (2011) D-Asp: a new player in reproductive endocrinology of the amphibian *Rana esculenta*. J Chromatogr B Analyt Technol Biomed Life Sci 879 (29):3268–3276. doi:10.1016/j.jchromb.2011.04.007

Raucci F, Fiore MMD (2009) The reproductive activity in the testis of Podarcis s. sicula involves D-aspartic acid: a study on c-kit receptor protein, tyrosine kinase activity and PCNA protein during annual sexual cycle. Gen Comp Endocrinol 161(3):373–383. doi:10.1016/j.ygcen.2009.02.002

Raucci F, Assisi L, D'Aniello S, Spinelli P, Botte V, Di Fiore MM (2004) Testicular endocrine activity is upregulated by D-aspartic acid in the green frog, *Rana esculenta*. J Endocrinol 182 (2):365–376

Raucci F, Santillo A, D'Aniello A, Chieffi P, Baccari GC (2005) D-aspartate modulates transcriptional activity in Harderian gland of frog, *Rana esculenta*: morphological and molecular evidence. J Cell Physiol 204(2):445–454. doi:10.1002/jcp.20316

Reissner K, Aswad D (2003) Deamidation and isoaspartate formation in proteins: unwanted alterations or surreptitious signals? Cell Mol Life Sci 60(7):1281–1295

Saitoh Y, Katane M, Kawata T, Maeda K, Sekine M, Furuchi T, Kobuna H, Sakamoto T, Inoue T, Arai H, Nakagawa Y, Homma H (2012) Spatiotemporal localization of D-amino acid oxidase and D-aspartate oxidases during development in Caenorhabditis elegans. Mol Cell Biol 32 (10):1967–1983. doi:10.1128/MCB.06513-11

Santillo A, Monteforte R, Raucci F, D'Aniello A, Baccari GC (2006) Occurrence of D-aspartate in the harderian gland of Podarcis s. sicula and its effect on gland secretion. J Exp Zool A Comp Exp Biol 305A(8):610–619. doi:10.1002/jez.a.301

Santillo A, Pinelli C, Burrone L, Chieffi Baccari G, Di Fiore MM (2013) D-aspartic acid implication in the modulation of frog brain sex steroid levels. Gen Comp Endocrinol 181:72–76. doi:10.1016/j.ygcen.2012.11.003

Saunders NR, Daneman R, Dziegielewska KM, Liddelow SA (2013) Transporters of the blood-brain and blood-CSF interfaces in development and in the adult. Mol Aspects Med 34 (2–3):742–752. doi:10.1016/j.mam.2012.11.006

Scanlan C, Shi T, Hatcher NG, Rubakhin SS, Sweedler JV (2010) Synthesis, accumulation, and release of D-aspartate in the Aplysia californica CNS. J Neurochem 115(5):1234–1244. doi:10.1111/j.1471-4159.2010.07020.x

Schell MJ, Cooper OB, Snyder SH (1997) D-aspartate localizations imply neuronal and neuroendocrine roles. Proc Natl Acad Sci U S A 94(5):2013–2018

Schieber A, Brückner H, Rupp-Classen M, Pecht W, Nowitzki-Gfimm S, Classen HG (1997) Evaluation of D-amino acid levels in rat by gas chromatography-selected ion monitoring mass spectrometry: no evidence for subacute toxicity of orally fed D-proline and D-aspartic acid. J Chromatogr B Biomed Sci Appl 691(1):1–12, http://dx.doi.org/10.1016/S0378-4347(96)00378-7

Shibata K, Watanabe T, Yoshikawa H, Abe K, Takahashi S, Kera Y, Yamada RH (2003) Purification and characterization of aspartate racemase from the bivalve mollusk *Scapharca broughtonii*. Comp Biochem Physiol B: Biochem Mol Biol 134(2):307–314. doi:10.1016/s1096-4959(02)00267-1

Shikata Y, Watanabe T, Teramoto T, Inoue A, Kawakami Y, Nishizawa Y, Katayama K, Kuwada M (1995) Isolation and characterization of a peptide isomerase from funnel-web spider venom. J Biol Chem 270(28):16719–16723. doi:10.1074/jbc.270.28.16719

Shinbo T, Yamaguchi T, Nishimura K, Sugiura M (1987) Chromatographic separation of racemic amino acids by use of chiral crown ether-coated reversed-phase packings. J Chromatogr 405:145–153, http://dx.doi.org/10.1016/S0021-9673(01)81756-8

Simo C, Rizzi A, Barbas C, Cifuentes A (2005) Chiral capillary electrophoresis-mass spectrometry of amino acids in foods. Electrophoresis 26(7–8):1432–1441. doi:10.1002/elps.200406199

Skoog DA, Holler FJ, Crouch SR (2007) Principles of instrumental analysis, 6th edn. Thomson Brooks/Cole, Belmont

Song Y, Liang F, Liu Y-M (2007) Quantification of D-amino acids in the central nervous system of Aplysia californica by liquid chromatography/tandem mass spectrometry. Rapid Commun Mass Spectrom 21(1):73–77. doi:10.1002/rcm.2803

Song Y, Feng Y, Lu X, Zhao S, Liu CW, Liu YM (2008) D-amino acids in rat brain measured by liquid chromatography/tandem mass spectrometry. Neurosci Lett 445(1):53–57. doi:10.1016/j.neulet.2008.08.058

Soustelle L, Besson MT, Rival T, Birman S (2002) Terminal glial differentiation involves regulated expression of the excitatory amino acid transporters in the Drosophila embryonic CNS. Dev Biol 248(2):294–306

Spinelli P, Brown ER, Ferrandino G, Branno M, Montarolo PG, D'Aniello E, Rastogi RK, D'Aniello B, Baccari GC, Fisher G, D'Aniello A (2006) D-aspartic acid in the nervous system of Aplysia limacina: possible role in neurotransmission. J Cell Physiol 206(3):672–681. doi:10.1002/jcp.20513

Stephenson RC, Clarke S (1989) Succinimide formation from aspartyl and asparaginyl peptides as a model for the spontaneous degradation of proteins. J Biol Chem 264(11):6164–6170

Stevens BR (2010) Amino acid transport by epithelial membranes. In: Gerencser GA (ed) Epithelial transport physiology. Humana Press, Totowa, NJ, pp 353–378. doi:10.1007/978-1-60327-229-2_15

Still JL, Buell MV, Knox WE, Green DE (1949) Studies on the cyclophorase system; D-aspartic oxidase. J Biol Chem 179(2):831–837

Swartz ME, Mazzeo JR, Grover ER, Brown PR (1995) Separation of amino acid enantiomers by micellar electrokinetic capillary chromatography using synthetic chiral surfactants. Anal Biochem 231(1):65–71, http://dx.doi.org/10.1006/abio.1995.1504

Takahashi O (2014) Just three water molecules can trigger the undesired nonenzymatic reactions of aspartic acid residues: new insight from a quantum-chemical study. J Phys Conf Ser 490:012147. doi:10.1088/1742-6596/490/1/012147

Takigawa Y, Homma H, Lee JA, Fukushima T, Santa T, Iwatsubo T, Imai K (1998) D-aspartate uptake into cultured rat pinealocytes and the concomitant effect on L-aspartate levels and melatonin secretion. Biochem Biophys Res Commun 248(3):641–647. doi:10.1006/bbrc.1998.8971

Terabe S, Shibata M, Miyashita Y (1989) Chiral separation by electronkinetic chromatography while bile salt micelles. J Chromatogr 480:403–411, http://dx.doi.org/10.1016/S0021-9673(01)84309-0

Tivesten A, Lundqvist A, Folestad S (1997) Selective chiral determination of aspartic and glutamic acid in biological samples by capillary electrophoresis. Chromatographia 44(11–12):623–633. doi:10.1007/Bf02466666

Topo E, Soricelli A, Di Maio A, D'Aniello E, Di Fiore MM, D'Aniello A (2010) Evidence for the involvement of D-aspartic acid in learning and memory of rat. Amino Acids 38(5):1561–1569. doi:10.1007/s00726-009-0369-x

Tsesarskaia M, Galindo E, Szókán G, Fisher G (2009) HPLC determination of acidic D-amino acids and their N-methyl derivatives in biological tissues. Biomed Chromatogr 23(6):581–587. doi:10.1002/bmc.1156

Tverdislov VA, Yakovenko LV, Ivlieva AA, Tverdislova IL (2011) Ionic and chiral asymmetries as physical factors of biogenesis and ontogenesis. Mosc U Phys B+ 66(2):105–115. doi:10.3103/S0027134911020184

Ueda T, Kitamura F, Mitchell R, Metcalf T, Kuwana T, Nakamoto A (1991) Chiral separation of naphthalene-2,3-dicarboxaldehyde-labeled amino acid enantiomers by cyclodextrin-modified micellar electrokinetic chromatography with laser-induced fluorescence detection. Anal Chem 63(24):2979–2981. doi:10.1021/ac00024a033

Umesh A, Cohen BN, Ross LS, Gill SS (2003) Functional characterization of a glutamate/aspartate transporter from the mosquito Aedes aegypti. J Exp Biol 206(Pt 13):2241–2255

Villar-Cerviño V, Barreiro-Iglesias A, Rodicio MC, Anadón R (2010) D-serine is distributed in neurons in the brain of the sea lamprey. J Comp Neurol 518(10):1688–1710. doi:10.1002/cne.22296

Waldhier MC, Dettmer K, Gruber MA, Oefner PJ (2010) Comparison of derivatization and chromatographic methods for GC–MS analysis of amino acid enantiomers in physiological samples. J Chromatogr B Analyt Technol Biomed Life Sci 878(15–16):1103–1112, http://dx.doi.org/10.1016/j.jchromb.2010.03.021

Waldhier MC, Almstetter MF, Nürnberger N, Gruber MA, Dettmer K, Oefner PJ (2011) Improved enantiomer resolution and quantification of free D-amino acids in serum and urine by comprehensive two-dimensional gas chromatography–time-of-flight mass spectrometry. J Chromatogr 1218(28):4537–4544, http://dx.doi.org/10.1016/j.chroma.2011.05.039

Wan H, Blomberg LG (2000) Chiral separation of amino acids and peptides by capillary electrophoresis. J Chromatogr 875(1–2):43–88, http://dx.doi.org/10.1016/S0021-9673(99)01209-1

Wang LP, Ota N, Romanova EV, Sweedler JV (2011) A novel pyridoxal 5′-phosphate-dependent amino acid racemase in the Aplysia californica central nervous system. J Biol Chem 286(15):13765–13774. doi:10.1074/jbc.M110.178228

Welch CJ (1994) Evolution of chiral stationary phase design in the Pirkle laboratories. J Chromatogr 666(1–2):3–26, http://dx.doi.org/10.1016/0021-9673(94)80367-6

Yamamoto A, Tanaka H, Ishida T, Horiike K (2011) Immunohistochemical localization of D-aspartate oxidase in porcine peripheral tissues. Amino Acids 41(2):529–536. doi:10.1007/s00726-010-0785-y

Yamane H, Tsuneyoshi Y, Denbow DM, Furuse M (2009) N-methyl-D-aspartate and alpha-amino-3-hydroxy-5-methyl-4-isoxazolepropionate receptors involved in the induction of sedative effects under an acute stress in neonatal chicks. Amino Acids 37(4):733–739. doi:10.1007/s00726-008-0203-x

Yang L, Chen CJ, Liu X, Shi J, Wang GA, Zhu LD, Guo LP, Glennon JD, Scully NM, Doherty BE (2010) Use of cyclodextrin-modified gold nanoparticles for enantioseparations of drugs and amino acids based on pseudostationary phase-capillary electrochromatography. Electrophoresis 31(10):1697–1705. doi:10.1002/elps.200900541

Zaar K (1996) Light and electron microscopic localization of D-aspartate oxidase in peroxisomes of bovine kidney and liver: an immunocytochemical study. J Histochem Cytochem 44(9):1013–1019. doi:10.1177/44.9.8773567

Zaar K, Köst H-P, Schad A, Völkl A, Baumgart E, Fahimi HD (2002) Cellular and subcellular distribution of D-aspartate oxidase in human and rat brain. J Comp Neurol 450(3):272–282. doi:10.1002/cne.10320

Zampolli MG, Basaglia G, Dondi F, Sternberg R, Szopa C, Pietrogrande MC (2007) Gas chromatography–mass spectrometry analysis of amino acid enantiomers as methyl chloroformate derivatives: application to space analysis. J Chromatogr 1150(1–2):162–172. http://dx.doi.org/10.1016/j.chroma.2006.12.033

Chapter 13
Homeostasis of Free D-Aspartate in Mammalian Cells

Hiroshi Homma and Masumi Katane

Abstract Among the free D-amino acids found in mammals, D-aspartate (D-Asp) has been shown to play crucial roles in the central nervous, neuroendocrine, and endocrine systems. Here, we present an overview of recent studies on free D-Asp, focusing on homeostasis in mammalian cells, especially the molecular mechanisms necessary to synthesize, degrade, and release D-Asp.

Keywords D-Aspartate • Homeostasis • Biosynthesis • Degradation • Efflux

Although biosynthesis of D-Asp has been demonstrated in several cultured mammalian cells, the complete biosynthetic pathway of D-Asp in mammals remains to be elucidated. In contrast to the enigma of biosynthesis of D-Asp, however, D-Asp oxidase (DDO) has long been known as a degradative enzyme that stereospecifically acts on acidic D-amino acids and is the sole catabolic enzyme to act on D-Asp in mammals. Because recent studies have implicated D-Asp in the pathophysiology of *N*-methyl-D-Asp receptor-related diseases, such as depression and schizophrenia, DDO is considered an attractive therapeutic target, and DDO inhibitors that augment brain D-Asp levels are now considered as potential antipsychotic drugs for treatment of these diseases.

There appear to be distinct release pathways of D-Asp in mammalian cells. Two pathways for the efflux of D-Asp of cytoplasmic origin have been identified; one is a spontaneous and continuous release pathway, and the other is through volume-sensitive organic anion channels. An alternative pathway for the release of D-Asp via exocytotic discharge of vesicular D-Asp has also been identified. In mammalian tissues, D-Asp is present in specific cells, and the aforementioned mechanisms may be selectively employed in distinct cell types to regulate D-Asp homeostasis.

H. Homma (✉) • M. Katane
Laboratory of Biomolecular Science, Graduate School of Pharmaceutical Life Sciences, Kitasato University, 5-9-1 Shirokane, Minato-ku, Tokyo 108-8641, Japan
e-mail: hommah@pharm.kitasato-u.ac.jp

T. Yoshimura et al. (eds.), *D-Amino Acids*, DOI 10.1007/978-4-431-56077-7_13

13.1 Introduction

In 1986, Dunlop et al. first reported the existence of free D-aspartate (D-Asp) in mammals, including humans (Dunlop et al. 1986). Since then, substantial amounts of D-Asp have been found in a wide variety of mammalian tissues, particularly the central nervous, neuroendocrine, and endocrine systems. Using several different methods, including high-performance liquid chromatography (HPLC) analysis and immunohistochemical staining with anti-D-Asp antibody, the expression levels and localization of D-Asp in mammalian tissues and cells have been studied, and ontogenetic changes in its contents and localization have been determined in various tissues. The biological and physiological activities of D-Asp also have been examined in detail in selected neuroendocrine and endocrine tissues, such as the pineal gland, pituitary gland, hypothalamus, and testis. A variety of review articles on D-Asp have been recently published, including publications describing D-Asp quantification and localization in various tissues (D'Aniello 2007; Furuchi and Homma 2005; Homma 2002, 2007) and biological activities in endocrine tissues (D'Aniello 2007; Di Fiore et al. 2014; Furuchi and Homma 2005; Homma 2002; Katane and Homma 2011) and the nervous system (Errico et al. 2009, 2012; Ota et al. 2012). In this review, we focus on the homeostasis of D-Asp in mammalian cells, especially its biosynthesis and degradation in the cells and its efflux from and reuptake into the cells. In this chapter, the methodology used to analyze D-Asp (Chap. 12) and the biological activities of D-Asp in both nonmammalian (Chaps. 11 and 12) and mammalian cells (Chap. 13) are reviewed.

13.2 Biosynthesis of D-Asp

Since the discovery of free D-Asp in mammals (Dunlop et al. 1986), much attention has been paid to the origins and synthetic pathways of this D-amino acid. Substantial amounts of D-Asp have been detected in various lower organisms, and in mammals ingested D-Asp can be absorbed in the intestine and transported to several tissues (D'Aniello et al. 1993a). Therefore, mammalian D-Asp can be partially derived from enterobacterium and/or diet. On the other hand, Neidle and Dunlop observed that D-Asp levels were remarkably increased in fertilized chicken eggs during incubation (Neidle and Dunlop 1990). Since the egg is a closed system without dietary supplementation, this observation strongly suggested that the biosynthesis of D-Asp occurs in chicken embryos and therefore perhaps also in the mammalian body. Subsequent studies utilizing several different methods, including HPLC analysis, immunocytochemical staining, and enzymatic assays based on D-Asp-specific oxidase reactivity, showed that rat pheochromocytoma PC12 cells contain D-Asp and that D-Asp levels, both intracellular and in the cell culture medium, increase over time (Fig. 13.1A; Long et al. 1998). Since extracellular D-Asp is not taken up by cells, and the cells do not express the L-Glu transporter, this suggests

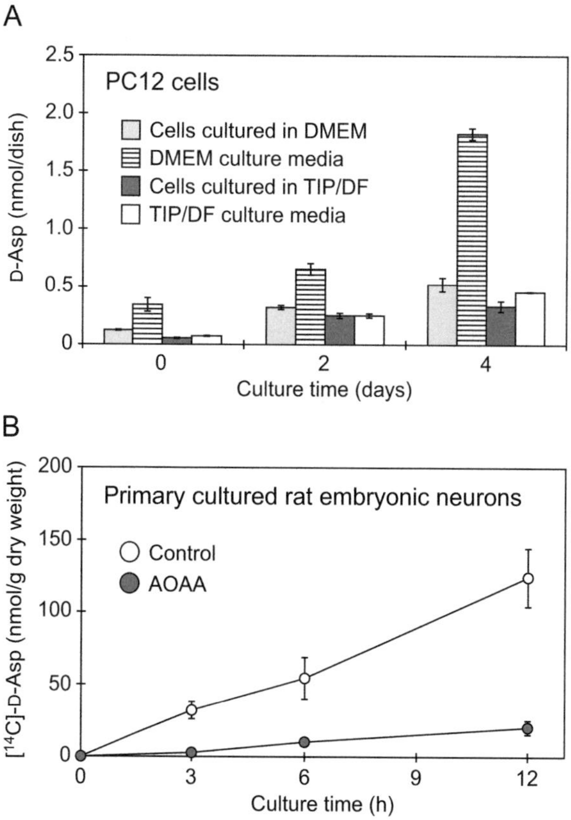

Fig. 13.1 Biosynthesis of D-Asp in cultured mammalian cells. (**A**) PC12 cells were cultured in DMEM or TIP/DF medium. The level of D-Asp in cells cultured in DMEM (*light gray bars*), DMEM culture media (*striped bars*), cells cultured in TIP/DF (*dark gray bars*), and TIP/DF culture media (*open bars*) were determined at the indicated culture times. Data are presented as the means from one triplicate assay ± standard deviation. For details, see text and Long et al. (1998). (**B**) Purified rat embryonic neurons were cultured with [^{14}C]-L-Asp in the absence (*open circles*) or presence (*gray circles*) of AOAA. The [^{14}C]-D-Asp contents in the cells were determined at the indicated culture times. Data are presented as the mean ± standard error of the mean ($n = 4$). For details, see text and Wolosker et al. (2000) (Reproduced with permission from Elsevier (Long et al. 1998; Wolosker et al. 2000))

that intracellular biosynthesis of D-Asp occurs in these cells. D-Asp synthesis has also been observed in a subclone of PC12 cells (Long et al. 2002), rat pituitary tumor GH_3 cells (Long et al. 2000), and recently in human cervical adenocarcinoma HeLa cells (Matsuda et al. 2015). In PC12 cells, the highest proportion of D-Asp (D-Asp/[D-Asp + L-Asp] × 100 %) was approximately 14 %, whereas biosynthesis of D-Asp was not observed in mouse fibroblast Swiss 3T3 cells or human neuroblastoma NB-1 cells (Long et al. 1998). Furthermore, in primary cultures of rat embryonic neurons, high levels of endogenous D-Asp were detected, and [^{14}C]-D-Asp biosynthesis was demonstrated using [^{14}C]-L-Asp as the precursor molecule (Fig. 13.1B;

Wolosker et al. 2000). Treatment of these neuronal cultures with amino-oxyacetic acid (AOAA) markedly inhibited the accumulation of [^{14}C]-D-Asp (Wolosker et al. 2000). Since AOAA is a potent inhibitor of pyridoxal phosphate (PLP)-dependent enzymes, this result suggests that a PLP-dependent enzyme is involved in the conversion of L-Asp to D-Asp. A likely candidate for this enzyme is a racemase, and in fact, an Asp-specific, PLP-dependent racemase (EC 5.1.1.13) from the bivalve *Scapharca broughtonii* has been cloned and characterized (Abe et al. 2006; Shibata et al. 2003a, b; Watanabe et al. 1998). A second Asp racemase has also been cloned from the sea slug *Aplysia californica* (Wang et al. 2011), and Asp-specific racemase-like activity (conversion of L-Asp to D-Asp) has also been reported in other animals, including the cephalopod (D'Aniello et al. 2005), frog (Raucci et al. 2005), lizard (Assisi et al. 2001), and mollusk (Spinelli et al. 2006). Recently, the gene that encodes Asp racemase (*Got 1-like 1*, *Got1l1*) was identified in mouse (Kim et al. 2010), and the recombinant form of this enzyme has been shown to exhibit PLP-dependent Asp racemase activity. Notably, the deduced amino acid sequence of the mouse Asp racemase is more homologous to that of mammalian Asp aminotransferase than to mammalian Ser racemases and the *S. broughtonii* or *A. californica* Asp racemases, whereas the *S. broughtonii* and *A. californica* Asp racemases show relatively high amino acid sequence identity with the mammalian Ser racemases. Therefore, it appears that the Asp racemases of mouse, *S. broughtonii*, and *A. californica* have evolved convergently from different ancestral proteins to acquire a similar catalytic mechanism.

In mice, Asp racemase is abundant in the brain, heart, and testis, followed by the adrenal gland, and is expressed at negligible levels in the liver, lung, kidney, and spleen (Kim et al. 2010). Immunohistochemical analysis revealed coincident localization of Asp racemase and D-Asp in the mouse brain, pituitary gland, hippocampus, pineal gland, adrenal medulla, and testis. Interestingly, depletion of Asp racemase in the adult mouse hippocampus by retrovirus-mediated expression of Asp racemase-targeting short-hairpin RNA elicited profound defects in dendritic development and neuron survival (Kim et al. 2010). Thus, the current data suggest that Asp racemase plays an essential role in neuronal development, consistent with the previously proposed role for D-Asp in neurogenesis and development of the brain (Sakai et al. 1998; Wolosker et al. 2000).

In the rat and human genomes, the genes annotated as homologous to the mouse Asp racemase are designated *Got1l1* and *GOT1L1*, respectively, and the proteins encoded by these genes are presumed to be responsible for the biosynthesis of D-Asp. However, the lengths of the predicted amino acid sequences of mouse, rat, and human GOT1L1s (404, 315, and 421 amino acids, respectively) are not conserved. In our recent study, we investigated whether *Got1l1* and/or *GOT1L1* are involved in Asp synthesis. First, the correlation between cellular D-Asp content and the expression level of GOT1L1 mRNA was examined in several rat and human cell lines. The expression levels of mRNA encoding D-Asp oxidase (DDO, also known as DASPO; EC 1.4.3.1), the sole catabolic enzyme that acts on free D-Asp in mammals (Katane and Homma 2010; Ohide et al. 2011), were also measured in these cell lines. Second, the effect of knockdown of the *Got1l1* gene on D-Asp biosynthesis

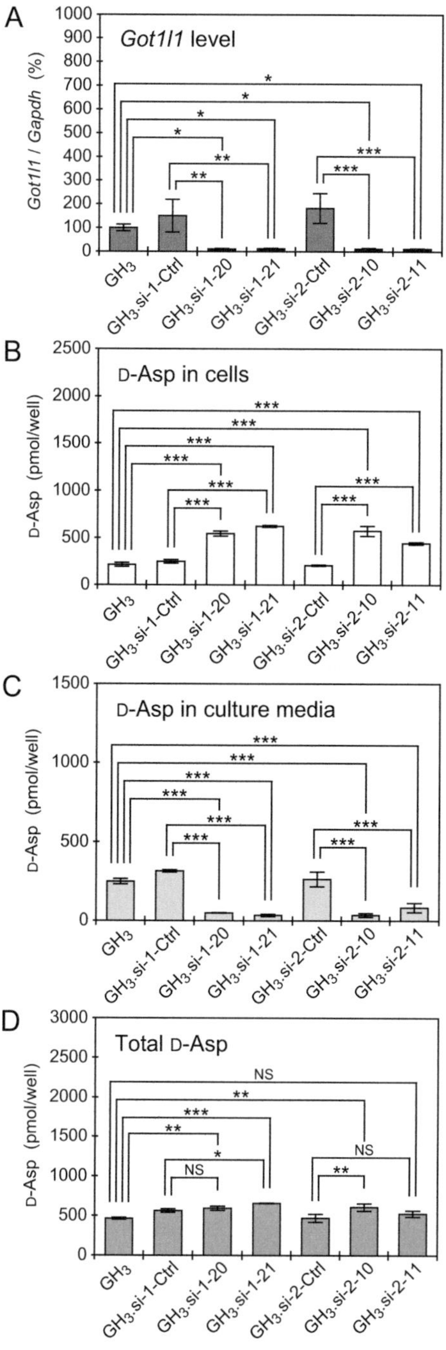

Fig. 13.2 The effect of *Got1l1* gene knockdown on D-Asp synthesis in GH_3 cells. (**A**) Real-time PCR analyses of *Got1l1* mRNA levels normalized to those of mRNA encoding glyceraldehyde-3-phosphate dehydrogenase (GAPDH) in wild-type GH_3 cells and GH_3 cells treated with *Got1l1*-specific (GH_3.si-1-20, GH_3.si-1-21, GH_3.si-2-10, and GH_3.si-2-11 cells) or scrambled control siRNAs (GH_3.si-1-Ctrl and GH_3.si-2-Ctrl cells). The *Got1l1* expression levels in each cell type are expressed relative to that in wild-type GH_3 cells. Data are presented as the mean $\pm$ standard deviation ($n = 3$). (B–D) HPLC analyses of D-Asp synthesis by wild-type GH_3 cells and GH_3 cells treated with *Got1l1*-specific or scrambled control siRNAs. Cells were seeded at a density of 1×10^6 into 6-well plates and cultured. The following day, the medium was replaced and the cells were cultured for 24 h. The D-Asp contents of the cells (**B**) and the culture medium (**C**), and the total D-Asp content (D-Asp contents in the cells plus media) (**D**), are shown. Data are presented as the mean $\pm$ standard deviation ($n = 3$). $^*P < 0.05$, $^{**}P < 0.01$, and $^{***}P < 0.001$ based on the Tukey–Kramer multiple comparison test. NS, not significant ($P > 0.05$)

Table 13.1 Tissue levels of D- and L-Asp in mouse brain

Brain region	Genotype	D-Asp (pmol/mg tissue)	L-Asp (nmol/mg tissue)
Frontal cortex	Wild type	8.15 ± 1.64	0.59 ± 0.13
	Ser racemase KO	2.82 ± 0.27 **	0.44 ± 0.02
Hippocampus	Wild type	9.00 ± 0.55	0.66 ± 0.04
	Ser racemase KO	4.27 ± 0.41 ***	0.66 ± 0.04
Striatum	Wild type	7.77 ± 0.69	0.39 ± 0.02
	Ser racemase KO	2.85 ± 0.69 ***	0.39 ± 0.02
Cerebellum	Wild type	1.89 ± 0.05	0.54 ± 0.04
	Ser racemase KO	1.86 ± 0.08	0.55 ± 0.02

Data are from Horio et al. (2013). For details, see text and Horio et al. (2013). Reproduced with permission from Elsevier (Horio et al. 2013)
Each value shown is the mean ± standard error of the mean ($n = 10$). ** $P < 0.01$ and *** $P < 0.001$ (Student's t-test) compared to wild type. Tissue levels of D- and L-Ala in the frontal cortex, hippocampus, striatum, and cerebellum were comparable between wild-type and Ser racemase KO mice (Horio et al. 2013)

was examined. The results failed to show any correlation between the expression levels of *Got1l1*/*GOT1L1* and D-Asp contents in either the rat or human cell lines (Matsuda et al. 2015). Furthermore, during the culture of the cells that are known to synthesize D-Asp (Long et al. 2000), knockdown of *Got1l1* did not reduce the total contents of D-Asp (D-Asp in the cells plus D-Asp in the media) (Fig. 13.2), confirming that GOT1L1 contributes little, if at all, to the biosynthesis of D-Asp in rat and human cells (Matsuda et al. 2015).

In the other report (Tanaka-Hayashi et al. 2015), a *Got1l1* knockout (KO) mouse was generated, and the amino acid contents in the wild-type and KO mice were compared. No significant differences were observed in the amino acid contents in the hippocampus or testis. In the report, the recombinant protein encoded by mouse *Got1l1* was also produced, although it was reported to be difficult to obtain large amounts of soluble form of the protein. However, the recombinant protein failed to produce D-Asp from L-Asp. Instead, the protein exhibited L-Asp aminotransferase activity. Thus, the whole picture of the biosynthetic pathway of D-Asp in mammals remains to be elucidated.

It is noteworthy that *A. californica* Asp racemase shows racemase activity toward Asp as well as Ser (Wang et al. 2011), while *S. broughtonii* Asp racemase was reported to exhibit no racemase activity toward Ser (Abe et al. 2006; Shibata et al. 2003b). Ser racemase in mammals might have some activity toward Asp to produce D-Asp, and a Ser racemase KO mouse exhibited reduced levels of D-Asp in some regions of the brain (Table 13.1; Horio et al. 2013). Thus, D-Asp synthesis may be closely related to D-Ser metabolism.

An alternative pathway for D-Asp synthesis in mammals involving D-amino acid aminotransferase (EC 2.6.1.21) may be possible. This enzyme transfers an amino group from a D-amino acid to an oxaloacetic acid, resulting in the production of D-Asp. In fact, the cDNA of D-amino acid aminotransferase has been cloned from *Arabidopsis thaliana*, and the recombinant form of this enzyme has been shown to catalyze transamination from D-Glu to oxaloacetic acid to produce D-Asp (Funakoshi et al. 2008). However, no D-amino acid aminotransferase has been identified in mammals. On the other hand, free D-Asp can be derived from D-aspartyl residue-containing proteins. It is known that L-asparaginyl and L-aspartyl residues in proteins undergo spontaneous isomerization under physiological conditions to generate D-aspartyl and/or isoaspartyl residues (Fujii 2005; Fujii et al. 2010). In addition, D-aspartyl endopeptidase, a protease that can generate free D-Asp through the degradation of D-aspartyl residue-containing proteins, has been identified in rabbit and mouse (Kinouchi et al. 2004).

13.3 Degradation of D-Asp

DDO has long been known as a degradative enzyme that stereospecifically acts on D-Asp. In 1949, Still et al. first reported the enzymatic activity of DDO in rabbit kidney extracts (Still et al. 1949). This enzyme is stereospecific for acidic D-amino acids and degrades not only D-Asp but also D-Glu and *N*-methyl-D-Asp (NMDA) (D'Aniello et al. 1993b; Katane et al. 2011, 2015). DDO has been identified in a wide variety of eukaryotes, including humans, but not in bacteria or plants. D-Amino acid oxidase (DAO, also abbreviated DAAO, EC 1.4.3.3), the other degradative enzyme stereospecific for D-amino acids, acts on several neutral and basic D-amino acids, but not on acidic D-amino acids (Krebs 1935). DDO is a flavin adenine dinucleotide (FAD)-containing flavoprotein in which one molecule of FAD is bound noncovalently to one molecule of DDO. DDO catalyzes the oxidative deamination of D-amino acids with oxygen to generate the corresponding 2-oxo acids, along with hydrogen peroxide and ammonia (in the case of NMDA, methylamine is produced instead of ammonia) (Katane and Homma 2010; Ohide et al. 2011).

In mammals, DDO activity is highest in the kidney, followed by the liver and brain, and is low in other peripheral tissues (Van Veldhoven et al. 1991; Yamada et al. 1988). Immunohistochemical analysis revealed that DDO is localized to the epithelial cells of proximal renal tubules in the kidney and hepatocytes in the liver (Zaar 1996). The intracellular localization of DDO to the peroxisome has been demonstrated (Beard 1990; Van Veldhoven et al. 1991; Zaar 1996; Zaar et al. 1989), where catalase degrades toxic hydrogen peroxide, one of the enzymatic reaction products. The DDO contents of the kidney and liver are relatively low at birth and rapidly increase thereafter (D'Aniello et al. 1993a; Kera et al. 1993; Van Veldhoven et al. 1991). This is in contrast to the D-Asp contents in these tissues, which decrease as development proceeds (Dunlop et al. 1986; D'Aniello et al. 1993a). This is also true in the brain, in which the D-Asp contents in the cerebrum rapidly decrease as development proceeds (Hashimoto et al. 1995), while the DDO contents rapidly increase (Van Veldhoven et al. 1991). Enzyme histochemical analysis of rat brain tissues revealed that DDO is present in several regions, including the choroid plexus, ependyma, hippocampus, dentate gyrus, olfactory bulb, pituitary gland, granule cell layer, and white matter of the cerebellum (Schell et al. 1997). Notably, DDO is concentrated in nerve cells, such as the hippocampal pyramidal neurons and olfactory epithelial neurons, tissues in which D-Asp immunoreactivity is not observed. Thus, the location and activity levels of DDO correlate inversely with the presence of D-Asp, suggesting that, in mammals, DDO degrades endogenous D-Asp as a physiological substrate, thereby regulating the cellular levels of this D-amino acid.

Recent findings obtained from studies using DDO-deficient mice collectively suggest that D-Asp acts as a signaling molecule in the nervous system by binding to the NMDA receptor and plays an important role in the regulation of brain functions (Errico et al. 2008a, b, 2011a, b, 2014; Weil et al. 2006). In neurological assessments such as the Morris water maze and contextual fear conditioning, altered behaviors are observed in DDO-deficient mice (Errico et al. 2008b). D-Asp contents in the hippocampus of DDO-deficient mice are approximately 13-fold higher than in wild-type mice (Errico et al. 2008a). Furthermore, in DDO-deficient mice, significant suppression is observed in the synaptic depotentiation following long-term potentiation (LTP) induction of the hippocampus (Fig. 13.3A), a process in which the NMDA receptor is believed to play a pivotal role. Similarly, high-frequency stimulation of corticostriatal fibers induces long-term depression (LTD) in corticostriatal slices of wild-type mice but not of DDO-deficient mice (Fig. 13.3b; Errico et al. 2008b). In the corticostriatal system, it is believed that enhanced NMDA receptor signaling blocks the induction of LTD, and D-Asp is known to stimulate NMDA receptor activity by binding to the L-Glu-binding site (Fagg and Matus 1984; Olverman et al. 1988). Thus, it is likely that NMDA receptor-mediated neurotransmission is enhanced by elevated concentrations of D-Asp in DDO-deficient mice.

DDO-deficient mice exhibit a reduced immobility time in the Porsolt forced-swim test, a model of depression, suggesting that the genetic ablation of DDO has a specific antidepressant effect (Weil et al. 2006). In addition, deficits in prepulse inhibition induced by amphetamine, a potent dopamine releaser, or by MK-801, a noncompetitive antagonist of the NMDA receptor, are attenuated in DDO-deficient mice (Errico et al. 2008b). These results suggest that increased levels of D-Asp in the brain protect against sensorimotor gating deficits, a defect that is observed in schizophrenic patients. Schizophrenia is a chronic, severe, and disabling mental disorder in which excessive dopaminergic transmission is believed to be involved. Although the mechanism of onset of this disease has yet to be fully clarified, it has been recently accepted that reduced neurotransmission via the NMDA receptor may be involved in the pathophysiology of schizophrenia. Therefore, increased attention has been focused on DDO as a therapeutic target, and DDO inhibitors able to augment D-Asp levels in the brain are being investigated as antipsychotic drugs for the treatment of schizophrenia.

We screened 257 different compounds, including biologically active compounds of microbial origin and pre-existing drugs, for their ability to inhibit the enzymatic activity of mouse DDO on D-Asp. This screening identified thiolactomycin (TLM) as a candidate DDO inhibitor (Katane et al. 2010). TLM, an antibiotic isolated

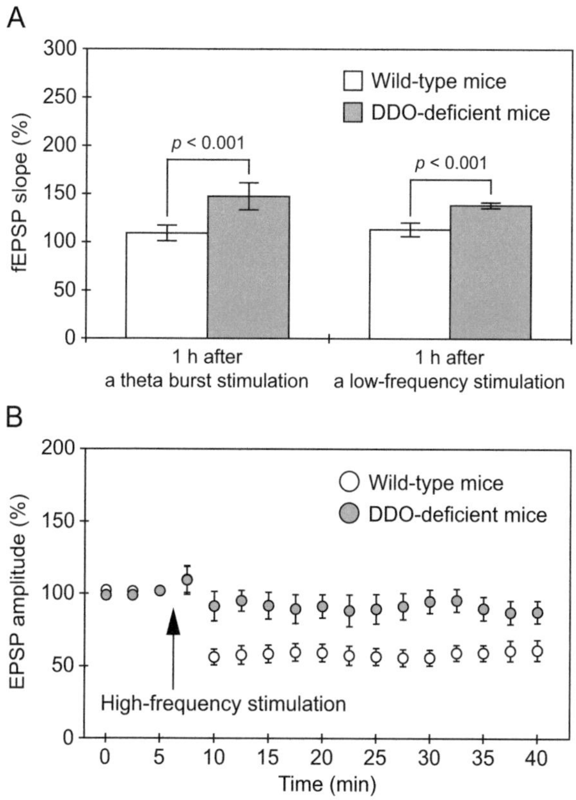

Fig. 13.3 Alterations in long-term synaptic plasticity in DDO-deficient mice. (**A**) The vertical axis represents the percentage of the field excitatory postsynaptic potential (fEPSP) slopes relative to the baselines. fEPSP slopes were measured 1 h after a theta burst stimulation (*left*). Potentiation recovers almost to baseline in hippocampal slices from wild-type mice (*open bar*), whereas the induced LTP is maintained in slices from DDO-deficient mice (*gray bar*). fEPSP slopes were measured 1 h after low-frequency stimulation following the induction of LTP (*right*). Synaptic depotentiation is observed in slices from wild-type mice (*open bar*) but not in slices from DDO-deficient mice (*gray bar*). Data are presented as the mean ± standard error of the mean ($n = 4$–6). The figure was drawn based on data from Errico et al. (2008a). For details, see text and Errico et al. (2008a). (**B**) The vertical axis represents the percentage of the EPSP amplitudes relative to the amplitude recorded before high-frequency stimulation. High-frequency stimulation of corticostriatal fibers induces LTD in corticostriatal slices from wild-type mice (*open circles*) but not in slices from DDO-deficient mice (*gray circles*). Data are presented as the mean ± standard error of the mean ($n = 8$–9). For details, see text and Errico et al. (2008b) (Reproduced with permission from the Society of Neuroscience (Errico et al. 2008b))

initially from a species of *Nocardia*, is known to inhibit type II fatty acid synthases in bacteria. The dose-dependent effect of TLM on the enzymatic activity of mouse DDO was examined, and the 50 % inhibitory concentration (IC_{50}) was determined and compared to that of malonate, a known classical competitive inhibitor of mammalian DDO. Both malonate and TLM inhibited DDO activity in a dose-

Table 13.2 Effect of FAD concentrations on the inhibitory activity of TLM against mouse DDO

FAD concentration (μM)	IC_{50} (μM) Malonate	TLM
60	5305 ± 409	1957 ± 124
3	5418 ± 134	510 ± 61

Data are from Katane et al. (2010). For details, see text and Katane et al. (2010). Reproduced with permission from Elsevier (Katane et al. 2010).
IC_{50} values were determined using 20 mM D-Asp as the substrate. Each value shown is the mean ± standard deviation ($n = 3$).

dependent manner. Interestingly, decreasing the concentration of FAD added to the reaction mixture markedly enhanced the inhibitory activity of TLM against DDO, whereas the inhibitory activity of malonate against DDO remained constant regardless of FAD concentration (Table 13.2). These results suggest that TLM inhibits both the binding of FAD to DDO and the binding of the substrate to the enzyme. Indeed, additional experiments using the holo-form and apo-form of DDO revealed that TLM inhibits the activity of DDO by competing with both the substrate and the FAD cofactor (Katane et al. 2010).

To gain insight into the mechanism of binding of TLM to DDO, structural models of the apo-form of DDO complexed with TLM were constructed using Glide software (Schrödinger Suite 2009; Friesner et al. 2004; Halgren et al. 2004). Since the 3D structure of DDO has not been determined, a structural model of mouse DDO was proposed using the 3D structure of human DAO, which was solved by X-ray crystallographic analysis (PDB ID: 2DU8) (Kawazoe et al. 2006), as the template structure. Since TLM can exist in two tautomeric forms in solution, models of mouse DDO complexed with each of the TLM tautomers were constructed. In one model, the keto and hydroxyl groups of TLM interact via hydrogen bonds with the backbone amide of Ala-48 and the side-chain guanidino group of Arg-237, respectively (Fig. 13.4A). In the other model, the keto group of TLM interacts with the backbone amides of Gly-49 and Met-50, and the hydroxyl group of TLM interacts with the side-chain guanidino groups of Arg-237 and Arg-278 (Fig. 13.4B). The structurally equivalent residues of Ala-48, Gly-49, and Met-50 in mouse DDO are Ala-49, Gly-50, and Leu-51, respectively, in human and pig DAOs, and the backbone amides of these residues are presumed to interact with the flavin ring of FAD (Kawazoe et al. 2006; Mattevi et al. 1996). In addition, Arg-237 and Arg-278 are presumed to be important for substrate binding and catalytic activity of DDO (Katane et al. 2007, 2011). Collectively, it appears that TLM binds to the active site of DDO and interferes with the proper orientation of both the substrate and FAD in the active site, resulting in the inhibition of enzymatic activity. Indeed, introduction of an Arg-237-to-Ala mutation in mouse DDO reduces its sensitivity to TLM (Katane et al. 2010). Interestingly, TLM also inhibits not only the activity of human DDO, which shares a high amino acid sequence identity with mouse DDO (80 %), but also the activity of mammalian (human, pig, and mouse) DAOs, which possess moderate sequence identity to mouse DDO (41 %). Among the aforementioned amino acid residues, Ala-48, Gly-49, and

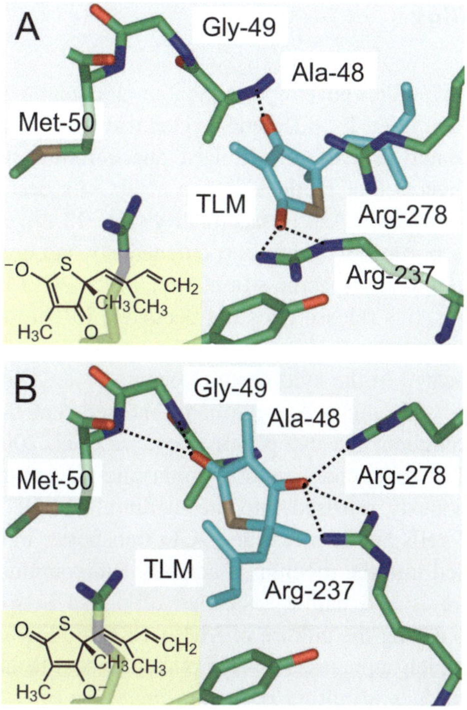

Fig. 13.4 Structural models of mouse DDO complexed with TLM. Since TLM can exist in two tautomeric forms in solution, two models of mouse DDO complexed with each of the TLM tautomers were constructed and are shown in A and B. The chemical structures of the two tautomers of TLM, which were used for the construction of the respective structural models, are also shown in the lower left in each of the panels. The carbon atoms in TLM and the side chains of amino acid residues are colored *cyan* and *green*, respectively. Other atoms are colored as follows: nitrogen, *blue*; oxygen, *red*; and sulfur, *khaki*. Dotted lines denote possible hydrogen bonds. The figure was modified from Katane et al. (2010). For details, see text and Katane et al. (2010)

Arg-278 of mouse DDO are conserved in these DAOs. However, the structurally equivalent residues of Met-50 and Arg-237 in mouse DDO are Leu and Tyr residues, respectively. Therefore, it is possible that the orientation of TLM bound to the active sites of these DAOs differs from that in DDO. Additional studies are required to clarify the mode of binding of TLM to DAO. Recently, we identified several candidate inhibitors based on in silico screening of approximately 4 million commercially available compounds for their capacity to bind the active site of human DDO as well as DAO (Katane et al. 2013 and our unpublished observations). Characterization of these compounds and in vitro screening of their inhibitory activity are currently in progress.

13.4 D-Asp Efflux

Although a number of studies have reported that intracellular D-Asp can be released into the extracellular milieu by different mechanisms, the details of this process remain to be elucidated. Therefore, we used pheochromocytoma PC12 cells to investigate the pathways that mediate D-Asp efflux. As described above, these cells are able to synthesize D-Asp intracellularly (Fig. 13.1A; Adachi et al. 2004; Long et al. 1998), a portion of which is spontaneously and continuously released into the medium without any specific stimulation. Since PC12 cells do not express the L-Glu transporter, this release does not occur via L-Glu transporter-mediated reverse transport. Subcellular fractionation assays revealed that endogenous D-Asp is predominantly located in the cytoplasm (Koyama et al. 2006), and immunocytochemical staining with anti-D-Asp antibody showed that D-Asp is distributed homogeneously throughout the cytoplasm (Koyama et al. 2006). These findings indicated that in PC12 cells, endogenous cytoplasmic D-Asp appears to be spontaneously and continuously released into the medium. In MPT1 cells, which are derived from PC12 cells and express the L-Glu transporter in their plasma membrane, D-Asp released into the medium is actively and continuously taken up into the cells; thus there is a dynamic exchange of D-Asp between the intra- and extracellular spaces during the culture of MPT1 cells. Following inhibition of the L-Glu transporter, which suppresses D-Asp uptake, the spontaneous and continuous efflux of D-Asp continues, resulting in the accumulation of D-Asp in the medium (Koyama et al. 2005).

This spontaneous, continuous release pathway is different from the exocytotic release of dopamine observed in PC12 cells, in which dopamine accumulates in large dense-core vesicles (LDCVs). The kinetic profile of the exocytotic release of dopamine is distinct from the spontaneous and continuous release of D-Asp. In addition, dopamine release is significantly suppressed by a voltage-gated Ca^{2+} channel blocker, and by knockdown of the SNARE protein, SNAP-25, which is an essential component of the exocytosis, whereas the release of D-Asp is not inhibited under these conditions (Fig. 13.5; Koyama et al. 2006). These results indicate that the spontaneous, continuous release of endogenous D-Asp occurs through efflux and that the vesicular exocytotic pathway is not substantially involved. However, the molecular entity that mediates this spontaneous efflux remains to be identified.

Another pathway for the transmembrane efflux of D-Asp in PC12 cells is via volume-sensitive organic anion channels (VSOCs), which connect the cytoplasm and extracellular space. When PC12 cells are exposed to hypotonic medium, significant amounts of endogenous D-Asp are released through VSOCs, which open in response to hypotonic stimulation (Koyama et al. 2006). This D-Asp efflux is substantially suppressed by treatment with putative VSOC blockers, such as 5-nitro-2-(3-phenylpropylamino)benzoic acid (NPPB) and 1,9-dideoxyforskolin (Fig. 13.6). However, these VSOC blockers do not inhibit the spontaneous release of D-Asp from the cells, suggesting that this pathway is distinct from that of D-Asp

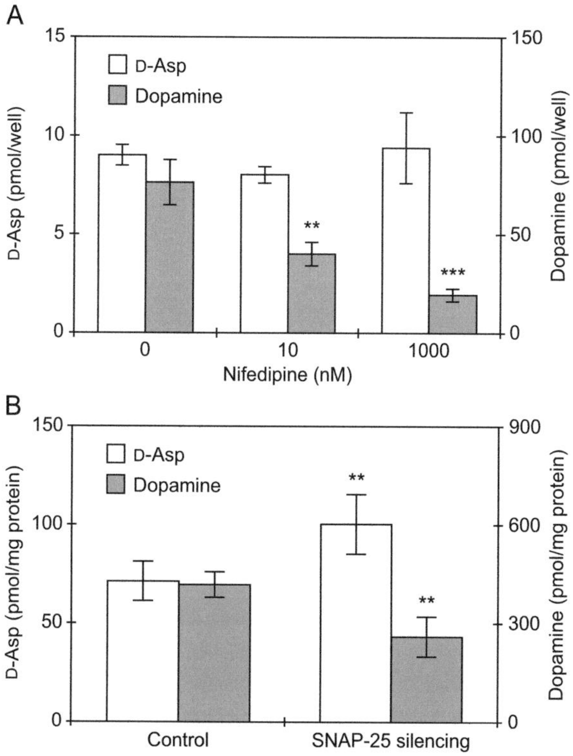

Fig. 13.5 Spontaneous and continuous release of cytoplasmic D-Asp in PC12 cells. (**A**) PC12 cells were cultured in the presence of the indicated concentrations of nifedipine, a voltage-gated Ca^{2+} channel blocker, and D-Asp (*open bars*) and dopamine (*gray bars*) efflux was determined after 20 min. Data are presented as the mean ± standard deviation ($n=3$). $^{**}P<0.01$ and $^{***}P<0.001$ (Student's *t*-test) compared to D-Asp or dopamine levels determined in the absence (0 nM) of nifedipine. For details, see text and Koyama et al. (2006). (**B**) PC12 cells were transfected with the SNAP-25-specific siRNA expression plasmid (SNAP-25 silencing) or the parental plasmid (control), and D-Asp (*open bars*) and dopamine (*gray bars*) efflux were then determined. Data are presented as the mean ± standard deviation ($n=3$). $^{**}P<0.01$ (Student's *t*-test) compared to D-Asp or dopamine levels determined in control cells. For details, see text and Koyama et al. (2006) (Reproduced with permission from Elsevier (Koyama et al. 2006))

efflux through VSOCs. Interestingly, treatment of PC12 cells with apoptosis-inducing agents, including staurosporine, tumor necrosis factor-α, hydrogen peroxide, and C2-ceramide, evokes the release of D-Asp (Furuchi et al. 2009). This increase in D-Asp release is almost completely inhibited by the putative VSOC blockers NPPB and 4,4′-diisothiocyanostilbene-2,2′-sulphonic acid, indicating that this D-Asp efflux is conceivably mediated through the VSOC pathway (Furuchi et al. 2009). An essential component of VSOC was identified recently, (Qiu et al. 2014; Voss et al. 2014), which is expected to facilitate further progress in the characterization of this efflux pathway.

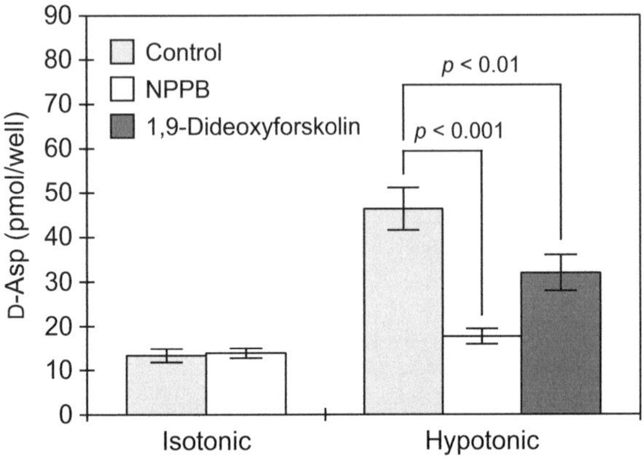

Fig. 13.6 Hypotonic stimulus-induced efflux of endogenous D-Asp in PC12 cells. PC12 cells were incubated in isotonic (316 mOsM) or hypotonic (216 mOsM) solutions in the absence (*light gray bars*) or presence of NPPB (*open bars*) or 1,9-dideoxyforskolin (*dark gray bars*), and D-Asp efflux was determined after 20 min. For details, see text and Koyama et al. (2006) (Reproduced with permission from Elsevier (Koyama et al. 2006))

The physiological significance of the enhanced efflux of D-Asp observed upon the induction of apoptosis has been discussed in our previous publications (Furuchi et al. 2009; Katane and Homma 2011): our studies revealed a feedback mechanism in the testis involving D-Asp-mediated stimulation of testosterone production by Leydig cells. In the testis, insufficient production of testosterone by the Leydig cells results in the acceleration of germ cell apoptosis in the seminiferous tubules, thereby enhancing D-Asp efflux from spermatogenic germ cells. In turn, this efflux stimulates testosterone production in the interstitial Leydig cells to support the normal manuration of the germ cells in the seminiferous tubules.

Another pathway for D-Asp efflux involves vesicular exocytosis (D'Aniello et al. 2011). Depolarization induced by a high concentration of KCl or treatment with acetylcholine results in the release of a significant amount of endogenous D-Asp from rat adrenal tissue slices (Fig. 13.7A; Wolosker et al. 2000). L-Asp is also released under these conditions, but the magnitude of release is much smaller. In these cells, D-Asp release is Ca^{2+} dependent and is completely inhibited by the addition of extracellular EGTA. It is of note that EGTA also significantly suppresses the basal release of D-Asp in the absence of stimulation (Wolosker et al. 2000). In our previous report, we showed that D-Asp is associated with granule-like structures around the nuclei in 2068 cells, a cell line derived from PC12 cells. In cells preloaded with [^{14}C]-D-Asp, D-Asp efflux was Ca^{2+} dependent and stimulated by depolarization with high concentrations of KCl or a Ca^{2+} ionophore (Fig. 13.7B; Long et al. 2001). These results suggest that D-Asp is released from 2068 cells via an exocytotic pathway, which is markedly different than the results observed in PC12 cells. However, the exocytotic release of D-Asp has been reported in PC12 cells (Nakatsuka et al. 2001). In this study, D-Asp immunoreactivity was associated with particulate structures, and D-Asp was

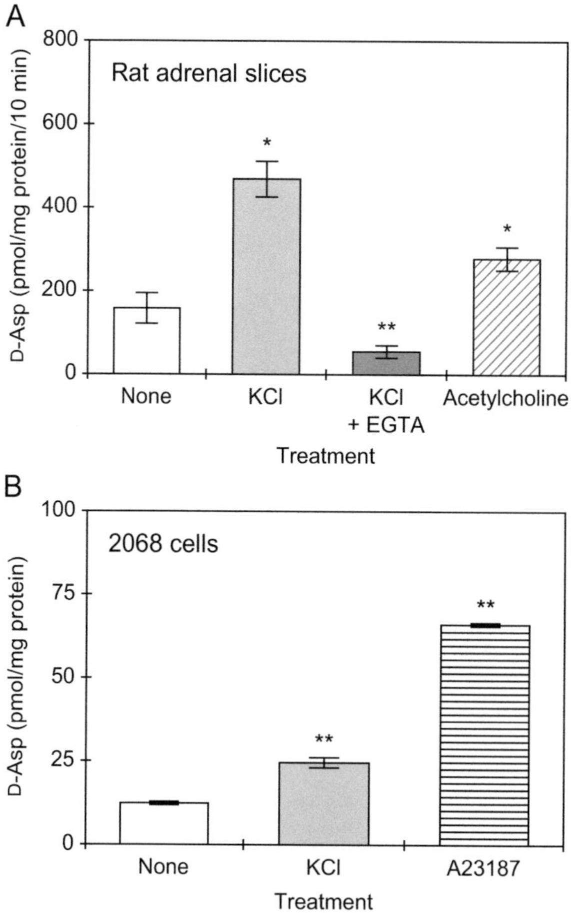

Fig. 13.7 Exocytotic discharge of vesicular D-Asp. (**A**) Rat adrenal slices were freshly prepared and equilibrated with oxygenated Krebs–Henseleit buffer for 30 min at 37 °C. After equilibration, release of D-Asp was elicited by treatment with 55 mM KCl (*light gray bar*), 55 mM KCl plus 0.2 mM EGTA (*dark gray bar*), or 100 μM acetylcholine (*hatched bar*), and D-Asp efflux was then determined. Data are presented as the mean ± standard error of the mean ($n = 6$–8). * $P < 0.05$ and ** $P < 0.01$ compared to the D-Asp level determined in the absence of any treatment (*open bar*). The figure was drawn based on data from Wolosker et al. (2000). For details, see text and Wolosker et al. (2000). (**B**) 2068 cells were preloaded with [^{14}C]-D-Asp. After contaminating extracellular [^{14}C]-D-Asp was washed out, the cells were incubated in the presence of 55 mM KCl (*light gray bar*) or 5 μM of the Ca^{2+} ionophore A23187 (*striped bar*) for 30 min, and the radioactivity released into the medium was then measured. Data are presented as the mean ± standard deviation ($n = 3$). ** $P < 0.01$ compared to the D-Asp level determined in the absence of any treatment (*open bar*). For details, see text and Long et al. (2001) (Reproduced with permission from Bioimaging Society (Long et al. 2001))

shown to co-localize with dopamine in LDCVs. In addition, D-Asp was released in response to high concentrations of KCl or exposure to neurotoxin in a manner similar to dopamine. These studies contradict the results obtained with PC12 cells in our studies. One possible explanation for these differences is that the PC12 cells employed in these studies are a spontaneously occurring flat variant of PC12 cells that express molecular component(s) (such as vesicular transporters) that concentrate D-Asp in the intracellular vesicles, allowing for exocytotic release.

There appear to be distinct release pathways for D-Asp in different cell lines. For example, D-Asp of cytoplasmic origin can be released via either a spontaneous and continuous release pathway or through VSOC channels. A third alternative pathway for the efflux of D-Asp is via exocytotic discharge. It is conceivable that these pathways are selectively utilized depending on the functional state of the cells in vivo. There are several proposed efflux pathways for D-Ser, a signaling molecule that gates the activity of synaptic NMDA receptor. Although D-Ser homeostasis in the synapse of the brain is still not fully delineated, two pathways for its release have been proposed: one through vesicular exocytosis from astrocytes, where D-Ser is stored in synaptic-like vesicles (Kang et al. 2013; Henneberger et al. 2010), and another pathway that mediates the release of cytoplasmic D-Ser through the Asc-1 transporter in the plasma membrane of neuronal cells (Maucler et al. 2013; Rosenberg et al. 2013). Further studies are necessary to fully elucidate the mechanisms of D-Asp efflux and their physiological significance.

References

Abe K, Takahashi S, Muroki Y, Kera Y, Yamada R (2006) Cloning and expression of the pyridoxal 5′-phosphate-dependent aspartate racemase gene from the bivalve mollusk *Scapharca broughtonii* and characterization of the recombinant enzyme. J Biochem 139:235–244

Adachi M, Koyama H, Long Z, Sekine M, Furuchi T, Imai K, Nimura N, Shimamoto K, Nakajima T, Homma H (2004) L-Glutamate in the extracellular space regulates endogenous D-aspartate homeostasis in rat pheochromocytoma MPT1 cells. Arch Biochem Biophys 424:89–96

Assisi L, Botte V, D'Aniello A, Di Fiore MM (2001) Enhancement of aromatase activity by D-aspartic acid in the ovary of the lizard *Podarcis s. sicula*. Reproduction 121:803–808

Beard ME (1990) D-Aspartate oxidation by rat and bovine renal peroxisomes: an electron microscopic cytochemical study. J Histochem Cytochem 38:1377–1381

D'Aniello A (2007) D-Aspartic acid: an endogenous amino acid with an important neuroendocrine role. Brain Res Rev 53:215–234

D'Aniello A, D'Onofrio G, Pischetola M, D'Aniello G, Vetere A, Petrucelli L, Fisher GH (1993a) Biological role of D-amino acid oxidase and D-aspartate oxidase. Effects of D-amino acids. J Biol Chem 268:26941–26949

D'Aniello A, Vetere A, Petrucelli L (1993b) Further study on the specificity of D-amino acid oxidase and D-aspartate oxidase and time course for complete oxidation of D-amino acids. Comp Biochem Physiol B 105:731–734

D'Aniello S, Spinelli P, Ferrandino G, Peterson K, Tsesarskia M, Fisher G, D'Aniello A (2005) Cephalopod vision involves dicarboxylic amino acids: D-aspartate, L-aspartate and L-glutamate. Biochem J 386:331–340

D'Aniello S, Somorjai I, Garcia-Fernàndez J, Topo E, D'Aniello A (2011) D-Aspartic acid is a novel endogenous neurotransmitter. FASEB J 25:1014–1027

Di Fiore MM, Santillo A, Baccari GC (2014) Current knowledge of D-aspartate in glandular tissues. Amino Acids 46:1805–1818

Dunlop DS, Neidle A, McHale D, Dunlop DM, Lajtha A (1986) The presence of free D-aspartic acid in rodents and man. Biochem Biophys Res Commun 26:27–32

Errico F, Nisticò R, Palma G, Federici M, Affuso A, Brilli E, Topo E, Centonze D, Bernardi G, Bozzi Y, D'Aniello A, Di Lauro R, Mercuri NB, Usiello A (2008a) Increased levels of D-aspartate in the hippocampus enhance LTP but do not facilitate cognitive flexibility. Mol Cell Neurosci 37:236–246

Errico F, Rossi S, Napolitano F, Catuogno V, Topo E, Fisone G, D'Aniello A, Centonze D, Usiello A (2008b) D-aspartate prevents corticostriatal long-term depression and attenuates schizophrenia-like symptoms induced by amphetamine and MK-801. J Neurosci 28:10404–10414

Errico F, Napolitano F, Nisticò R, Centonze D, Usiello A (2009) D-aspartate: an atypical amino acid with neuromodulatory activity in mammals. Rev Neurosci 20:429–440

Errico F, Nisticò R, Napolitano F, Bonito-Oliva A, Romano R, Barbieri F, Florio T, Russo C, Mercuri NB, Usiello A (2011a) Persistent increase of D-aspartate in D-aspartate oxidase mutant mice induces a precocious hippocampal age-dependent synaptic plasticity and spatial memory decay. Neurobiol Aging 32:2061–2074

Errico F, Bonito-Oliva A, Bagetta V, Vitucci D, Romano R, Zianni E, Napolitano F, Marinucci S, Di Luca M, Calabresi P, Fisone G, Carta M, Picconi B, Gardoni F, Usiello A (2011b) Higher free D-aspartate and *N*-methyl-D-aspartate levels prevent striatal depotentiation and anticipate L-DOPA-induced dyskinesia. Exp Neurol 232:240–250

Errico F, Napolitano F, Nisticò R, Usiello A (2012) New insights on the role of free D-aspartate in the mammalian brain. Amino Acids 43:1861–1871

Errico F, Nisticò R, Di Giorgio A, Squillace M, Vitucci D, Galbusera A, Piccinin S, Mango D, Fazio L, Middei S, Trizio S, Mercuri NB, Teule MA, Centonze D, Gozzi A, Blasi G, Bertolino A, Usiello A (2014) Free D-aspartate regulates neuronal dendritic morphology, synaptic plasticity, gray matter volume and brain activity in mammals. Transl Psychiatry 4, e417

Fagg GE, Matus A (1984) Selective association of *N*-methyl aspartate and quisqualate types of L-glutamate receptor with brain postsynaptic densities. Proc Natl Acad Sci U S A 81:6876–6880

Friesner RA, Banks JL, Murphy RB, Halgren TA, Klicic JJ, Mainz DT, Repasky MP, Knoll EH, Shelley M, Perry JK, Shaw DE, Francis P, Shenkin PS (2004) Glide: a new approach for rapid, accurate docking and scoring. 1. Method and assessment of docking accuracy. J Med Chem 47:1739–1749

Fujii N (2005) D-Amino acid in elderly tissues. Biol Pharm Bull 28:1585–1589

Fujii N, Kaji Y, Fujii N, Nakamura T, Motoie R, Mori Y, Kinouchi T (2010) Collapse of homochirality of amino acids in proteins from various tissues during aging. Chem Biodiversity 7:1389–1397

Funakoshi M, Sekine M, Katane M, Furuchi T, Yohda M, Yoshikawa T, Homma H (2008) Cloning and functional characterization of *Arabidopsis thaliana* D-amino acid aminotransferase—D-aspartate behavior during germination. FEBS J 275:1188–1200

Furuchi T, Homma H (2005) Free D-aspartate in mammals. Biol Pharm Bull 28:1566–1570

Furuchi T, Suzuki T, Sekine M, Katane M, Homma H (2009) Apoptotic inducers activate the release of D-aspartate through a hypotonic stimulus-triggered mechanism in PC12 cells. Arch Biochem Biophys 490:118–128

Halgren TA, Murphy RB, Friesner RA, Beard HS, Frye LL, Pollard WT, Banks JL (2004) Glide: a new approach for rapid, accurate docking and scoring. 2. Enrichment factors in database screening. J Med Chem 47:1750–1759

Hashimoto A, Oka T, Nishikawa T (1995) Anatomical distribution and postnatal changes in endogenous free D-aspartate and D-serine in rat brain and periphery. Eur J Neurosci 7:1657–1663

Henneberger C, Papouin T, Oliet SHR, Rusakov DA (2010) Long-term potentiation depends on release of D-serine from astrocytes. Nature 463:232–236

Homma H (2002) D-Aspartate in the mammalian body. Viva Origino 30:204–215 (available online at http://www.origin-life.gr.jp/3004/3004204/3004204.pdf)

Homma H (2007) Biochemistry of D-aspartate in mammalian cells. Amino Acids 32:3–11

Horio M, Ishima T, Fujita Y, Inoue R, Mori H, Hashimoto K (2013) Decreased levels of free D-aspartic acid in the forebrain of serine racemase (*Srr*) knock-out mice. Neurochem Int 62:843–847

Kang N, Peng H, Yu Y, Stanton PK, Guilarte TR, Kang J (2013) Astrocytes release D-serine by a large vesicle. Neuroscience 240:243–257

Katane M, Homma H (2010) D-Aspartate oxidase: the sole catabolic enzyme acting on free D-aspartate in mammals. Chem Biodiversity 7:1435–1449

Katane M, Homma H (2011) D-Aspartate—an important bioactive substance in mammals: a review from an analytical and biological point of view. J Chromatogr B 879:3108–3121

Katane M, Furuchi T, Sekine M, Homma H (2007) Molecular cloning of a cDNA encoding mouse D-aspartate oxidase and functional characterization of its recombinant proteins by site-directed mutagenesis. Amino Acids 32:69–78

Katane M, Saitoh Y, Hanai T, Sekine M, Furuchi T, Koyama N, Nakagome I, Tomoda H, Hirono S, Homma H (2010) Thiolactomycin inhibits D-aspartate oxidase: a novel approach to probing the active site environment. Biochimie 92:1371–1378

Katane M, Saitoh Y, Maeda K, Hanai T, Sekine M, Furuchi T, Homma H (2011) Role of the active site residues arginine-216 and arginine-237 in the substrate specificity of mammalian D-aspartate oxidase. Amino Acids 40:467–476

Katane M, Osaka N, Matsuda S, Maeda K, Kawata T, Saitoh Y, Sekine M, Furuchi T, Doi I, Hirono S, Homma H (2013) Identification of novel D-amino acid oxidase inhibitors by *in silico* screening and their functional characterization *in vitro*. J Med Chem 56:1894–1907

Katane M, Kawata T, Nakayama K, Saitoh Y, Kaneko Y, Matsuda S, Saitoh Y, Miyamoto T, Sekine M, Homma H (2015) Characterization of the enzymatic and structural properties of human D-aspartate oxidase and comparison with those of the rat and mouse enzymes. Biol Pharm Bull 38:298–305

Kawazoe T, Tsuge H, Pilone MS, Fukui K (2006) Crystal structure of human D-amino acid oxidase: context-dependent variability of the backbone conformation of the VAAGL hydrophobic stretch located at the *si*-face of the flavin ring. Protein Sci 15:2708–2717

Kera Y, Nagasaki H, Iwashima A, Yamada R (1993) Gender dependence of D-aspartate oxidase activity in rat tissues. Comp Biochem Physiol B 104:739–742

Kim PM, Duan X, Huang AS, Liu CY, Ming G-L, Song H, Snyder SH (2010) Aspartate racemase, generating neuronal D-aspartate, regulates adult neurogenesis. Proc Natl Acad Sci U S A 107:3175–3179

Kinouchi T, Ishiura S, Mabuchi Y, Urakami-Manaka Y, Nishio H, Nishiuchi Y, Tsunemi M, Takada K, Watanabe M, Ikeda M, Matsui H, Tomioka S, Kawahara H, Hamamoto T, Suzuki K, Kagawa Y (2004) Mammalian D-aspartyl endopeptidase: a scavenger for noxious racemized proteins in aging. Biochem Biophys Res Commun 314:730–736

Koyama H, Sekine M, Furuchi T, Katane M, Nimura N, Shimamoto K, Nakajima T, Homma H (2005) A novel L-glutamate transporter inhibitor reveals endogenous D-aspartate homeostasis in rat pheochromocytoma MPT1 cells. Life Sci 76:2933–2944

Koyama H, Adachi M, Sekine M, Katane M, Furuchi T, Homma H (2006) Cytoplasmic localization and efflux of endogenous D-aspartate in pheochromocytoma 12 cells. Arch Biochem Biophys 446:131–139

Krebs HA (1935) Metabolism of amino-acids: deamination of amino-acids. Biochem J 29:1620–1644

Long Z, Homma H, Lee J-A, Fukushima T, Santa T, Iwatsubo T, Yamada R, Imai K (1998) Biosynthesis of D-aspartate in mammalian cells. FEBS Lett 434:231–235

Long Z, Lee J-A, Okamoto T, Nimura N, Imai K, Homma H (2000) D-Aspartate in a prolactin-secreting clonal strain of rat pituitary tumor cells (GH_3). Biochem Biophys Res Commun 276:1143–1147

Long Z, Sekine M, Nimura N, Lee J-A, Imai K, Iwatsubo T, Homma H (2001) Immunocytochemical study of D-aspartate in the 2068 rat pheochromocytoma cell line. Bioimages 9:61–67

Long Z, Sekine M, Adachi M, Furuchi T, Imai K, Nimura N, Homma H (2002) Cell density inversely regulates D- and L-aspartate levels in rat pheochromocytoma MTP1 cells. Arch Biochem Biophys 404:92–97

Matsuda S, Katane M, Maeda K, Kaneko Y, Saitoh Y, Miyamoto T, Sekine M, Homma H (2015) Biosynthesis of D-aspartate in mammals—the rat and human homologues of mouse aspartate racemase are not responsible for the biosynthesis of D-aspartate. Amino Acids (in press). doi:10.1007/s00726-015-1926-0

Mattevi A, Vanoni MA, Todone F, Rizzi M, Teplyakov A, Coda A, Bolognesi M, Curti B (1996) Crystal structure of D-amino acid oxidase: a case of active site mirror-image convergent evolution with flavocytochrome b_2. Proc Natl Acad Sci U S A 93:7496–7501

Maucler C, Pernot P, Vasylieva N, Pollegioni L, Marinesco S (2013) *In vivo* D-serine hetero-exchange through alanine-serine-cysteine (ASC) transporters detected by microelectrode biosensors. ACS Chem Neurosci 4:772–781

Nakatsuka S, Hayashi M, Muroyama A, Otsuka M, Kozaki S, Yamada H, Moriyama Y (2001) D-Aspartate is stored in secretory granules and released through a Ca^{2+}-dependent pathway in a subset of rat pheochromocytoma PC12 cells. J Biol Chem 276:26589–26596

Neidle A, Dunlop DS (1990) Developmental changes in free D-aspartic acid in the chicken embryo and in the neonatal rat. Life Sci 46:1517–1522

Ohide H, Miyoshi Y, Maruyama R, Hamase K, Konno R (2011) D-Amino acid metabolism in mammals: biosynthesis, degradation and analytical aspects of the metabolic study. J Chromatogr B 879:3162–3168

Olverman HJ, Jones AW, Mewett KN, Watkins JC (1988) Structure/activity relations of *N*-methyl-D-aspartate receptor ligands as studied by their inhibition of [^{3}H]D-2-amino-5-phosphonopentanoic acid binding in rat brain membranes. Neuroscience 26:17–31

Ota N, Shi T, Sweedler JV (2012) D-aspartate acts as a signaling molecule in nervous and neuroendocrine systems. Amino Acids 43:1873–1886

Qiu Z, Dubin AE, Mathur J, Tu B, Reddy K, Miraglia LJ, Reinhardt J, Orth AP, Patapoutian A (2014) SWELL1, a plasma membrane protein, is an essential component of volume-regulated anion channel. Cell 157:447–458

Raucci F, Santillo A, D'Aniello A, Chieffi P, Baccari GC (2005) D-aspartate modulates transcriptional activity in Harderian gland of frog, *Rana esculenta*: morphological and molecular evidence. J Cell Physiol 204:445–454

Rosenberg D, Artoul S, Segal AC, Kolodney G, Radzishevsky I, Dikopoltsev E, Foltyn VN, Inoue R, Mori H, Billard J-M, Wolosker H (2013) Neuronal D-serine and glycine release via the Asc-1 transporter regulates NMDA receptor-dependent synaptic activity. J Neurosci 33:3533–3544

Sakai K, Homma H, Lee J-A, Fukushima T, Santa T, Tashiro K, Iwatsubo T, Imai K (1998) Emergence of D-aspartic acid in the differentiating neurons of the rat central nervous system. Brain Res 808:65–71

Schell MJ, Cooper OB, Snyder SH (1997) D-aspartate localizations imply neuronal and neuroendocrine roles. Proc Natl Acad Sci U S A 94:2013–2018

Schrödinger Suite 2009. Schrödinger, LLC, New York, NY, USA

Shibata K, Watanabe T, Yoshikawa H, Abe K, Takahashi S, Kera Y, Yamada R (2003a) Purification and characterization of aspartate racemase from the bivalve mollusk *Scapharca broughtonii*. Comp Biochem Physiol B Biochem Mol Biol 134:307–314

Shibata K, Watanabe T, Yoshikawa H, Abe K, Takahashi S, Kera Y, Yamada R (2003b) Nucleotides modulate the activity of aspartate racemase of *Scapharca broughtonii*. Comp Biochem Physiol B Biochem Mol Biol 134:713–719

Spinelli P, Brown ER, Ferrandino G, Branno M, Montarolo PG, D'Aniello E, Rastogi RK, D'Aniello B, Baccari GC, Fisher G, D'Aniello A (2006) D-aspartic acid in the nervous system of *Aplysia limacina*: possible role in neurotransmission. J Cell Physiol 206:672–681

Still JL, Buell MV, Knox WE, Green DE (1949) Studies on the cyclophorase system. VII. D-aspartic oxidase. J Biol Chem 179:831–837

Tanaka-Hayashi A, Hayashi S, Inoue R, Ito K, Konno K, Yoshida T, Watanabe M, Yoshimura T, Mori H (2015) Is D-aspartate produced by glutamic-oxaloacetic transaminase-1 like 1 (Got1l1): a putative aspartate racemase? Amino Acids 47:79–86

Van Veldhoven PP, Brees C, Mannaerts GP (1991) D-aspartate oxidase, a peroxisomal enzyme in liver of rat and man. Biochim Biophys Acta 1073:203–208

Voss FK, Ullrich F, Münch J, Lazarow K, Lutter D, Mah N, Andrade-Navarro MA, von Kries JP, Stauber T, Jentsch TJ (2014) Identification of LRRC8 heteromers as an essential component of the volume-regulated anion channel VRAC. Science 344:634–638

Wang L, Ota N, Romanova EV, Sweedler JV (2011) A novel pyridoxal 5′-phosphate-dependent amino acid racemase in the *Aplysia californica* central nervous system. J Biol Chem 286:13765–13774

Watanabe T, Shibata K, Kera Y, Yamada R (1998) Occurrence of free D-aspartate and aspartate racemase in the blood shell *Scapharca broughtonii*. Amino Acids 14:353–360

Weil ZM, Huang AS, Beigneux A, Kim PM, Molliver ME, Blackshaw S, Young SG, Nelson RJ, Snyder SH (2006) Behavioural alterations in male mice lacking the gene for D-aspartate oxidase. Behav Brain Res 171:295–302

Wolosker H, D'Aniello A, Snyder SH (2000) D-aspartate disposition in neuronal and endocrine tissues: ontogeny, biosynthesis and release. Neuroscience 100:183–189

Yamada R, Nagasaki H, Wakabayashi Y, Iwashima A (1988) Presence of D-aspartate oxidase in rat liver and mouse tissues. Biochim Biophys Acta 965:202–205

Zaar K (1996) Light and electron microscopic localization of D-aspartate oxidase in peroxisomes of bovine kidney and liver: an immunocytochemical study. J Histochem Cytochem 44:1013–1019

Zaar K, Völkl A, Fahimi HD (1989) D-aspartate oxidase in rat, bovine and sheep kidney cortex is localized in peroxisomes. Biochem J 261:233–238

Chapter 14
Neuromodulatory Activity of D-Aspartate in Mammals

Francesco Errico and Alessandro Usiello

Abstract D-aspartate is transiently present in the mammalian brain since it substantially occurs in embryonic phases and strongly decreases after birth, due to the postnatal activity of the catabolizing enzyme D-aspartate oxidase (DDO). Pharmacological evidence indicates that D-aspartate binds to and activates NMDA receptors (NMDARs). To decipher the function of D-aspartate in mammals, genetic and pharmacological mouse models with nonphysiological high levels of this D-amino acid have been generated. Their characterization has evidenced that increased D-aspartate enhances hippocampal NMDAR-dependent synaptic plasticity, dendritic morphology, and spatial memory. In line with the hypothesis of a NMDAR hypofunction in the pathogenesis of schizophrenia, it has been also shown that increased D-aspartate produces corticostriatal adaptations resembling those observed after chronic haloperidol treatment, and protects against prepulse inhibition deficits induced by psychotomimetic drugs. Moreover, a study in healthy humans has demonstrated that genetic variation in *DDO* gene, predicting potential increase in D-aspartate levels in *postmortem* prefrontal cortex, is associated with greater prefrontal gray matter and activity during working memory. Interestingly, a significant reduction of D-aspartate content has been detected in the *postmortem* brain of patients with schizophrenia. In line with a dichotomous effect of NMDAR stimulation on brain physiology, constitutive elevation of D-aspartate levels is also associated with detrimental effects, including precocious hippocampal age-related deterioration. Based on the agonistic role of D-aspartate on NMDARs and on its

F. Errico (✉)
Laboratory of Behavioural Neuroscience, Ceinge Biotecnologie Avanzate, Naples, Via G. Salvatore 486-80145, Naples, Italy

Department of Molecular Medicine and Medical Biotechnology, University of Naples "Federico II", 80138 Naples, Italy
e-mail: erricof@ceinge.unina.it

A. Usiello
Laboratory of Behavioural Neuroscience, Ceinge Biotecnologie Avanzate, Naples, Via G. Salvatore 486-80145, Naples, Italy

Department of Environmental, Biological and Pharmaceutical Sciences and Technologies, The Second University of Naples (SUN), 81100 Caserta, Italy

T. Yoshimura et al. (eds.), *D-Amino Acids*, DOI 10.1007/978-4-431-56077-7_14

abundance during prenatal life, future studies will be crucial to address the effect of this molecule on developmental processes controlled by NMDARs.

Keywords D-aspartate • D-aspartate oxidase • NMDA receptor • Mouse models

14.1 Introduction

Until a few decades ago, the presence of D-amino acids was believed to be restricted only to bacteria and low animal species (Corrigan 1969; D'Aniello and Giuditta 1977). Accumulating results in the years 80–90 have surprisingly shown that D-forms of amino acids are evolutionary well conserved in mammals, including humans, in which they are present in several tissue types. In particular, significant amounts of free D-aspartate (D-Asp) and D-serine (D-Ser) were found in the brain where they follow a peculiar regional and temporal pattern of emergence (Dunlop et al. 1986; Hamase et al. 1997; Hashimoto et al. 1993; Neidle and Dunlop 1990; Hashimoto et al. 1995). Such discovery, together with the long-established identification of flavoenzymes responsible for the oxidative deamination of dicarboxylic and neutral D-amino acids (Krebs 1935; Still et al. 1949), greatly supported the hypothesis that these atypical molecules might have a defined biological role in mammals. Today, it is well established that D-Ser is a physiological endogenous co-agonist of NMDA receptors (NMDARs) at central excitatory synapses of mammalian brain (Martineau et al. 2006; Wolosker and Radzishevsky 2013; Billard 2012; Van Horn et al. 2013) where it influences development (Kim et al. 2005), synaptic transmission and plasticity (Mothet et al. 2000; Papouin et al. 2012; Rosenberg et al. 2013; Li et al. 2013; Yang et al. 2003; Fossat et al. 2012), and behaviors such as cognition and social interaction (Labrie et al. 2009; DeVito et al. 2011; Basu et al. 2009). On the other side, altered levels of D-Ser seem to produce a disturbed NMDAR-dependent signaling and be causative for several pathological states including a psychiatric disorder like schizophrenia (Javitt et al. 2012; Labrie et al. 2012; Coyle and Tsai 2004b). Differently from D-Ser, the neurobiological role of D-Asp is becoming clearer only in the last years. In this chapter, we will overview the main advances on D-Asp, starting from the first pharmacological and neurochemical findings that have paved the way to the most recent knowledge based on mouse models with abnormal endogenous levels of this D-amino acid.

14.2 Occurrence of Free D-Aspartate in the Mammalian Brain

Experimental methods of D-Asp detection in the mammalian brain have shown that this D-amino acid transiently occurs in the rodents and humans since it is abundant at developmental stages and drastically decreases after birth (Dunlop et al. 1986; Hashimoto et al. 1993, 1995; Neidle and Dunlop 1990; Sakai et al. 1998; Wolosker et al. 2000). Interestingly, HPLC analysis performed on human prefrontal cortex (PFC) homogenates unveiled that D-Asp amount at gestational week 14 exceeds the levels of the corresponding L-form (Hashimoto et al. 1993). In the rat brain, two different immunohistochemical studies have investigated the presence of D-Asp in the rat brain at embryonic (Sakai et al. 1998) and postnatal (Wolosker et al. 2000) developmental phases. In the prenatal brain, D-Asp appears at embryonic day 12 (E12) in the ventrocaudal forebrain, midbrain, and hindbrain in the cytoplasm of neuroblasts, which have already ceased proliferative activity, but not in mitotic cells (Sakai et al. 1998). In the ventrocaudal forebrain, D-Asp appears in cell bodies of neuroblasts that migrate toward the outer layer of neural epithelium. When the migration process is completed and the layer has been established, D-Asp shifts to axons. Between E14 and E20, D-Asp occurrence increases and extends to the whole brain, including the cerebral cortex. In this area, D-Asp is observed in both cytoplasm and processes of neuroblasts of the intermediate zone and in the marginal zone, which mainly consists of axons. At postnatal day 0 (P0), D-Asp immunoreactivity is reduced (Sakai et al. 1998). Differently from Sakai et al, in newborn rats D-Asp is found in considerable amount by Wolosker et al. in forebrain regions including the cerebral cortex, olfactory bulbs, thalamus, and hypothalamus and in part of the midbrain. At P2, the staining also extends caudally to the hindbrain and cerebellum. At this stage D-Asp is concentrated in zones actively involved in developmental processes, such as the subventricular zone and the cortical plate of the cerebral cortex, hippocampal pyramidal neurons of the CA1-CA3 subfield and dentate gyrus, and the granule cells of the external granular layer of the cerebellum, that have not yet reached their final localization. Starting from P7, D-Asp levels substantially decrease. At P28, D-Asp is visible only in restricted areas of the brain (external plexiform layer of the olfactory bulbs, supraoptic and paraventricular nuclei of the hypothalamus and the medial habenula) (Schell et al. 1997). Besides neuronal cells, no evidence for D-Asp staining in glia has been so far collected (Schell et al. 1997; Wolosker et al. 2000).

14.3 The Metabolism of D-Aspartate

Early hypotheses to explain the endogenous occurrence of D-Asp in mammalian tissues were based on the potential derivation of D-Asp from food, intestinal bacterial flora, and degradation of metabolically stable proteins. However, transient occurrence of D-Asp and changes in its localization in the developing brain imply that the endogenous levels of this D-amino acid must be regulated by a dedicated biochemical mechanism controlling both biosynthesis and degradation of this D-amino acid. In support of an endogenous biosynthesis of D-Asp, in the chicken embryo brain, the levels of this D-amino acid increase with the incubation time of the egg that is a closed system (Neidle and Dunlop 1990). However, for a long time aspartate racemase activities, generating D-Asp, have been well documented only in bacteria (Kochhar and Christen 1992; Long et al. 2001; Yamauchi et al. 1992). Synthesis of D-Asp was first demonstrated in vitro by time-dependent accumulation of this D-amino acid in cultured rat pheochromocytoma (PC12) cells in a D-Asp-free environment (Long et al. 1998). A few years ago, a mouse pyridoxal 5′-phosphate (PLP)-dependent glutamate-oxaloacetate transaminase 1-like 1 (Got1l1), which substantially converts L-Asp to D-Asp and colocalizes with D-Asp in the adult brain, has been identified as the main source of endogenous D-Asp in this organ (Kim et al. 2010). On the other side, a recent publication has demonstrated that Got1l1 mRNA is undetectable in the brain. Moreover, the recombinant enzyme shows L-Asp aminotransferase activity but lacks Asp racemase activity. In agreement with biochemical results, knockout mice for *Got1l1* display D-Asp levels comparable to those of wild-type animals (Tanaka-Hayashi et al. 2015), suggesting there might be an as yet unknown enzyme for D-Asp synthesis. Interestingly, another study has evidenced that D-Asp levels are reduced in the forebrain of serine racemase knockout mice (*Srr*-KO) (Horio et al. 2013), thus suggesting a novel potential pathway for the generation of D-Asp, which would interact with that of D-Ser.

If the mechanism of D-Asp biosynthesis is still controversial in mammals, the existence of an enzyme catabolizing free D-Asp, D-aspartate oxidase (DDO, EC 1.4.3.1), has long been established (Still et al. 1949). DDO is a flavin adenine dinucleotide (FAD)-containing enzyme (Van Veldhoven et al. 1991) which oxidizes D-Asp in presence of H_2O and O_2, producing α-oxaloacetate, H_2O_2, and NH_4^+ ions (D'Aniello et al. 1993). DDO also oxidizes other dicarboxylic D-amino acids in vitro, such as D-glutamate and N-methyl D-aspartate (NMDA), while it is inactive toward basic and neutral D-amino acids, including D-Ser (Setoyama and Miura 1997), that are degraded by the D-amino acid oxidase (DAAO, EC 1.4.3.3), a flavoenzyme homologous to DDO (Pollegioni et al. 2007; Sacchi et al. 2012; Negri et al. 1992). The DDO sequence possesses a functional C-terminal tripeptide for the targeting to peroxisomes (Amery et al. 1998; Setoyama and Miura 1997), where this enzyme is supposed to oxidize D-Asp and safely release its toxic catabolite, H_2O_2 (Katane and Homma 2010; Beard 1990). In the brain, DDO activity strongly increases from birth until 6 weeks of life (Van Veldhoven et al. 1991) and is predominantly localized in neuronal population (Zaar et al. 2002) with an expression pattern reciprocal to the localization of its physiological substrate, D-Asp (Schell et al. 1997). The substantial increase of D-Asp levels in knockout mice for *Ddo* gene (*Ddo*$^{-/-}$) can be considered as the definitive proof that DDO is the enzyme responsible for physiological degradation of D-Asp (Errico et al. 2006; Huang et al. 2006).

14.4 Pharmacological Influence of D-Aspartate on NMDAR-Dependent Transmission

In the 1980s, neuropharmacological studies aimed at finding novel agonists or antagonists for ionotropic glutamate receptors revealed that D-Asp binds to the glutamate site of NMDA receptors (NMDARs) (Fagg and Matus 1984; Monahan and Michel 1987; Ogita and Yoneda 1988; Olverman et al. 1988; Ransom and Stec 1988). In line with this observation, recent evidence revealed that local applications of D-Asp on adult mouse brain slices trigger inward currents both in CA1 pyramidal neurons of the hippocampus (Fig. 14.1a) and in the GABAergic striatal medium spiny neurons (Fig. 14.1b, c), antagonized by competitive and noncompetitive NMDAR blockers, like D-AP5 and MK801, respectively (Errico et al. 2008a, b). Inward currents triggered by D-Asp are associated with a transient increase of intracellular Ca^{2+} in hippocampal pyramidal neurons (Errico et al. 2011c), an event expected in case of NMDAR activation. Unlike other neurotransmitter receptors, stimulation of NMDARs requires co-activation by two ligands: glutamate on agonist site of GluN2 and either D-serine or glycine as co-agonist (Li et al. 2013; Mothet et al. 2000; Rosenberg et al. 2013). Neuropharmacological evidence has shown that the D-Asp-mediated activation of NMDARs occurs via interaction with the glutamate site on each of the GluN2 subunits (Errico et al. 2011b). Interestingly, residual D-Asp-dependent currents still persist after the simultaneous perfusion of selective antagonists of GluN2A, 2B, and 2C-D subunits or after the application of high concentrations of D-AP5 or MK-801 (Errico et al. 2008a, b, 2011b, c). This effect suggests that D-Asp may also trigger NMDAR-independent currents. In this respect, it has been previously shown that D-Asp can inhibit kainate-induced AMPAR currents in acutely isolated rat hippocampal

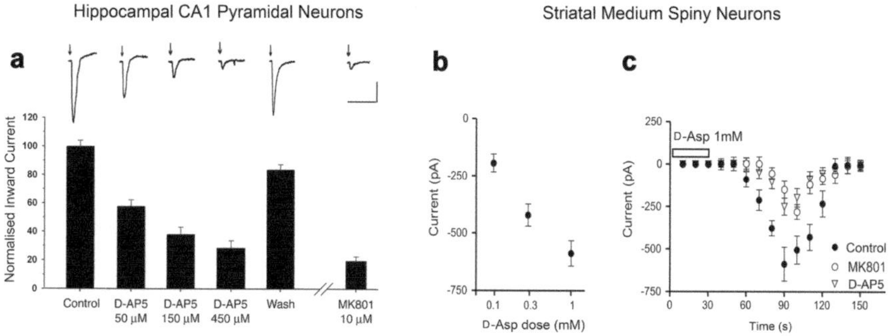

Fig. 14.1 D-aspartate stimulates NMDARs. (**a**) Inward currents recorded from mouse hippocampal CA1 pyramidal neurons following local application of D-Asp (*arrows*) are reduced by D-AP5 in a concentration-dependent and reversible manner and persistently diminished by MK801. Calibration bars: 100 pA mV, 30 s. (**b**) Bath application of D-Asp induce dose-dependent inward currents in mouse striatal neurons. (**c**) Preincubation of striatal slices with D-AP5 or MK801 causes a marked inhibition of D-Asp-induced currents ((**a**) Modified from Errico et al. 2008a. (**b**, **c**) Modified from Errico et al 2008b)

neurons (Gong et al. 2005) or activate mGlu5 receptors, coupled to polyphosphoinositide hydrolysis, in neonate rat hippocampal and cortical slices (Molinaro et al. 2010).

In the light of the above evidence, it could be hypothesized that endogenous D-Asp may have an effect on NMDAR-dependent transmission influencing synaptic communication within neuronal networks. If this is the case, dedicated mechanisms should allow the hypothetical release and subsequent removal of this D-amino acid from synapses. Different works using mammalian tissue slices and cells or synaptosomal preparations have suggested that D-Asp can be released through vesicular Ca^{2+}-mediated exocytotic processes (Davies and Johnston 1976; Malthe-Sorenssen et al. 1979; Nakatsuka et al. 2001; Wolosker et al. 2000; D'Aniello et al. 2010) which would presume that secretory organelles in subset of cells are able to actively store D-Asp (Nakatsuka et al. 2001). In adrenal slices, for example, D-Asp release evoked by K^+ depolarization is completely suppressed by the Ca^{2+}-selective chelating agent EGTA (Wolosker et al. 2000). Ca^{2+}-dependent mechanisms of release have been observed also in vivo. Indeed, K^+ injection into the mouse cerebral cortex elicits pronounced D-Asp staining in the choroid plexus (Schell et al. 1997). In addition to Ca^{2+}-dependent process, other studies showed that intracellular D-Asp may also be spontaneously released (Adachi et al. 2004; Koyama et al. 2006; Wolosker et al. 2000) or exchanged through a mechanism involving L-Glu transporters (Anderson et al. 2001; Bak et al. 2003).

Intracellular uptake of neuronal D-Asp may occur through L-Glu/L-Asp transporter, a carrier system that utilizes the Na^+/K^+ electrochemical gradient to move excitatory amino acids against their concentration gradient. Indeed, this carrier system is stereoselective for L-enantiomer of Glu but recognizes and transports both L- and D- Asp with the same efficiency (Palacin et al. 1998). Experimental approaches using [^{3}H]D-Asp autoradiography (Taxt and Storm-Mathisen 1984) or immunostaining with D-Asp antibody (Gundersen et al. 1993) have demonstrated that D-Asp, preloaded on rat hippocampal slices, shows a laminar distribution identical to L-Glu, corresponding to the terminal areas of the main excitatory fiber pathways of the hippocampus (Gundersen et al. 1993; Taxt and Storm-Mathisen 1984). Reuptake of D-Asp seems to involve both nerve terminals of asymmetrical synapses and glia, likely due to regional and subtype heterogeneity of the transporter system (Gundersen et al. 1993; Garthwaite and Garthwaite 1985).

14.5 Mouse Models with Increased Levels of D-Aspartate

In the attempt to clarify the neurobiological in vivo role of D-Asp, in the last years two knockout mice for *Ddo* gene have been independently generated ($Ddo^{-/-}$) (Errico et al. 2006; Huang et al. 2006). HPLC analysis revealed a strong increase of D-Asp amount in the brain and in peripheral organs of both $Ddo^{-/-}$ mouse lines, compared to their respective wild-type littermates, while no difference between genotypes was found in the content of dicarboxylic L-amino acids, L-Asp (Errico

Table 14.1 Free D-aspartate levels and fold increase in the brain of $Ddo^{-/-}$ and D-aspartate-treated mice

Brain Area	Age (month)	D-Asp content (nmol/g tissue) $Ddo^{+/+}$ mice	Ddo^{-do} mice	Fold increase	References
Hippocampus	4–5	119 ± 12	1847 ± 180	15.5	Errico et al., (2011a)
	9–10	15 ± 7	1810 ± 129	15.7	
	13–14	140 ± 13	2048 ± 276	14.6	
Striatum	4	31 ± 6	325 ± 23	10.5	Errico et al., (2008b, 2011a)
Cortex		43 ± 6	817 ± 40	19.0	
Cerebellum		34 ± 8	447 ± 22	13.1	
Olfactory bulbs		33 ± 4	319 ± 34	9.6	

Brain area	20 mM D-Asp oral administration. Delivery time (month)	Age (month)	D-Asp content (nmol/g tissue) H_2O-treated mice	D-Asp-treated mice	Fold increase	References
Hippocampus	1	2	22 ± 3	102 ± 11	4.6	Errico et al. (2008a)
	3	4.5	41 ± 2	99 ± 5	2.4	Errico et al. (2011b)
	12	13.5	71 ± 11	158 ± 29	2.2	
Striatum	2	3.5	17 ± 4	91 ± 14	5.3	Errico et al. (2008b)
Cortex			58 ± 4	153 ± 11	2.6	
Cerebellum			18 ± 1	81 ± 12	4.5	

et al. 2006; Huang et al. 2006), and L-Glu (Huang et al. 2006). Neurochemical evaluation in specific brain regions like the hippocampus, striatum, cortex, cerebellum, and olfactory bulbs show an approximately 10–20-fold increase of D-Asp levels in $Ddo^{-/-}$ animals, compared to wild-type controls (Errico et al. 2008a, b, 2011a, c) (Table 14.1). As a likely consequence of yet unknown methylation events, a significant augmentation of NMDA levels was also found in $Ddo^{-/-}$ brains (Errico et al. 2006, 2011c). In order to increase the endogenous levels of D-Asp without gene targeting approach, chronic oral administration of D-Asp (20 mM) to C57BL/6 J mice in tap water has also been pursued. D-Asp administration produces an approximately two to fivefold increase in the hippocampus, cortex, striatum, and cerebellum, compared to the same brain areas of untreated mice (Errico et al. 2008a, b, 2011b) (Table 14.1).

14.6 Elevated Levels of D-Aspartate in the Mouse Brain Affect NMDA Receptor-Dependent Synaptic Plasticity, Dendritic Morphology, and Cognition

In agreement with the ability of D-Asp to stimulate glutamatergic transmission, increased levels of D-Asp in both adult *Ddo*$^{-/-}$ and D-Asp-treated mice are associated with enhanced NMDAR-dependent early-phase long-term potentiation (E-LTP) in the hippocampal CA1 area of mice (Errico et al. 2008a, 2011b, c). Experiments of intermittent oral administration of D-Asp to C57BL/6 J animals indicate that the magnitude of NMDAR-dependent E-LTP in the hippocampus is regulated by changes in the brain levels of this D-amino acid. Indeed, 3 weeks of interruption after chronic D-Asp administration are able to wash out the excess of D-Asp and, in turn, to normalize E-LTP amplitude at physiological levels. Finally, further one-month treatment with D-Asp, after three-week withdrawal, re-establishes synaptic plasticity at previously potentiated levels (Errico et al. 2011b). Elevated D-Asp levels can influence also long-lasting forms of hippocampal synaptic plasticity. Indeed, the execution of a paradigm for E-LTP induction, that causes a decaying LTP in wild-type slices, is sufficient to induce stable late-phase LTP (L-LTP) in slices from *Ddo*$^{-/-}$ and D-Asp-treated animals (Errico et al. 2014) (Fig. 14.2a, b). In both animal models, D-Asp-dependent L-LTP is insensible to rapamycin but is fully prevented by cytochalasin D administration (Errico et al. 2014). The D-Asp-dependent enhancement in LTP may occur through direct activation of NMDARs by D-Asp and/or by indirect influence of D-Asp on the release of endogenous L-Glu following tetanic stimulation. In line with a role for D-Asp on NMDAR-dependent transmission, mice chronically treated with D-Asp also reveal increased frequency of NMDAR-mediated miniature excitatory postsynaptic currents (mEPSCs) in pyramidal neurons of the medial PFC layer II/III (Errico et al. 2014). The enhancement of NMDAR-dependent transmission in D-Asp-treated mice is mirrored by greater basal metabolic activity in fronto-hippocampal areas, as assessed with basal cerebral blood volume (bCBV)-weighted functional magnetic resonance imaging (fMRI) (Errico et al. 2014).

In line with enhanced NMDAR-dependent transmission and facilitated induction of the late phase of synaptic plasticity, elevation of D-Asp levels is also associated with modifications in neuronal cytoarchitecture since both *Ddo*$^{-/-}$ and D-Asp-treated mice display increased dendritic length and spine density (Fig. 14.2c, d), and greater complexity of dendrites arborization in pyramidal neurons of the PFC and hippocampus (Errico et al. 2014). Coherently with D-Asp-dependent potentiation in structural and functional synaptic plasticity, increased levels of D-Asp at adulthood improve spatial cognitive abilities of *Ddo*$^{-/-}$ mice, when tested in NMDAR-dependent tasks involving hippocampal circuits, such as the hidden-platform version of the Morris water maze and the contextual fear conditioning (Errico et al. 2008b, 2011c).

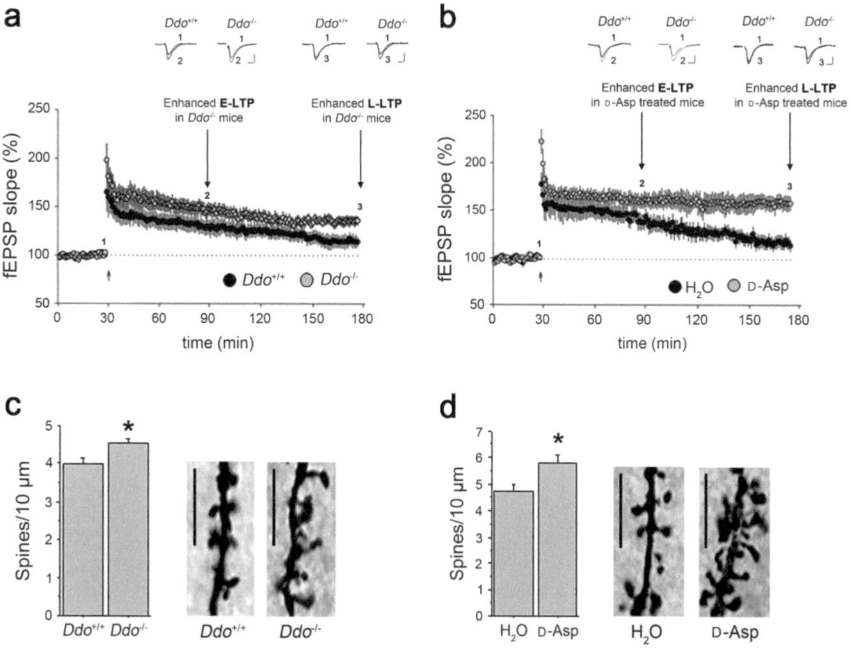

Fig. 14.2 Facilitatory role of increased D-aspartate on hippocampal NMDAR-dependent early- and late-phase LTP. Time plot of hippocampal field excitatory postsynaptic potential (fEPSP) responses showing that an early-phase LTP (E-LTP) stimulation paradigm (1 s, 100 Hz tetanus) elicits increased E-LTP in (**a**) *Ddo*$^{-/-}$ and (**b**) D-Asp-treated mice, compared to their relative controls (*Ddo*$^{+/+}$ and H_2O-treated mice, respectively). Interestingly, the same protocol of E-LTP stimulation, that causes a decaying LTP in control animals, is sufficient to induce stable late-phase LTP (L-LTP) in (**a**) *Ddo*$^{-/-}$ and (**b**) D-Asp-treated mice. (**a**, **b**) Insets show fEPSPs from representative experiments during a baseline interval and following LTP induction. Scale: *vertical bar*, 0.5 mV; *horizontal bar*, 10 ms. (**c**, **d**) Spine density (number of spines per 10 μm) is increased in the hippocampal CA1 area of (**c**) *Ddo*$^{-/-}$ and (**d**) D-Asp-treated mice, compared to their relative controls (*Ddo*$^{+/+}$ and H_2O-treated mice, respectively). The right panels show representative dendrite comparison between (**c**) genotypes or (**d**) treatments. $^*P < 0.05$, compared with (**c**) *Ddo*$^{+/+}$ or (**d**) H_2O-treated animals (Student's *t*-test). *Scale bar*, 5 μm (Modified from Errico et al 2014)

14.7 Abnormal Levels of D-Aspartate Affect Mouse NMDA Receptor-Dependent Processes and Human Prefrontal Phenotypes Relevant to Schizophrenia

The hypothesis of a developmental hypofunction of NMDARs in the pathogenesis of schizophrenia is today supported by a large body of evidence (Moghaddam and Javitt 2012; Coyle 2012; Coyle and Tsai 2004a; Sawa and Snyder 2003). In this regard, several studies suggest that reduced levels of the co-agonist D-Ser may result in disturbed NMDAR neurotransmission and be thus involved in the pathophysiology of this psychiatric disorder (Coyle and Tsai 2004b; Javitt et al. 2012; Labrie et al. 2012). In the light of the influence played by D-Asp on NMDAR-dependent transmission and of its neurodevelopmental occurrence, it has been recently hypothesized that this D-amino acid may act as a potential endogenous mediator for NMDAR-related processes involved in schizophrenia. To evaluate in vivo this hypothesis, mouse models with increased levels of D-Asp have been tested in the prepulse inhibition (PPI) paradigm, a cross species operational measure of sensorimotor gating regarded today as an endophenotypic trait of schizophrenia (Gottesman and Gould 2003; Geyer 2006). Results showed that chronic exposure to higher D-Asp levels does not affect the basal sensorimotor filtering of adult *Ddo*$^{-/-}$ and D-Asp-treated mice but substantially reduces the inhibitory deficits induced in these mice by acute treatment with psychotomimetic drugs like amphetamine and MK801 (Errico et al. 2008b) (Fig. 14.3a). Electrophysiological experiments are in line with a potential influence of D-Asp in schizophrenia-related processes dependent by NMDAR transmission. Indeed, increased levels of D-Asp are able to inhibit the induction of long-term depression (LTD) at corticostriatal synapses of both *Ddo*$^{-/-}$ and D-Asp-treated mice (Errico et al. 2008b) (Fig. 14.3b), an effect that is similarly obtained in conditions of enhanced NMDAR signaling and after chronic treatment with the typical antipsychotic haloperidol (Centonze et al. 2004).

Results collected in preclinical animal models provide a promising background to evaluate whether also in humans D-Asp may have an impact on phenotypes relevant to schizophrenia. In this regard, the detection of the levels of endogenous free D-Asp and NMDA in a small cohort of *postmortem* prefrontal cortex (PFC) and striatum samples of schizophrenic patients revealed that the content of both D-amino acids is consistently reduced when compared to that found in healthy subjects (Errico et al. 2013). Interestingly, the decline of endogenous D-Asp and NMDA correlates with a selective reduction in the levels of the NMDAR subunits GluN1, GluN2A, and GluN2B in the prefrontal cortex of patients with schizophrenia (Errico et al. 2013). Another work has examined the association of *DDO* gene variants with a series of complex prefrontal phenotypes (Errico et al. 2014). Analysis of data from 268 brains of nonpsychiatric individuals obtained from the *postmortem* collection bank BrainCloud (http://braincloud.jhmi.edu) (Colantuoni et al. 2011) has evidenced that the C allele of rs3757351 is significantly associated with reduced expression of *DDO* mRNA, when compared to the T allele. This result predicts a potential increase of endogenous D-Asp levels in this brain region. Then, to evaluate the functional effect of this single-nucleotide polymorphism (SNP), healthy individuals were subjected to voxel-based morphometry ($n = 152$) and to BOLD fMRI prefrontal activity during performance of the 1- and 2-back working memory task ($n = 143$). Results evidenced that subjects with the C allele also display augmented prefrontal gray matter volume and greater prefrontal activity, when compared to individuals with the T allele.

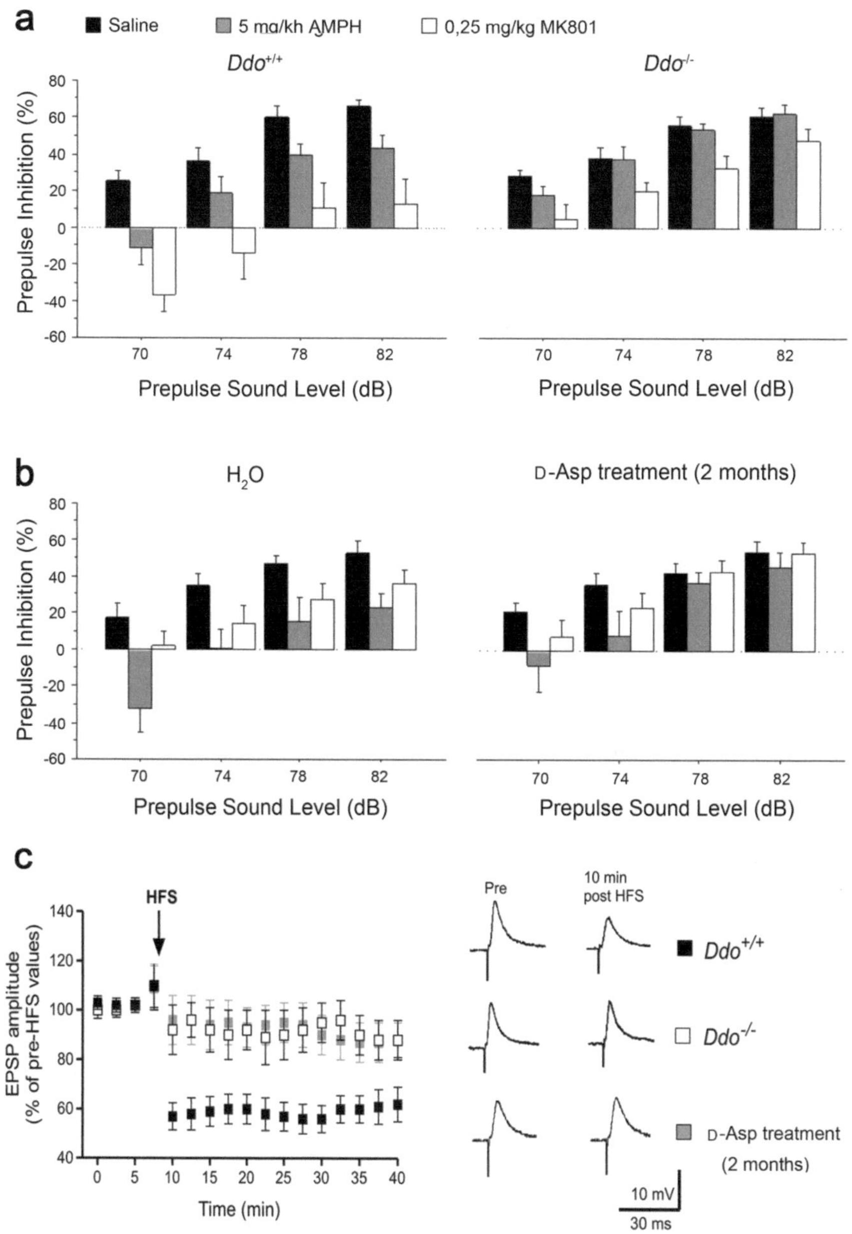

Fig. 14.3 Increased levels of D-aspartate attenuate schizophrenia-like symptoms induced by psychotomimetic drugs. The dopamine releaser amphetamine (5 mg/kg) and the NMDAR antagonist MK801 (0.25 mg/kg) are both able to induce a significant decreased PPI in (**a**) *Ddo*$^{+/+}$ and (**b**) H_2O-treated mice but do not affect the inhibitory responses of (**a**) *Ddo*$^{-/-}$ and (**b**) D-Asp-treated mice. Percentage of PPI is tested using prepulse sounds of 70, 74, 78, and 82 dB, coupled to pulse sounds of 120 dB on a background noise level of 65 dB (Errico et al. 2008a). (**c**) Corticostriatal LTD is absent in slices from *Ddo*$^{-/-}$ mice and from mice chronically exposed to D-Asp. The physiological traces on the right are examples of EPSPs recorded before and after high-frequency stimulation (HFS) of corticostriatal fibers in slices from *Ddo*$^{+/+}$, *Ddo*$^{-/-}$, and

14.8 Elevated Levels of D-Aspartate Can Produce Detrimental Effects in the Brain Through an Excessive Stimulation of NMDA Receptors

As an endogenous agonist of NMDARs, D-Asp is potentially able to influence all the functions regulated by this receptor subclass and to guide the direction of events that depend on it. In the previous paragraphs, we have shown that higher levels of D-Asp in mouse models are able to enhance functional and structural synaptic plasticity, improve spatial memory, and protect against sensorimotor gating deficits induced by psychotomimetic drugs, probably via activation of NMDARs. These ameliorating effects are in line with the view that stimulation of NMDARs, within a certain limit, is crucial for promoting synaptic strength, connectivity, and the formation of learning and memory. By contrast, it is well known that intense and/or chronic activation of NMDARs can lead to neuronal death and be, therefore, harmful for brain functioning (Hardingham and Bading 2003; Lancelot and Beal 1998). In this regard, increased D-Asp levels in mice can prove detrimental during aging due to continuous overstimulation of NMDARs. In fact, while increased levels of endogenous D-Asp enhance the NMDAR-dependent E-LTP in 4-/5-month-old *Ddo*$^{-/-}$ mice, the persistent upregulation of this D-amino acid accelerates the age-related decay of synaptic plasticity in 9-/10- and, even more, in 13-/14-month-old animals (Errico et al. 2011c) (Fig. 14.4). Consistently, the spatial memory improvement found in 4-/5-month-old *Ddo*$^{-/-}$ mice turns in a drastic worsening of learning and memory abilities at 13–14 months of age (Errico et al. 2011c) (Fig. 14.4). In line with results obtained in knockout mice, long-term treatment with D-Asp for 12 months to C57BL/6 J mice is able to significantly reduce E-LTP at CA1 synapses, compared to non-treated mice (Errico et al. 2011b). The direct and reversible effect of D-Asp on hippocampus-dependent E-LTP is further highlighted by the fact that interruption of its administration for 3 weeks, after 12-month continuous treatment, can restore hippocampal synaptic plasticity at control levels (Errico et al. 2011b). Changes in NMDAR-dependent LTP induced by D-Asp are not associated with deregulated expression of NMDARs and AMPARs. In fact, in hippocampal homogenates from both *Ddo*$^{-/-}$ and D-Asp-treated mice, protein levels of the NMDAR subunits, GluN1, GluN2A, GluN2B, and of AMPAR subunits, GluR1 and GluR2/3, are comparable between genotypes or treatments (Errico et al. 2011b, c). Moreover, in both knockout and D-Asp-treated animals, the increase in D-Asp levels is constant over time since the amount of D-Asp is comparable in *Ddo*$^{-/-}$ animals of 4–5, 9–10, and 13–14 months of age (Errico et al. 2011c) as well as in mice treated with D-Asp for 3 or 12 months (Errico et al. 2011b) (Table 14.1). These observations further highlight the hypothesis that

---

Fig. 14.3 (continued) d-Asp-treated animals. This effect resembles that produced in wild-type mice by chronic treatment with the atypical antipsychotic haloperidol (see text) (Modified from Errico et al 2008b)

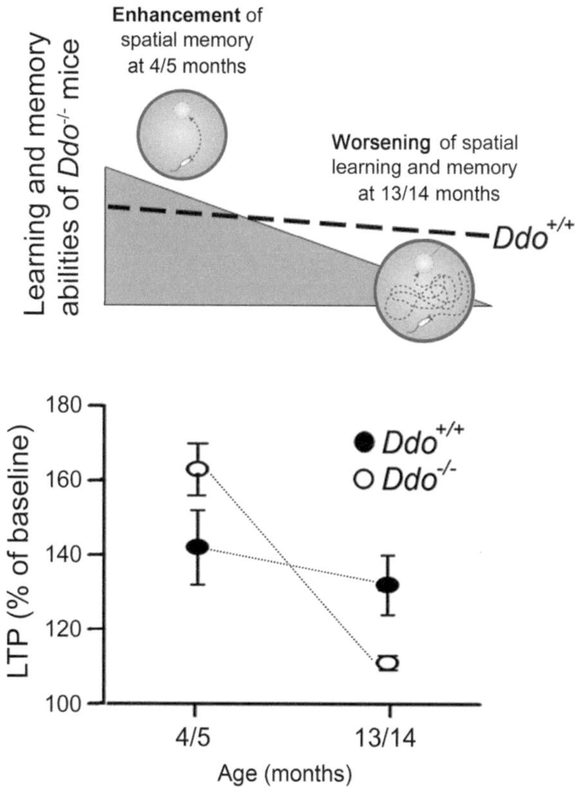

Fig. 14.4 Schematic representation of the opposite, age-related effects of increased D-Asp levels on hippocampus-dependent spatial memory (*top panel*) and E-LTP (*bottom panel*) of *Ddo*$^{-/-}$ mice during adult life. Spatial memory is evaluated in the probe test performed during the hidden-platform version of the Morris water maze paradigm. In the probe test, animals are allowed to swim for 60 s in the absence of the platform, in order to evaluate their memory retention after the learning phase (Errico et al. 2011c). E-LTP values refer to data reported in Errico et al (2011c)

excessive activation of NMDARs by increased D-Asp levels may produce beneficial or detrimental effects on hippocampal age-dependent processes depending on the time window and length of the stimulation. This is confirmed by experiments of short-term D-Asp administration in elderly mice. Indeed, one-month treatment with D-Asp to 12-month-old C57BL/6 J females is able to increase their E-LTP at levels even higher than those measured in two-month-old untreated controls (Errico et al. 2011b).

Detrimental effect of D-Asp has been described not only in aged animals chronically exposed to increased levels of D-Asp but also in young-adult *Ddo*$^{-/-}$ mice (approximately 4 months old) under particular experimental conditions mimicking pathological states. In this regard, exaggerated D-Asp levels induce a hyperexcitability state of nociceptive specific neurons in the dorsal horn of the spinal cord. This phenotype is associated with a reduced pain threshold in inflammatory and neuropathic pain condition (Boccella et al. 2015). Another work has evidenced that the pre-existing striatal hyper-glutamatergic tone produced by excessive D-Asp levels is able to trigger a precocious expression of levodopa-induced dyskinesia in 6-hydroxydopamine-lesioned hemiparkinsonian *Ddo*$^{-/-}$ mice (Errico et al. 2011a). Finally, it has been hypothesized that genetically enhanced NMDAR transmission can exacerbate synaptic transmission defects and accelerate the clinical manifestation of motor deficits scored in $Ddo^{-/-}$ mice under an experimental autoimmune encephalomyelitis pathological state (Grasselli et al. 2013).

14.9 Conclusions

Prompted by the first pharmacological and localization studies, the results obtained in the last years in animal models with increased levels of D-Asp have substantially improved our understanding on the neurobiological role of this atypical amino acid. A molecule thought to be a vestigial residue inherited from primordial organisms is today believed to contribute to higher brain functions in mammals. Nevertheless, it should be remarked that the results so far collected derive from a nonphysiological condition in which the levels of D-Asp are maintained forcedly high at adulthood, and, therefore, they cannot provide comprehensive information to understand the physiological role of the D-Asp in the brain. Generation of novel animal models with reduced levels of D-Asp since embryonic stage may help to answer this question, thus providing an explanation for the massive presence of D-Asp in the developing brain. Knowledge coming from new animal models could in turn clarify the potential involvement of D-Asp in schizophrenia. Indeed, it is well known that the etiology of this mental illness may involve pathological processes, caused by both genetic and environmental factors, that begin in utero and develop until adolescence or young adulthood when they lead to the emergence of positive and/or negative symptoms (Fatemi and Folsom 2009; Owen et al. 2011; Lu et al. 2011). One hypothesis could be that a putative reduction in D-Asp content may have substantial clinical relevance if it starts during embryonic phases, when critical neurodevelopmental processes like neurogenesis, survival, migration, and generation of neuronal circuitry are under the regulatory control of NMDARs (Ikonomidou et al. 2001; Nacher and McEwen 2006; Ritter et al. 2002). Future studies will help to clarify the still unclear role of this atypical amino acid in the mammalian brain.

References

Adachi M, Koyama H, Long Z, Sekine M, Furuchi T, Imai K, Nimura N, Shimamoto K, Nakajima T, Homma H (2004) L-Glutamate in the extracellular space regulates endogenous D-aspartate homeostasis in rat pheochromocytoma MPT1 cells. Arch Biochem Biophys 424 (1):89–96. doi:10.1016/j.abb.2004.01.016 S0003986104000529 [pii]

Amery L, Brees C, Baes M, Setoyama C, Miura R, Mannaerts GP, Van Veldhoven PP (1998) C-terminal tripeptide Ser-Asn-Leu (SNL) of human D-aspartate oxidase is a functional peroxisome-targeting signal. Biochem J 336(Pt 2):367–371

Anderson CM, Bridges RJ, Chamberlin AR, Shimamoto K, Yasuda-Kamatani Y, Swanson RA (2001) Differing effects of substrate and non-substrate transport inhibitors on glutamate uptake reversal. J Neurochem 79(6):1207–1216

Bak LK, Schousboe A, Waagepetersen HS (2003) Characterization of depolarization-coupled release of glutamate from cultured mouse cerebellar granule cells using DL-threo-beta-benzyloxyaspartate (DL-TBOA) to distinguish between the vesicular and cytoplasmic pools. Neurochem Int 43(4–5):417–424, doi:S0197018603000305 [pii]

Basu AC, Tsai GE, Ma CL, Ehmsen JT, Mustafa AK, Han L, Jiang ZI, Benneyworth MA, Froimowitz MP, Lange N, Snyder SH, Bergeron R, Coyle JT (2009) Targeted disruption of serine racemase affects glutamatergic neurotransmission and behavior. Mol Psychiatry 14 (7):719–727. doi:10.1038/mp.2008.130

Beard ME (1990) D-aspartate oxidation by rat and bovine renal peroxisomes: an electron microscopic cytochemical study. J Histochem Cytochem 38(9):1377–1381

Billard JM (2012) D-Amino acids in brain neurotransmission and synaptic plasticity. Amino Acids 43(5):1851–1860. doi:10.1007/s00726-012-1346-3

Boccella S, Vacca V, Errico F, Marinelli S, Squillace M, Guida F, Di Maio A, Vitucci D, Palazzo E, De Novellis V, Maione S, Pavone F, Usiello A (2015) D-aspartate modulates nociceptive-specific neuron activity and pain threshold in inflammatory and neuropathic pain condition in mice. BioMed Res Int 2015:905906. doi:10.1155/2015/905906

Centonze D, Usiello A, Costa C, Picconi B, Erbs E, Bernardi G, Borrelli E, Calabresi P (2004) Chronic haloperidol promotes corticostriatal long-term potentiation by targeting dopamine D2L receptors. J Neurosci 24(38):8214–8222. doi:10.1523/JNEUROSCI.1274-04.2004 24/38/8214 [pii]

Colantuoni C, Lipska BK, Ye T, Hyde TM, Tao R, Leek JT, Colantuoni EA, Elkahloun AG, Herman MM, Weinberger DR, Kleinman JE (2011) Temporal dynamics and genetic control of transcription in the human prefrontal cortex. Nature 478(7370):519–523. doi:10.1038/nature10524

Corrigan JJ (1969) D-amino acids in animals. Science 164(876):142–149

Coyle JT (2012) NMDA receptor and schizophrenia: a brief history. Schizophr Bull 38 (5):920–926. doi:sbs076 [pii] 10.1093/schbul/sbs076

Coyle JT, Tsai G (2004a) NMDA receptor function, neuroplasticity, and the pathophysiology of schizophrenia. Int Rev Neurobiol 59:491–515. doi:10.1016/S0074-7742(04)59019-0 S0074774204590190 [pii]

Coyle JT, Tsai G (2004b) The NMDA receptor glycine modulatory site: a therapeutic target for improving cognition and reducing negative symptoms in schizophrenia. Psychopharmacology (Berl) 174(1):32–38. doi:10.1007/s00213-003-1709-2

D'Aniello A, Giuditta A (1977) Identification of D-aspartic acid in the brain of Octopus vulgaris Lam. J Neurochem 29(6):1053–1057

D'Aniello A, Vetere A, Petrucelli L (1993) Further study on the specificity of D-amino acid oxidase and D-aspartate oxidase and time course for complete oxidation of D-amino acids. Comp Biochem Physiol 105(3–4):731–734

D'Aniello S, Somorjai I, Garcia-Fernandez J, Topo E, D'Aniello A (2010) D-Aspartic acid is a novel endogenous neurotransmitter. FASEB J 25(3):1014–1027. doi:fj.10-168492 [pii] 10.1096/fj.10-168492

Davies LP, Johnston GA (1976) Uptake and release of D- and L-aspartate by rat brain slices. J Neurochem 26(5):1007–1014

DeVito LM, Balu DT, Kanter BR, Lykken C, Basu AC, Coyle JT, Eichenbaum H (2011) Serine racemase deletion disrupts memory for order and alters cortical dendritic morphology. Genes Brain Behav 10(2):210–222. doi:10.1111/j.1601-183X.2010.00656.x

Dunlop DS, Neidle A, McHale D, Dunlop DM, Lajtha A (1986) The presence of free D-aspartic acid in rodents and man. Biochem Biophys Res Commun 141(1):27–32

Errico F, Pirro MT, Affuso A, Spinelli P, De Felice M, D'Aniello A, Di Lauro R (2006) A physiological mechanism to regulate D-aspartic acid and NMDA levels in mammals revealed by D-aspartate oxidase deficient mice. Gene 374:50–57

Errico F, Nistico R, Palma G, Federici M, Affuso A, Brilli E, Topo E, Centonze D, Bernardi G, Bozzi Y, D'Aniello A, Di Lauro R, Mercuri NB, Usiello A (2008a) Increased levels of D-aspartate in the hippocampus enhance LTP but do not facilitate cognitive flexibility. Mol Cell Neurosci 37(2):236–246. doi:S1044-7431(07)00223-0 [pii] 10.1016/j.mcn.2007.09.012

Errico F, Rossi S, Napolitano F, Catuogno V, Topo E, Fisone G, D'Aniello A, Centonze D, Usiello A (2008b) D-aspartate prevents corticostriatal long-term depression and attenuates schizophrenia-like symptoms induced by amphetamine and MK-801. J Neurosci 28 (41):10404–10414. doi:28/41/10404 [pii] 10.1523/JNEUROSCI.1618-08.2008

Errico F, Bonito-Oliva A, Bagetta V, Vitucci D, Romano R, Zianni E, Napolitano F, Marinucci S, Di Luca M, Calabresi P, Fisone G, Carta M, Picconi B, Gardoni F, Usiello A (2011a) Higher free D-aspartate and N-methyl-D-aspartate levels prevent striatal depotentiation and anticipate L-DOPA-induced dyskinesia. Exp Neurol 232(2):240–250. doi:S0014-4886(11)00321-9 [pii] 10.1016/j.expneurol.2011.09.013

Errico F, Nistico R, Napolitano F, Mazzola C, Astone D, Pisapia T, Giustizieri M, D'Aniello A, Mercuri NB, Usiello A (2011b) Increased D-aspartate brain content rescues hippocampal age-related synaptic plasticity deterioration of mice. Neurobiol Aging 32(12):2229–2243. doi:10.1016/j.neurobiolaging.2010.01.002

Errico F, Nistico R, Napolitano F, Oliva AB, Romano R, Barbieri F, Florio T, Russo C, Mercuri NB, Usiello A (2011c) Persistent increase of D-aspartate in D-aspartate oxidase mutant mice induces a precocious hippocampal age-dependent synaptic plasticity and spatial memory decay. Neurobiol Aging 32(11):2061–2074. doi:S0197-4580(09)00399-6 [pii] 10.1016/j.neurobiolaging.2009.12.007

Errico F, Napolitano F, Squillace M, Vitucci D, Blasi G, de Bartolomeis A, Bertolino A, D'Aniello A, Usiello A (2013) Decreased levels of d-aspartate and NMDA in the prefrontal cortex and striatum of patients with schizophrenia. J Psychiatr Res doi:S0022-3956(13)00196-9 [pii] 10.1016/j.jpsychires.2013.06.013

Errico F, Nistico R, Di Giorgio A, Squillace M, Vitucci D, Galbusera A, Piccinin S, Mango D, Fazio L, Middei S, Trizio S, Mercuri NB, Teule MA, Centonze D, Gozzi A, Blasi G, Bertolino A, Usiello A (2014) Free D-aspartate regulates neuronal dendritic morphology, synaptic plasticity, gray matter volume and brain activity in mammals. Transl Psychiatry 4, e417. doi:10.1038/tp.2014.59 tp201459 [pii]

Fagg GE, Matus A (1984) Selective association of N-methyl aspartate and quisqualate types of L-glutamate receptor with brain postsynaptic densities. Proc Natl Acad Sci U S A 81 (21):6876–6880

Fatemi SH, Folsom TD (2009) The neurodevelopmental hypothesis of schizophrenia, revisited. Schizophr Bull 35(3):528–548. doi:10.1093/schbul/sbn187

Fossat P, Turpin FR, Sacchi S, Dulong J, Shi T, Rivet JM, Sweedler JV, Pollegioni L, Millan MJ, Oliet SH, Mothet JP (2012) Glial D-serine gates NMDA receptors at excitatory synapses in prefrontal cortex. Cereb Cortex 22(3):595–606. doi:10.1093/cercor/bhr130

Garthwaite G, Garthwaite J (1985) Sites of D-[^{3}H]aspartate accumulation in mouse cerebellar slices. Brain Res 343(1):129–136

Geyer MA (2006) The family of sensorimotor gating disorders: comorbidities or diagnostic overlaps? Neurotox Res 10(3–4):211–220

Gong XQ, Frandsen A, Lu WY, Wan Y, Zabek RL, Pickering DS, Bai D (2005) D-aspartate and NMDA, but not L-aspartate, block AMPA receptors in rat hippocampal neurons. Br J Pharmacol 145(4):449–459

Gottesman II, Gould TD (2003) The endophenotype concept in psychiatry: etymology and strategic intentions. Am J Psychiatry 160(4):636–645

Grasselli G, Rossi S, Musella A, Gentile A, Loizzo S, Muzio L, Di Sanza C, Errico F, Musumeci G, Haji N, Fresegna D, Sepman H, De Chiara V, Furlan R, Martino G, Usiello A,

Mandolesi G, Centonze D (2013) Abnormal NMDA receptor function exacerbates experimental autoimmune encephalomyelitis. Br J Pharmacol 168:502–517. doi:10.1111/j.1476-5381.2012.02178.x

Gundersen V, Danbolt NC, Ottersen OP, Storm-Mathisen J (1993) Demonstration of glutamate/aspartate uptake activity in nerve endings by use of antibodies recognizing exogenous D-aspartate. Neuroscience 57(1):97–111, doi:0306-4522(93)90114-U [pii]

Hamase K, Homma H, Takigawa Y, Fukushima T, Santa T, Imai K (1997) Regional distribution and postnatal changes of D-amino acids in rat brain. Biochim Biophys Acta 1334 (2–3):214–222, doi:S0304-4165(96)00095-5 [pii]

Hardingham GE, Bading H (2003) The Yin and Yang of NMDA receptor signalling. Trends Neurosci 26(2):81–89. doi:10.1016/S0166-2236(02)00040-1

Hashimoto A, Kumashiro S, Nishikawa T, Oka T, Takahashi K, Mito T, Takashima S, Doi N, Mizutani Y, Yamazaki T et al (1993) Embryonic development and postnatal changes in free D-aspartate and D-serine in the human prefrontal cortex. J Neurochem 61(1):348–351

Hashimoto A, Oka T, Nishikawa T (1995) Anatomical distribution and postnatal changes in endogenous free D-aspartate and D-serine in rat brain and periphery. Eur J Neurosci 7 (8):1657–1663

Horio M, Ishima T, Fujita Y, Inoue R, Mori H, Hashimoto K (2013) Decreased levels of free D-aspartic acid in the forebrain of serine racemase (Srr) knock-out mice. Neurochem Int 62 (6):843–847. doi:10.1016/j.neuint.2013.02.015

Huang AS, Beigneux A, Weil ZM, Kim PM, Molliver ME, Blackshaw S, Nelson RJ, Young SG, Snyder SH (2006) D-aspartate regulates melanocortin formation and function: behavioral alterations in D-aspartate oxidase-deficient mice. J Neurosci 26(10):2814–2819

Ikonomidou C, Bittigau P, Koch C, Genz K, Hoerster F, Felderhoff-Mueser U, Tenkova T, Dikranian K, Olney JW (2001) Neurotransmitters and apoptosis in the developing brain. Biochem Pharmacol 62(4):401–405

Javitt DC, Zukin SR, Heresco-Levy U, Umbricht D (2012) Has an angel shown the way? Etiological and therapeutic implications of the PCP/NMDA model of schizophrenia. Schizophr Bull 38(5):958–966. doi:10.1093/schbul/sbs069 sbs069 [pii]

Katane M, Homma H (2010) D-aspartate oxidase: the sole catabolic enzyme acting on free D-aspartate in mammals. Chem Biodivers 7(6):1435–1449. doi:10.1002/cbdv.200900250

Kim PM, Aizawa H, Kim PS, Huang AS, Wickramasinghe SR, Kashani AH, Barrow RK, Huganir RL, Ghosh A, Snyder SH (2005) Serine racemase: activation by glutamate neurotransmission via glutamate receptor interacting protein and mediation of neuronal migration. Proc Natl Acad Sci U S A 102(6):2105–2110. doi:0409723102 [pii] 10.1073/pnas.0409723102

Kim PM, Duan X, Huang AS, Liu CY, Ming GL, Song H, Snyder SH (2010) Aspartate racemase, generating neuronal D-aspartate, regulates adult neurogenesis. Proc Natl Acad Sci U S A 107 (7):3175–3179. doi:10.1073/pnas.0914706107 0914706107 [pii]

Kochhar S, Christen P (1992) Mechanism of racemization of amino acids by aspartate aminotransferase. Eur J Biochem 203(3):563–569

Koyama H, Adachi M, Sekine M, Katane M, Furuchi T, Homma H (2006) Cytoplasmic localization and efflux of endogenous D-aspartate in pheochromocytoma 12 cells. Arch Biochem Biophys 446(2):131–139. doi:S0003-9861(05)00513-8 [pii] 10.1016/j.abb.2005.12.008

Krebs HA (1935) Metabolism of amino-acids: deamination of amino-acids. Biochem J 29 (7):1620–1644

Labrie V, Duffy S, Wang W, Barger SW, Baker GB, Roder JC (2009) Genetic inactivation of D-amino acid oxidase enhances extinction and reversal learning in mice. Learn Mem 16 (1):28–37. doi:10.1101/lm.1112209 16/1/28 [pii]

Labrie V, Wong AH, Roder JC (2012) Contributions of the D-serine pathway to schizophrenia. Neuropharmacology 62 (3):1484–1503. doi:S0028-3908(11)00041-4 [pii] 10.1016/j.neuropharm.2011.01.030

Lancelot E, Beal MF (1998) Glutamate toxicity in chronic neurodegenerative disease. Prog Brain Res 116:331–347

Li Y, Sacchi S, Pollegioni L, Basu AC, Coyle JT, Bolshakov VY (2013) Identity of endogenous NMDAR glycine site agonist in amygdala is determined by synaptic activity level. Nat Commun 4:1760. doi:10.1038/ncomms2779

Long Z, Homma H, Lee JA, Fukushima T, Santa T, Iwatsubo T, Yamada R, Imai K (1998) Biosynthesis of D-aspartate in mammalian cells. FEBS Lett 434(3):231–235

Long Z, Lee JA, Okamoto T, Sekine M, Nimura N, Imai K, Yohda M, Maruyama T, Sumi M, Kamo N, Yamagishi A, Oshima T, Homma H (2001) Occurrence of D-Amino Acids and a pyridoxal 5′-phosphate-dependent aspartate racemase in the acidothermophilic archaeon, Thermoplasma acidophilum. Biochem Biophys Res Commun 281(2):317–321. doi:10.1006/bbrc.2001.4353

Lu L, Mamiya T, Koseki T, Mouri A, Nabeshima T (2011) Genetic animal models of schizophrenia related with the hypothesis of abnormal neurodevelopment. Biol Pharm Bull 34 (9):1358–1363

Malthe-Sorenssen D, Skrede KK, Fonnum F (1979) Calcium-dependent release of D-[^{3}H]aspartate evoked by selective electrical stimulation of excitatory afferent fibres to hippocampal pyramidal cells in vitro. Neuroscience 4(9):1255–1263, doi:0306-4522(79)90155-6 [pii]

Martineau M, Baux G, Mothet JP (2006) D-serine signalling in the brain: friend and foe. Trends Neurosci 29 (8):481–491. doi:S0166-2236(06)00122-6 [pii] 10.1016/j.tins.2006.06.008

Moghaddam B, Javitt D (2012) From revolution to evolution: the glutamate hypothesis of schizophrenia and its implication for treatment. Neuropsychopharmacology 37(1):4–15. doi:10.1038/npp.2011.181 npp2011181 [pii]

Molinaro G, Pietracupa S, Di Menna L, Pescatori L, Usiello A, Battaglia G, Nicoletti F, Bruno V (2010) D-aspartate activates mGlu receptors coupled to polyphosphoinositide hydrolysis in neonate rat brain slices. Neurosci Lett 478(3):128–130

Monahan JB, Michel J (1987) Identification and characterization of an N-methyl-D-aspartate-specific L-[^{3}H]glutamate recognition site in synaptic plasma membranes. J Neurochem 48 (6):1699–1708

Mothet JP, Parent AT, Wolosker H, Brady RO Jr, Linden DJ, Ferris CD, Rogawski MA, Snyder SH (2000) D-serine is an endogenous ligand for the glycine site of the N-methyl-D-aspartate receptor. Proc Natl Acad Sci U S A 97(9):4926–4931

Nacher J, McEwen BS (2006) The role of N-methyl-D-asparate receptors in neurogenesis. Hippocampus 16(3):267–270. doi:10.1002/hipo.20160

Nakatsuka S, Hayashi M, Muroyama A, Otsuka M, Kozaki S, Yamada H, Moriyama Y (2001) D-Aspartate is stored in secretory granules and released through a Ca(2+)-dependent pathway in a subset of rat pheochromocytoma PC12 cells. J Biol Chem 276(28):26589–26596. doi:10.1074/jbc.M011754200

Negri A, Ceciliani F, Tedeschi G, Simonic T, Ronchi S (1992) The primary structure of the flavoprotein D-aspartate oxidase from beef kidney. J Biol Chem 267(17):11865–11871

Neidle A, Dunlop DS (1990) Developmental changes in free D-aspartic acid in the chicken embryo and in the neonatal rat. Life Sci 46(21):1517–1522

Ogita K, Yoneda Y (1988) Disclosure by triton X-100 of NMDA-sensitive [^{3}H] glutamate binding sites in brain synaptic membranes. Biochem Biophys Res Commun 153(2):510–517

Olverman HJ, Jones AW, Mewett KN, Watkins JC (1988) Structure/activity relations of N-methyl-D-aspartate receptor ligands as studied by their inhibition of [^{3}H]D-2-amino-5-phosphonopentanoic acid binding in rat brain membranes. Neuroscience 26(1):17–31

Owen MJ, O'Donovan MC, Thapar A, Craddock N (2011) Neurodevelopmental hypothesis of schizophrenia. British J Psychiatry 198(3):173–175. doi:10.1192/bjp.bp.110.084384

Palacin M, Estevez R, Bertran J, Zorzano A (1998) Molecular biology of mammalian plasma membrane amino acid transporters. Physiol Rev 78(4):969–1054

Papouin T, Ladepeche L, Ruel J, Sacchi S, Labasque M, Hanini M, Groc L, Pollegioni L, Mothet JP, Oliet SH (2012) Synaptic and extrasynaptic NMDA receptors are gated by different endogenous coagonists. Cell 150(3):633–646. doi:10.1016/j.cell.2012.06.029 S0092-8674 (12)00786-6 [pii]

Pollegioni L, Piubelli L, Sacchi S, Pilone MS, Molla G (2007) Physiological functions of D-amino acid oxidases: from yeast to humans. Cell Mol Life Sci 64(11):1373–1394. doi:10.1007/s00018-007-6558-4

Ransom RW, Stec NL (1988) Cooperative modulation of [^{3}H]MK-801 binding to the N-methyl-D-aspartate receptor-ion channel complex by L-glutamate, glycine, and polyamines. J Neurochem 51(3):830–836

Ritter LM, Vazquez DM, Meador-Woodruff JH (2002) Ontogeny of ionotropic glutamate receptor subunit expression in the rat hippocampus. Brain Res Dev Brain Res 139(2):227–236

Rosenberg D, Artoul S, Segal AC, Kolodney G, Radzishevsky I, Dikopoltsev E, Foltyn VN, Inoue R, Mori H, Billard JM, Wolosker H (2013) Neuronal D-serine and glycine release via the Asc-1 transporter regulates NMDA receptor-dependent synaptic activity. J Neurosci 33 (8):3533–3544. doi:10.1523/JNEUROSCI.3836-12.2013

Sacchi S, Caldinelli L, Cappelletti P, Pollegioni L, Molla G (2012) Structure-function relationships in human D-amino acid oxidase. Amino Acids 43(5):1833–1850. doi:10.1007/s00726-012-1345-4

Sakai K, Homma H, Lee JA, Fukushima T, Santa T, Tashiro K, Iwatsubo T, Imai K (1998) Emergence of D-aspartic acid in the differentiating neurons of the rat central nervous system. Brain Res 808(1):65–71

Sawa A, Snyder SH (2003) Schizophrenia: neural mechanisms for novel therapies. Mol Med 9 (1–2):3–9

Schell MJ, Cooper OB, Snyder SH (1997) D-aspartate localizations imply neuronal and neuroendocrine roles. Proc Natl Acad Sci U S A 94(5):2013–2018

Setoyama C, Miura R (1997) Structural and functional characterization of the human brain D-aspartate oxidase. J Biochem 121(4):798–803

Still JL, Buell MV et al (1949) Studies on the cyclophorase system; D-aspartic oxidase. J Biol Chem 179(2):831–837

Tanaka-Hayashi A, Hayashi S, Inoue R, Ito T, Konno K, Yoshida T, Watanabe M, Yoshimura T, Mori H (2015) Is D-aspartate produced by glutamic-oxaloacetic transaminase-1 like 1 (Got1l1): a putative aspartate racemase? Amino Acids 47:79–86. doi:10.1007/s00726-014-1847-3

Taxt T, Storm-Mathisen J (1984) Uptake of D-aspartate and L-glutamate in excitatory axon terminals in hippocampus: autoradiographic and biochemical comparison with gamma-aminobutyrate and other amino acids in normal rats and in rats with lesions. Neuroscience 11(1):79–100

Van Horn MR, Sild M, Ruthazer ES (2013) D-serine as a gliotransmitter and its roles in brain development and disease. Front Cell Neurosci 7:39. doi:10.3389/fncel.2013.00039

Van Veldhoven PP, Brees C, Mannaerts GP (1991) D-aspartate oxidase, a peroxisomal enzyme in liver of rat and man. Biochim Biophys Acta 1073(1):203–208

Wolosker H, Radzishevsky I (2013) The serine shuttle between glia and neurons: implications for neurotransmission and neurodegeneration. Biochem Soc Trans 41(6):1546–1550. doi:10.1042/BST20130220 BST20130220 [pii]

Wolosker H, D'Aniello A, Snyder SH (2000) D-aspartate disposition in neuronal and endocrine tissues: ontogeny, biosynthesis and release. Neuroscience 100(1):183–189

Yamauchi T, Choi SY, Okada H, Yohda M, Kumagai H, Esaki N, Soda K (1992) Properties of aspartate racemase, a pyridoxal 5′-phosphate-independent amino acid racemase. J Biol Chem 267(26):18361–18364

Yang Y, Ge W, Chen Y, Zhang Z, Shen W, Wu C, Poo M, Duan S (2003) Contribution of astrocytes to hippocampal long-term potentiation through release of D-serine. Proc Natl Acad Sci U S A 100(25):15194–15199. doi:10.1073/pnas.2431073100 2431073100 [pii]

Zaar K, Kost HP, Schad A, Volkl A, Baumgart E, Fahimi HD (2002) Cellular and subcellular distribution of D-aspartate oxidase in human and rat brain. J Comp Neurol 450(3):272–282

Part IV
D-Amino Acid Residues in Proteins

Chapter 15
D-Amino Acid Residues in Proteins Related to Aging and Age-Related Diseases and a New Analysis of the Isomers in Proteins

Noriko Fujii, Takumi Takata, Norihiko Fujii, Kenzo Aki, and Hiroaki Sakaue

Abstract Homochirality is essential for the development and maintenance of life. Until relatively recently, the homochirality of amino acids in living systems was believed to be maintained with the exception of the presence of D-amino acids in the cell wall of microorganisms. However, D-amino acids were recently found in various higher organisms in proteins and peptides and as free amino acids. In proteins, D-aspartate (Asp) residues have been detected in various tissues such as the eye lens, teeth, bone, aorta, ligament, brain, and skin of elderly individuals, and thus D-amino acids can no longer be considered as uncommon in living organisms. The presence of D-amino acids may change the higher-order structure of proteins, and this may be the cause of age-related diseases including cataract and Alzheimer's disease. D-Asp in aged tissues of living organisms is thought to result from the spontaneous racemization of the Asp residues. The racemization of Asp residues in proteins does not occur uniformly but does so at specific residues on the basis of the sequence context or structural considerations. Therefore, it is necessary to determine the nature of Asp residues at specific sites within particular proteins. However, the detection of D-amino acids in proteins to date has been complex and difficult. This review deals with 1) the presence of D-aspartate (Asp) residues in protein of living tissues, 2) the mechanism of D-Asp formation in protein under physiological conditions, 3) the influence of D-Asp on protein structure and function, and 4) recent advances in D-amino acid analysis in protein.

N. Fujii (✉) • T. Takata
Research Reactor Institute, Kyoto University, Kumatori, Osaka 590-0494, Japan
e-mail: nfujii@rri.kyoto-u.ac.jp

N. Fujii
Radioisotope Research Center, Teikyo University, Kaga Itabashi-ku, Tokyo 173-8605, Japan

K. Aki
Faculty of Pharmaceutical Sciences, Himeji Dokkyo University, Kamiohno, Himeji, Hyogo 670-8524, Japan

H. Sakaue
Division of Pharmacology, School of Pharmaceutical Sciences, International University of Health and Welfare Ohtawara, Tochigi 324-8501, Japan

T. Yoshimura et al. (eds.), *D-Amino Acids*, DOI 10.1007/978-4-431-56077-7_15

Keywords D-amino acid • Aging • Isoaspartate • Protein • Cataract • LC-MS/MS

15.1 Introduction

Amino acids contain one (or more) asymmetric tetrahedral carbon atoms. Therefore, the molecules can exist in two nonsuperimposable mirror image forms, that is, they can be right-handed (D-enantiomer) and left-handed (L-enantiomer) structures. It is thought that equal amounts of D- and L-amino acids existed on primeval earth before the emergence of life. However, during the stage of chemical evolution, only L-amino acids were selected for polymerization and formation of peptides and proteins after which life emerged. Although the chemical and physical properties of L-amino acids and D-amino acids are very similar with the exception of their optical character, the reasons for the elimination of D-amino acids and why all living organisms are now composed predominantly of L-amino acids are not well understood. However, it is clear that only one of the enantiomers could be selected because proteins, which consist of many amino acid diastereoisomers, would not be able to fold into proper structures in a manner similar to current proteins. Therefore, homochirality is essential for the development and maintenance of life. Once the L-amino acid world was established, D-amino acids were excluded from living systems. Consequently, there has been little study of the presence and function of D-amino acids in living organisms.

D-Amino acids, however, were recently found in various living higher organisms in the form of free amino acids, peptides, and proteins. Free D-aspartate (Asp) and D-serine (Ser) are present and may have important physiological function in mammals. Free D-Asp may play a role as a novel messenger in the maturation and differentiation of tissues (Katane and Homma* 2011), while free D-Ser is found in the brain (Hashimoto et al. 1992, 1993) and functions as a co-agonist of N-methyl-D-aspartate (NMDA) receptors (Hashimoto et al. 1993). Small peptides containing one D-amino acid have been found in various vertebrates and invertebrates. Dermorphin is the first D-amino acid-containing peptide found which was isolated from the skin of a frog (*Phyllomedusinae* from South and Middle America) and is an opioid peptide with the sequence Y [D-Ala]-FGYP (Montecucchi et al. 1981). The activity which is about 1000 times greater than that of morphine is lost by substituting L-Ala for D-Ala (Broccardo et al. 1981). Many small peptides that contain D-amino acids are described in another excellent review (Jolles 1998).

In proteins, D-Asp residues have been widely detected in various tissues such as eye lens (Masters et al. 1977; Fujii et al. 1994a, b; Fujii et al. 2011), teeth (Helfman and Bada 1976; Masuda et al. 2002), bone (Ritz et al. 1996; Cloos and Fledelius 2000), aorta (Powell et al. 1992), ligament (Ritz-Timme et al. 2003), brain (Roher et al. 1993), and skin (Fujii et al. 2002; Ritz-Timme et al. 2003) of elderly individuals. The presence of D-Asp in aged tissues of living organisms is a result of spontaneous racemization of the Asp residues in these particular proteins. Most researchers held the view that L-amino acids in proteins could never change to D-isomers under the physical conditions of the living body because proteins were believed to be difficult to modify chemically, since selection during evolution before the emergence of life worked to ensure such molecules had very stable properties. This general idea had no real basis in scientific fact but became established because D-amino acids could not be found. The racemization of Asp residues in proteins does not occur uniformly but does so at specific Asp residues on the basis of the sequence context or structural considerations that make the specific residues more susceptible to the reaction than others. It is therefore necessary to determine the nature of the Asp residues at specific sites within particular proteins.

Conventional enantioseparation of free amino acids by gas chromatography (GC) or reversed-phase high-performance liquid chromatography (RP-HPLC) is easier than looking for D-amino acids in the context of an intact protein. Identification of a very small quantity of D-amino acids at specific sites in proteins comprised almost entirely of L-amino acids is similar to looking for a needle in a haystack. In order to analyze the specific sites of protein-bound D-amino acids, several complex steps such as (1) purification of the protein, (2) enzymatic digestion of the protein, (3) separation and identification of the enzyme-digested peptides, (4) hydrolysis of the enzyme-digested peptides, (4) derivatization of the amino acids to diastereoisomers, (5) application of the diastereoisomers to reversed-phase high-performance liquid chromatography (RP-HPLC), and determination of the D/L ratio of amino acids by analysis of the respective peak areas. Recently, we proposed a new method of analysis for determining the Asp isomers at individual sites in a protein with decreased complexity using LC-MS systems.

This review deals with (1) the presence of D-amino acid residues in proteins, (2) the mechanism of how D-aspartate residues spontaneously occur in proteins under physiological conditions, (3) the influence of D-Asp on protein structure and function, and (4) the recent advances in D-amino acid analysis in proteins.

15.1.1 The Presence of D-Amino Acid Residues in Proteins

Although proteins consist exclusively of L-amino acids, D-amino acids have been detected in various tissues as described in the Introduction. Table 15.1 shows that D-amino acids have been found in many proteins from various tissues. It is therefore no longer uncommon to find D-amino acids in living organisms. Almost all D-amino acids found are D-Asp since Asp is the most easily racemizable amino acid. Earlier studies only showed that D-Asp accumulated in proteins of tissues with age. Because D-Asp was detected in homogenates of tissues, it could not be determined whether all of the aspartic acid in the protein was racemized uniformly or whether particular aspartic acid residues had a greater tendency to racemize in specific proteins.

Recent studies clearly indicate that Asp residues of proteins are not racemized uniformly but that D-Asp residues may be present at some specific sites in some particular proteins such as eye lens crystallins (Fujii et al. 1994a, b; Fujii et al. 2011, 2012), β-amyloid protein (Roher et al. 1993), histone H2B (Young et al. 2005), type I collagen (Cloos and Fledelius 2000), etc. In addition to these proteins, recent

Table 15.1 The presence of D-amino acids in proteins from various tissues

Tissue	Protein	Amino acid	Related disease	Specific sites
Tooth	Phosphophoryn	D-Asp	nd	nd
Bone	Osteocalcin	D-Asp	nd	nd
Bone	Type I collagen C-terminal telopeptide	D-Asp	Osteoporosis of Paget's disease	Asp 1211
Brain	Myelin	D-Asp	nd	nd
Brain	β-Amyloid	D-Asp	Alzheimer's disease	Asp 1, 7, Asp 23
Brain	β-Amyloid	D-Ser	Alzheimer's disease	Ser 8, 26
Brain	Histone H2B	D-Asp	nd	Asp 25
Erythrocyte		D-Asp		
Lens	αA-crystallin	D-Asp	Cataract	Asp-58, Asp-76, Asp-84, Asp-151
Lens	αB-crystallin	D-Asp	Cataract	Asp-36, Asp-62, Asp-96
Lens	βB2-crystallin	D-Asp	Cataract	Asp 4
Retina	nd	D-Asp	AMD	nd
Conjutiva	nd	D-Asp	Pinguecula	nd
Skin	Elastin	D-Asp	Elastosis	nd
Skin	Collagen	D-Asp		nd
Skin	Keratin	D-Asp		nd
Ligament	Elastin	D-Asp		nd
Lung	Elastin	D-Asp		nd
Aorta	Elastin	D-Asp	Arteriosclerosis	nd
	γ-Globulin	D-Asp, D-Cys		Asp 24 Cys 220
	Lysozyme	D-Asn		Asn 127

nd not determined

studies reported that Asn-127 in mouse lysozyme quickly racemizes after incubation (pH 7 and 37 °C) for 8 weeks, Cys220 in the hinge sequence of immunoglobulin gamma 1 (IgG1) quickly racemizes after storage for 6 months at 40 °C (Amano et al. 2011), and Asp-24 in the heavy chain peptide H5 is highly racemized (Zhang et al. 2011) (Table 15.1). These studies clearly indicate that racemization to D-amino acids occurs more easily than may have been expected.

15.1.2 How Do D-Aspartate Residues Spontaneously Occur at Specific Sites in Proteins Under Physiological Conditions?

We found several specific D-Asp sites in aged human lens proteins and have studied the mechanism of the spontaneous isomerization of proteins under physiological conditions. Human lens proteins are composed of three major structural proteins, namely, α-, β-, and γ-crystallins. α-Crystallin functions like a chaperone, binding to nonnative or unfolded proteins and protecting them against aggregation induced by heat, reduction, and chemical modification (Horwitz 1992). The chaperone-like activity of α-crystallin might play an important role in preventing the aggregation and insolubilization of other lenticular proteins, thereby maintaining the transparency of the eye lens. Alpha-crystallin is a polymer consisting of two subunits, αA and αB. We previously reported the presence of D-isomers at Asp-58, Asp-76, Asp-84, and Asp-151 in αA-crystallin (Fujii et al. 1994b, 2012); at Asp-36, Asp-62, and Asp-96 in αB-crystallin (Fujii et al. 1994a, 2012); and at Asp-4 in βB2-crystallin from aged human lenses (Fujii et al. 2011). D-Asp formation was accompanied by isomerization from the natural α-Asp to the abnormal β-Asp (Fujii et al. 1999). Racemization begins when the hydrogen atom attached to the α-carbon atom is released. Usually, this reaction proceeds with difficulty in mild conditions, such as those found in the body. However, Asp residues in proteins are susceptible to racemization because Asp has a carboxyl group in its side chain. Inversion and isomerization of Asp residues in proteins are considered to proceed via a succinimide intermediate. As shown in Fig. 15.1, the simultaneous formation of β- and D-Asp residues in the protein can be explained as follows: (1) When the lone-pair electron of the nitrogen atom of the amino acid residue following the Asp residue attacks the carboxyl group of the side chain of the Lα-Asp residue, L-succinimide is generated by intramolecular cyclization. (2) L-Succinimide can be converted to D-succinimide via an intermediate through keto–enol tautomerism. (3) D-Succinimide is then hydrolyzed at either side of its two carbonyl groups to form Dα- and Dβ-Asp; similarly, L-succinimide is hydrolyzed to form Lα- and Lβ-Asp. Thus, four isomers, Lα-Asp, Lβ-Asp, Dα-Asp, and Dβ-Asp, are simultaneously formed in the protein. The difference in abundance of the Asp isomers in the protein may be due to the rate constants for the formation of the isomers. The rate constant for hydrolysis from succinimidyl peptide to β-Asp peptide is about 5 times higher than the rate constant for hydrolysis from L-succinimidyl peptide to L-α-Asp peptide. Thus, of these Asp isomers, large amounts of D-β- and L-β-isomers are present, but the amount of D-α-isomer is not significant (Aki and Fujii 2013).

The rate of succinimide formation is considered to depend on the residue neighboring the Asp residue. When the neighboring amino acid has a small side chain, as found in glycine (Gly), alanine (Ala), or serine (Ser), succinimide forms easily because there is no steric hindrance (Geiger and Clarke 1987; Fujii et al. 1999). In fact, as shown in Table 15.2, many Gly, Ser, and Ala residues were found to be the residue adjacent to D-Asp sites. In addition to the effects of the

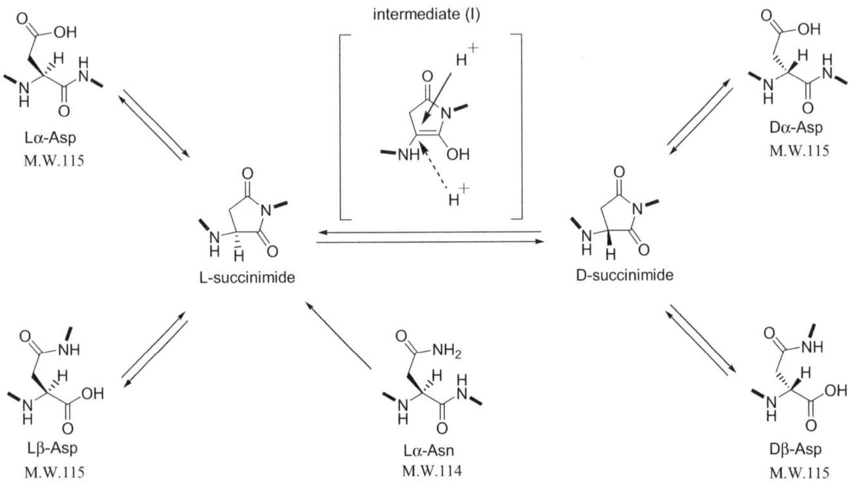

Fig. 15.1 Possible reaction pathways for spontaneous isomerization of Asp residues in protein

Table 15.2 Properties of D-Asp in various proteins

Protein	Site of D-Asp	D/L of Asp	Linkage	Next residue	Structure around Asp
αA-crystallin(1–173)80Y	58	3.10	β	Ser	nd
αA-crystallin(1–173)80Y	151	5.70	β	Ala	Flexible
αB-crystallin(1–175)	36	0.92	β	Leu	nd
αB-crystallin(1–175)	62	0.57	β	Thr	nd
βB2-crystallin 80Y	4	3.01	β	His	Flexible
β-Amyloid protein(1–42)	1	0.04	β	Ala	nd
β-Amyloid protein(1–42)	7	1.00	β	Ser	nd
Histone H2B(1–126)	25	0.14	nd	Gly	nd
Type I collagen C-terminal telopeptide (AHDGGR1209-1214)	1211	1.00	nd	Gly	nd
Light-chain peptide L2	12			Asp	Flexible
Heavy-chain peptide H5	24			Asp	Flexible

nd not determined

adjacent residues, Asp may also be susceptible to racemization when the residues are located in flexible regions as suggested in Table 15.2. These observations indicate that formation of succinimide in proteins depends both on the amino acids neighboring the Asp residues and on the higher-order structure of the protein.

15.1.3 *The Influence of D-Asp on Protein/Peptide Structure and Function*

15.1.3.1 D-β-Asp in Protein Promotes Massive and Heterogeneous Aggregation

The appearance of D-Asp isomers in a protein can cause major changes in the 3-D structure because the different side-chain orientation may induce an abnormal peptide backbone. In addition to D-Asp formation, the β-linkage of Asp may affect the quaternary structure because the main chain of the protein is elongated. Therefore, the presence of these isomers may be one of the triggers for abnormal aggregation. Moreover, these processes can induce partial unfolding of the corresponding proteins, leading to a disease state. In fact, samples of αA-crystallin containing large amounts of D-β-Asp obtained from donors of ~80 years of age have been shown to undergo abnormal aggregation to form massive and heterogeneous aggregates (Fujii et al. 2007). Specifically, αA-crystallin from normal young individuals (1-year-old, non-racemized samples) had an average sedimentation coefficient of 17 S at 37 °C, whereas the same protein from elderly individuals had an average sedimentation coefficient of 30 S (range, 20–60 S). Changes in the self-association of α-crystallin aggregates have also been correlated to changes in chaperone activity. α-Crystallin from young donors displays chaperone activity, but this activity is reduced by 60 % in aged α-crystallin aggregates (Fujii et al. 2007) (Table 15.3a).

This chaperone activity plays an important role in preventing the aggregation and insolubilization of other lenticular proteins. Hence, the loss of this activity adversely affects maintenance of the transparency of the eye lens.

15.1.3.2 A Single Substitution of an Asp Isomer in a Peptide Induces a Large Change in the Properties of the Peptide

As described in Sect. 3.1, the appearance of D- and β-Asp in a protein potentially induces large changes to its higher-order structure as well as to its function. However, it remains unclear whether the formation of the Asp isomer is the direct trigger for such a change. In order to clarify the effect of the inversion to D-isomers in a protein, we synthesized peptides corresponding to the 70–88 (KFVIFLDVKHFSPEDLTVK) fragment of human αA-crystallin, which is known to have chaperone function (Tanaka et al. 2008). The Lα-Asp corresponding to position 76 was replaced by diastereoisomers Lβ-Asp, Dα-Asp, and Dβ-Asp, and the biochemical properties of the four different peptides were then compared. The peptides containing abnormal isomers (Lβ-Asp, Dα-Asp, and Dβ-Asp) were more hydrophilic than the normal peptide (containing Lα-Asp) and adopted a random coil structure, rather than the normal β-sheet motif. The normal peptide promoted

Table 15.3a Influence of Asp isomers on protein properties

α-Crystallin	1 year	80 years
Asp isomers	ND	++++
Abnormal aggregates	ND	++++
Chaperone activity	100 %	40 %

ND not detected

Table 15.3b Influence of replacement of Lα-Asp-76 by Lβ-, Dα-, and Dβ-Asp isomers on the peptide αA-crystallin 70–88 (KFVIFLD[76]VKHFSPEDLTVK)

	Lα	Dα	Dβ	Lβ
Hydrophobicity	+++	++	+	-
Secondary structure	β-Sheet	Random	Random	Random
Chaperone activity	+	–	–	–

the aggregation of insulin, while the other three isomers suppressed its aggregation (Fujii et al. 2010).

This result clearly indicates that a single substitution of an Asp isomer in a peptide induces a large change in the properties of the peptide (Table 15.3b).

15.1.4 Recent Advances in D-Amino Acid Analysis in Proteins

The racemization of Asp residues in proteins does not occur uniformly but does so at specific Asp residues on the basis of the sequence context or structural considerations that make the specific residues more susceptible to reaction than others. Therefore, it is necessary to determine the nature of the Asp residues at specific sites within particular proteins.

The separation of the optical isomers of amino acids has previously been considered to be difficult because the physical and chemical properties of the optical isomers are the same. In addition to this, enantioseparation of the bound form of amino acids requires the hydrolysis of the protein/peptide before the analysis of the enantiomers. Conventional enantioseparation of amino acids has been performed using gas chromatography (GC) or reversed-phase high-performance liquid chromatography (RP-HPLC). GC analysis requires nanomole levels of sample, while picomole levels are required for reversed-phase high-performance liquid chromatography (RP-HPLC) analysis. GC analysis is direct enantioseparation through the use of a chiral capillary column, while reversed-phase high-performance liquid chromatography (RP-HPLC) analysis is indirect enantioseparation based on the analysis of the diastereoisomeric derivatives of the amino acid samples produced by chiral derivatizing reagents. Both methods require the appropriate amino acid derivatization or preparation in advance of the analysis, the former requiring changing the samples into the gaseous state before injection onto the GC and the latter requiring production of diastereoisomeric derivatives in the case of the non-chiral column. The process is very complex for free D-amino acid analysis. In addition, in order to analyze the specific sites of D-amino acids in protein, more complicated steps are required other than free D-amino acid analysis: (1) the protein is digested with an appropriate enzyme. (2) The resulting peptides are separated by reversed-phase high-performance liquid chromatography (RP-HPLC). (3) The peptides are identified by mass analysis and/or protein sequencing. (4) The α- or β-isomer of the identified peptides is determined by Edman degradation reaction. (5) The D/L ratio of the identified peptides is determined after hydrolysis with 6 N HCl and derivatization. (6) The diastereoisomers are analyzed by reversed-phase high-performance liquid chromatography (RP-HPLC) and the D/L ratio of amino acids determined by analysis of the respective peak areas. The resulting analysis of the isomerization of Asp residues in a protein can be accurate but it is a technically demanding process. Consequently, there has been little study of the presence and function of D-amino acids in living organisms.

In the review we propose a new accurate and quick LC-MS-based analysis for determining the specific sites having Asp isomers and quantifying the amounts of Asp isomers at the individual sites of all lens crystallins in the water-insoluble (WI) and the water-soluble (WS) fractions without the need for complicated purification from the lens tissues. Figure 15.2a, b shows a typical full LC-MS chromatogram of the tryptic peptides from the WI and WS lens proteins. Generally, each peptide would be expected to elute as one peak with one mass number; however, peptides which contain Asp isomers were separated into multiple peaks, and they eluted at different retention times during the LC-MS run even though they had entirely the same sequences. Using this property, we are able to identify peptides which have isomeric Asp residues. For example, the peptide predicted to correspond to positions 55–65 of αA-crystallin (αA 55–65; TVLDSGISEVR: $[M+2H]^{2+}=588.3$) as identified by the database was mainly separated into four

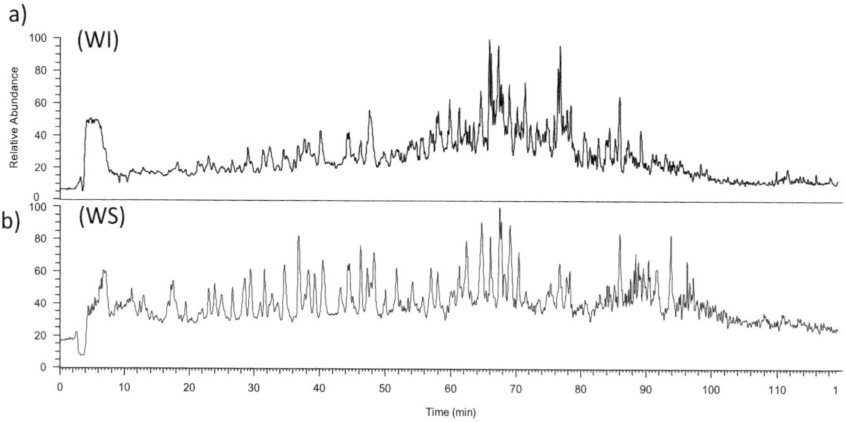

Fig. 15.2 LC-MS chromatogram of the tryptic peptides of water-insoluble (WI) and water-soluble (WS) fractions of lens proteins from elderly donor

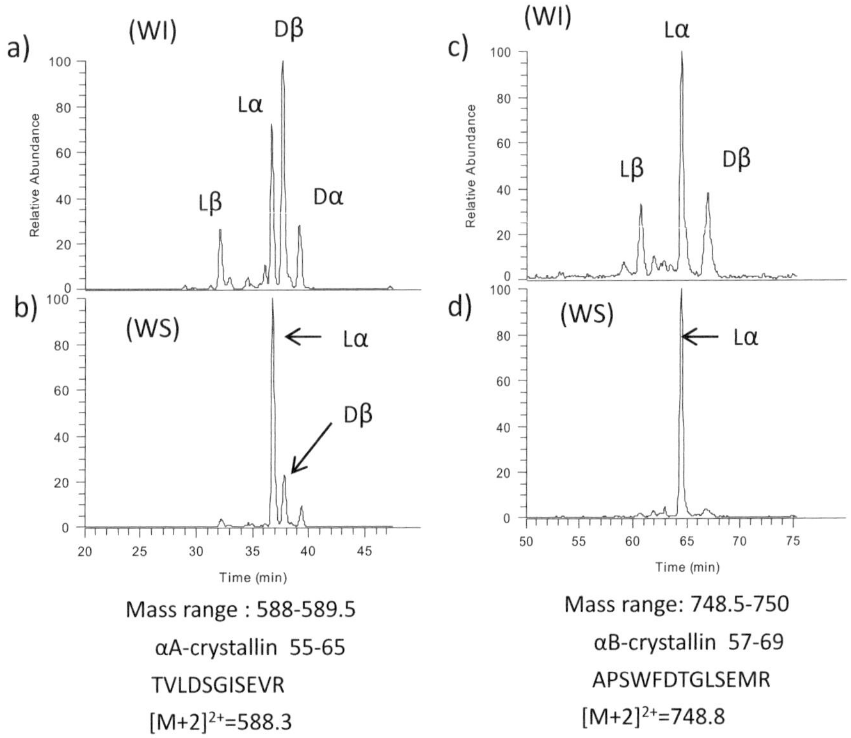

Fig. 15.3 LC-MS chromatogram of the tryptic peptides of water-insoluble (WI) and water-soluble (WS) fractions of lens proteins from elderly donor. (**a**) and (**b**): MS range 588–589.5 m/z (αA-crystallin 55–65) of WI and WS fractions, respectively. (**c**) and (**d**): mass range 748.5–750 m/z (αB-crystallin 57–69) of WI and WS fractions, respectively

different peaks which eluted at different times as shown in Fig. 15.3a, b. Figure 15.3c, d shows the LC-MS chromatogram of αB 57–69 (APSWFDTGLSEMR: $[M+2H]^{2+}$=748.8) from the WI and WS fractions, respectively. This peptide was also separated into several peaks.

The number of peptides from αA 55–65 and αB 57–69 was greater from WI protein (Fig. 15.3a, c) than from WS protein (Fig. 15.3b, c). A similar multiple separation of the various peptides containing Asp residues was obtained from all crystallins, that is, αA-, αB-, βA3-, βA4-, βB1-, βB2-, and γS-crystallin in both WI and WS proteins with the amounts of isomeric peptides in the WI fractions being greater than in the WS fractions. The results are summarized in Fig. 15.4. Figure 15.4 shows the amounts of the four Asp isomers of αA- and αB-crystallins from the WI and WS fractions. The amount of normal L-α-Asp is dramatically decreased, while the other isomeric ratios increased at all Asp sites in the WI fraction compared to the WS fraction.

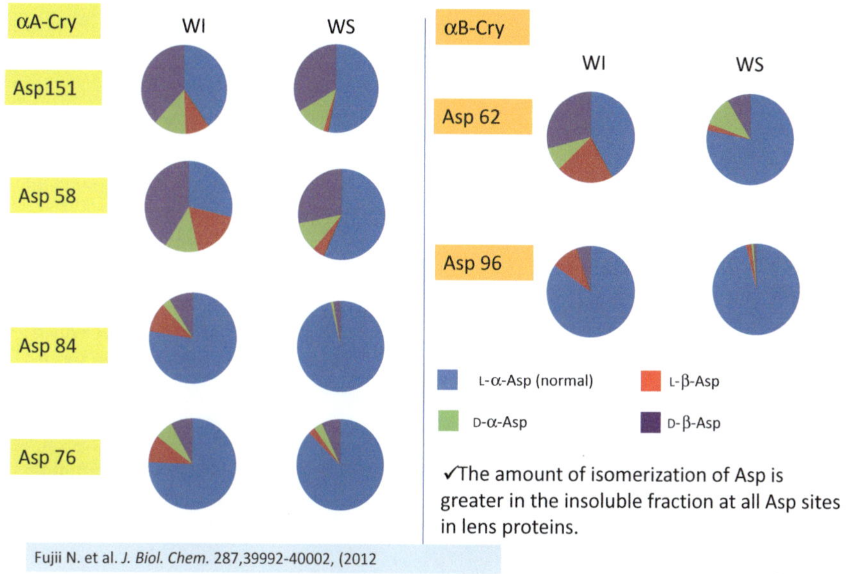

Fig. 15.4 Relative amounts of Asp isomers at various sites from WI and WS fractions of lenses from elderly donors

15.2 Prospects

Racemization and isomerization of amino acids in proteins can cause major changes in structure, since different side-chain orientations can induce an abnormal peptide backbone. Therefore, these posttranslational modifications can induce the partial unfolding of protein leading to a disease state. Thus, it is necessary to determine the levels of isomeric Asp residues at specific sites in any protein. Here, we describe a convenient and robust biochemical method for identifying the isomeric Asp sites in proteins using LC-MS systems (Fig. 15.5). There are many advantages to this new method: (1) No requirement for large amounts of sample proteins. (2) No requirement for the purification of lens proteins from the WI and WS fractions. (3) No requirement for complicated analytical steps which usually include the hydrolysis of the peptides followed by derivatization to the diastereoisomers of amino acids. This new method is able to search comprehensively for the Asp isomers in damaged or aged proteins from all living tissues and cells. Furthermore, the isomeric Asp sites can be determined, and the amounts of the Asp isomers can be quantified quickly and accurately at the femtomole level. This new method therefore improves the study of the isomerization of any amino acid which occurs spontaneously in living tissues or cells.

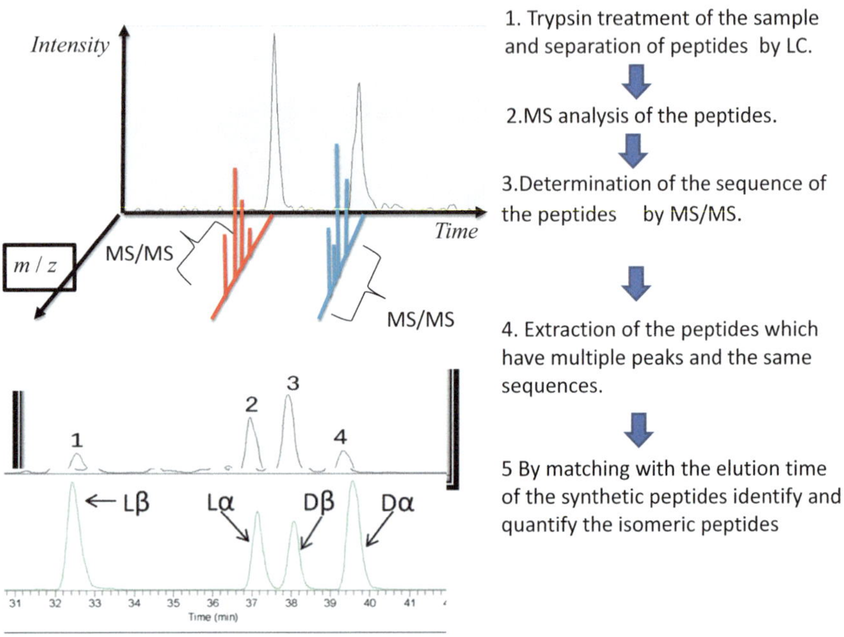

Fig. 15.5 How to identify peptides containing amino acid isomers?

References

Aki K, Fujii N (2013) Kinetics of isomerization and inversion of aspartate 58 of alphaA-crystallin peptide mimics under physiological conditions. PLoS One 8(3), e58515. doi:10.1371/journal.pone.0058515

Amano M, Hasegawa J, Kobayashi N, Kishi N, Nakazawa T, Uchiyama S, Fukui K (2011) Specific racemization of heavy-chain cysteine-220 in the hinge region of immunoglobulin gamma 1 as a possible cause of degradation during storage. Anal Chem 83(10):3857–3864

Broccardo M, Erspamer V, Falconieri Erspamer G, Improta G, Linari G, Melchiorri P, Montecucchi PC (1981) Pharmacological data on dermorphins, a new class of potent opioid peptides from amphibian skin. Br J Pharmacol 73(3):625–631

Cloos PA, Fledelius C (2000) Collagen fragments in urine derived from bone resorption are highly racemized and isomerized: a biological clock of protein aging with clinical potential. Biochem J 345(Pt 3):473–480

Fujii N, Ishibashi Y, Satoh K, Fujino M, Harada K (1994a) Simultaneous racemization and isomerization at specific aspartic acid residues in alpha B-crystallin from the aged human lens. Biochim Biophys Acta 1204(2):157–163

Fujii N, Satoh K, Harada K, Ishibashi Y (1994b) Simultaneous stereoinversion and isomerization at specific aspartic acid residues in alpha A-crystallin from aged human lens. J Biochem 116:663–669

Fujii N, Harada K, Momose Y, Ishii N, Akaboshi M (1999) D-amino acid formation induced by a chiral field within a human lens protein during aging. Biochem Biophys Res Commun 263 (2):322–326

Fujii N, Tajima S, Tanaka N, Fujimoto N, Takata T, Shimo-Oka T (2002) The presence of D-beta-aspartic acid-containing peptides in elastic fibers of sun-damaged skin: a potent marker for ultraviolet-induced skin aging. Biochem Biophys Res Commun 294(5):1047–1051

Fujii N, Shimmyo Y, Sakai M, Sadakane Y, Nakamura T, Morimoto Y, Kinouchi T, Goto Y, Lampi K (2007) Age-related changes of alpha-crystallin aggregate in human lens. Amino Acids 32(1):87–94

Fujii N, Fujii N, Kida M, Kinouchi T (2010) Influence of Lbeta-, Dalpha- and Dbeta-Asp isomers of the Asp-76 residue on the properties of alphaA-crystallin 70–88 peptide. Amino Acids 39 (5):1393–1399

Fujii N, Kawaguchi T, Sasaki H, Fujii N (2011) Simultaneous stereoinversion and isomerization at the Asp-4 residue in betaB2-crystallin from the aged human eye lenses. Biochemistry 50 (40):8628–8635

Fujii N, Sakaue H, Sasaki H (2012) A rapid, comprehensive liquid chromatography-mass spectrometry (LC-MS)-based survey of the Asp isomers in crystallins from human cataract lenses. J Biol Chem 287(47):39992–40002. doi:10.1074/jbc.M112.399972

Geiger T, Clarke S (1987) Deamidation, isomerization, and racemization at asparaginyl and aspartyl residues in peptides. Succinimide-linked reactions that contribute to protein degradation. J Biol Chem 262(2):785–794

Hashimoto A, Nishikawa T, Hayashi T, Fujii N, Harada K, Oka T, Takahashi K (1992) The presence of free D-serine in rat brain. FEBS Lett 296(1):33–36

Hashimoto A, Nishikawa T, Oka T, Takahashi K (1993) Endogenous D-serine in rat brain: N-methyl-D-aspartate receptor-related distribution and aging. J Neurochem 60(2):783–786

Helfman PM, Bada JL (1976) Aspartic acid racemisation in dentine as a measure of ageing. Nature 262(5566):279–281

Horwitz J (1992) Alpha-crystallin can function as a molecular chaperone. Proc Natl Acad Sci U S A 89(21):10449–10453

Jolles P (1998) D-amino acids in sequences of secreted peptides of multicellular organisms. Birkhauser Verlag, Basel, Switzerland, Basel, Boston, Berlin

Katane M, Homma* H (2011) D-Aspartate oxidase: the sole catabolic enzyme acting on free D-Aspartate in mammals. In: D-Amino acids in chemistry, life sciences, and biotechnology. Willey-VHC, Zürich

Masters PM, Bada JL, Zigler JS Jr (1977) Aspartic acid racemisation in the human lens during ageing and in cataract formation. Nature 268(5615):71–73

Masuda W, Nouso C, Kitamura C, Terashita M, Noguchi T (2002) D-Aspartic acid in bovine dentine non-collagenous phosphoprotein. Arch Oral Biol 47(11):757–762

Montecucchi PC, de Castiglione R, Piani S, Gozzini L, Erspamer V (1981) Amino acid composition and sequence of dermorphin, a novel opiate-like peptide from the skin of Phyllomedusa sauvagei. Int J Pept Protein Res 17(3):275–283

Powell JT, Vine N, Crossman M (1992) On the accumulation of D-aspartate in elastin and other proteins of the ageing aorta. Atherosclerosis 97(2–3):201–208

Ritz S, Turzynski A, Schutz HW, Hollmann A, Rochholz G (1996) Identification of osteocalcin as a permanent aging constituent of the bone matrix: basis for an accurate age at death determination. Forensic Sci Int 77(1–2):13–26

Ritz-Timme S, Laumeier I, Collins MJ (2003) Aspartic acid racemization: evidence for marked longevity of elastin in human skin. Br J Dermatol 149(5):951–959

Roher AE, Lowenson JD, Clarke S, Wolkow C, Wang R, Cotter RJ, Reardon IM, Zurcher-Neely HA, Heinrikson RL, Ball MJ, Greenberg BD (1993) Structural alterations in the peptide backbone of beta-amyloid core protein may account for its deposition and stability in Alzheimer's disease. J Biol Chem 268(5):3072–3083

Tanaka N, Tanaka R, Tokuhara M, Kunugi S, Lee YF, Hamada D (2008) Amyloid fibril formation and chaperone-like activity of peptides from alphaA-crystallin. Biochemistry 47(9):2961–2967

Young GW, Hoofring SA, Mamula MJ, Doyle HA, Bunick GJ, Hu Y, Aswad DW (2005) Protein L-isoaspartyl methyltransferase catalyzes in vivo racemization of Aspartate-25 in mammalian histone H2B. J Biol Chem 280(28):26094–26098

Zhang J, Yip H, Katta V (2011) Identification of isomerization and racemization of aspartate in the Asp-Asp motifs of a therapeutic protein. Anal Biochem 410(2):234–243

Part V
D-Amino Acid Metabolizing Enzyme

Chapter 16
D-Amino Acid-Metabolizing Enzyme

Tohru Yoshimura

Abstract To understand the roles of D-amino acids, information about their metabolizing enzymes is indispensable. This section provides an overview of important enzymes related to the synthesis and degradation of D-amino acids.

Keywords D-Amino acid oxidase • D-Amino acid aminotransferase • D-Asparate oxidase • Serine racemase • D-Serine dehydratase

16.1 Overview

16.1.1 Enzymes Involved in D-Amino Acid Synthesis

Free D-amino acids are synthesized by three pathways. The first one involves *N*-acyl-amino acid racemase and *N*-acyl-D-amino acid deacylase (Fig. 16.1). These bacterial enzymes convert L-amino acids to D-amino acids after *N*-acylation. This procedure was applied in the optical resolution of the racemic amino acids (Tokuyama and Hatano 1996), but it is not known to be employed in vivo.

The second pathway for D-amino acid synthesis is transamination catalyzed by D-amino acid aminotransferase. The enzyme requires pyridoxal 5′-phosphate (PLP) and catalyzes the mutual transfer of amino groups between various D-amino acids and keto acids (Fig. 16.2). Only D-amino acids can serve as an amino donor; therefore, D-amino acids are not synthesized de novo. The D-amino acid aminotransferases were found in bacteria, and the enzymes from thermophilic *Bacillus* species are well studied (Tanizawa and Asano 1989; Tanizawa and Masu 1989). Recently, the enzyme was also identified in *Arabidopsis thaliana* (Funakoshi et al. 2008). Its primary structure and properties are similar to those of the bacterial enzyme. Details about D-amino acid aminotransferases are described in Chap. 22.

The third route for D-amino acid synthesis is racemization of L-amino acids catalyzed by amino acid racemases (Yoshimura and Goto 2008). Racemization of

T. Yoshimura (✉)
Department of Applied Molecular Biosciences, Graduate School of Bioagricultural Sciences, Nagoya University, Chikusa-Ku, Nagoya, Aichi 464-8601, Japan
e-mail: yosimura@agr.nagoya-u.ac.jp

T. Yoshimura et al. (eds.), *D-Amino Acids*, DOI 10.1007/978-4-431-56077-7_16

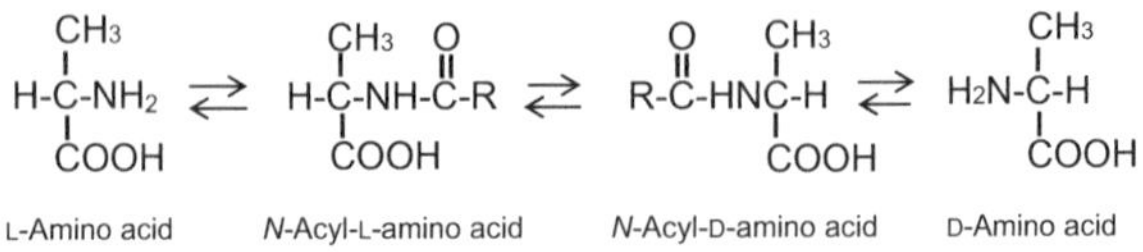

Fig. 16.1 Conversion of L-amino acid to D-amino acid via *N*-acyl-amino acid

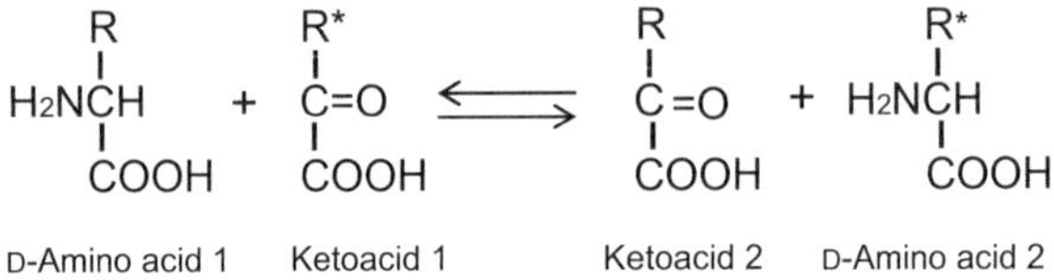

Fig. 16.2 Aminotransferase reaction

free amino acids is an entropy-driven reaction; thus, the reaction reaches the equilibrium of the 1:1 mixture between D- and L-amino acids. Two types of amino acid racemases are known: PLP-dependent enzymes and cofactor-independent enzymes. As described in previous chapters, D-serine and D-aspartate are physiologically important for mammals. In mammals, D-serine is synthesized by a PLP-dependent serine racemase (Wolosker and Mori 2012). The enzyme also catalyzes D- and L-serine dehydration to produce pyruvate and ammonia, which is characteristic of this enzyme. A sufficient amount of D-alanine is found in mammals (see Chap. 1); however, it is probably synthesized mainly by intestinal bacteria exploiting their alanine racemase, as the germ-free mice contain only low amount of D-alanine (Karakawa et al. 2013). The bacterial alanine racemase is a PLP-dependent enzyme, similarly to serine racemase, but its structure is completely different from that of the mammalian serine racemase, as shown in Fig. 16.3 (Yoshimura and Goto 2008). PLP enzymes are classified into seven evolutionarily differentiated groups, so-called fold types (B6 database: http://bioinformatics.unipr.it/cgi-bin/bioinformatics/B6db/home.pl). Enzymes belonging to different fold types are considered to have evolved from different ancestral proteins. Instead, enzymes in the same fold type evolved divergently; e.g., the fold-type I group contains enzymes with different functions, such as amino acid racemases, aminotransferases, and amino acid lyases, even though they have similar structures. Amino acid racemases are found in the fold-type groups I, II, and III, which suggests that there is no correlation between the fold type and reaction specificity. In this context, some PLP enzymes belonging to different fold types but catalyzing the same reactions evolved convergently. The mammalian serine racemase belongs to the fold-type II, and the bacterial alanine racemase belongs to the fold-type III (Fig. 16.3) (Yoshimura and Goto 2008). Serine racemase belonging to fold-type III was also reported. The enzyme was isolated from the vancomycin-resistant *Enterococcus gallinarum* strain BM4174, which owes its antibiotic resistance to this serine racemase (Arias et al. 1999).

Cofactor-independent racemases include glutamate racemase (Gallo and Knowles 1993), aspartate racemase (Yohda et al. 1991), proline racemase (Rudnick

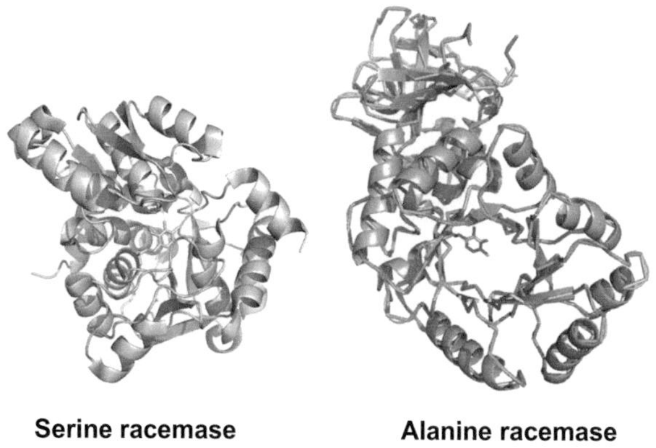

Fig. 16.3 Serine racemase and alanine racemase. Structures (monomer) of fold-type II serine racemase and fold-type III alanine racemase were depicted according to the PDB data of 1V71 and 1SFT, respectively

and Abeles 1975), and hydroxyproline epimerase (Adams and Frank 1980) and require cysteine residue(s) for the catalysis. The cofactor-independent aspartate racemase is described in more detail in Chap. 21.

16.1.2 Enzymes Involved in the D-Amino Acid Degradation

In mammals, D-serine is degraded by the FAD-dependent D-amino acid oxidase (see Chap. 19). D-Amino acid oxidases catalyze the oxidative deamination of various D-amino acids to the corresponding keto acids and ammonia (Fig. 16.4); e.g., D-serine is converted by the enzyme to hydroxypyruvate and ammonia. In the process of the D-amino acid oxidation, the enzyme catalyzes the substrate dehydrogenation to produce imino acid with the reduction of FAD. The imino acid is nonenzymatically hydrolyzed to α-keto acid and ammonia. The reduced FAD is oxidized by the molecular oxygen generating H_2O_2, and the enzyme returns to its initial state. The human D-amino acid oxidase is reviewed in more details in Chap. 19. D-Amino acid oxidase acts on various D-amino acids except for acidic amino acids. D-Aspartate is degraded in mammals by D-aspartate oxidase, a homologue of D-amino acid oxidase, as reviewed by Katane and Homma (2010). The mammalian D-aspartate oxidase is distributed mainly in the kidneys, liver, and brain. The mammalian enzyme acts on D-aspartate and *N*-methyl-D-aspartate, but its activity on D-glutamate is relatively low.

D-Serine is also degraded by D-serine dehydratase, which depends on PLP and catalyzes the dehydration of D-serine to pyruvate and ammonia (Fig. 16.5). There are two types of D-serine dehydratases, one found predominantly in bacteria (Dowhan and Snell 1970; Urusova et al. 2012) and the other present in both

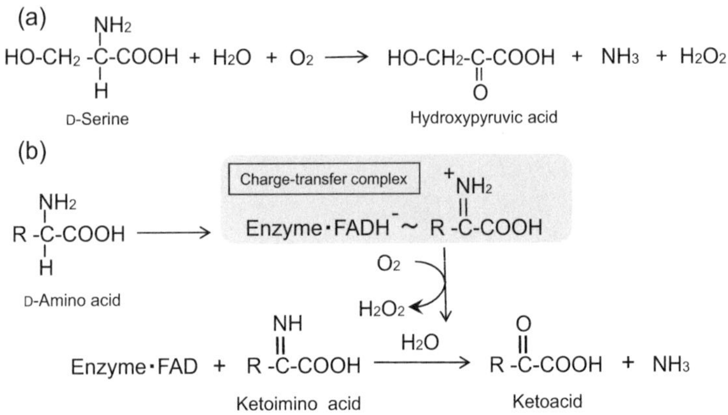

Fig. 16.4 Reaction catalyzed by D-amino acid oxidase. (**a**) D-Amino acid oxidase catalyzes oxidative deamination of D-serine to produce hydroxypyruvate, ammonia, and hydrogen peroxide. (**b**) Reaction mechanism of D-amino acid oxidase. Details are described in text

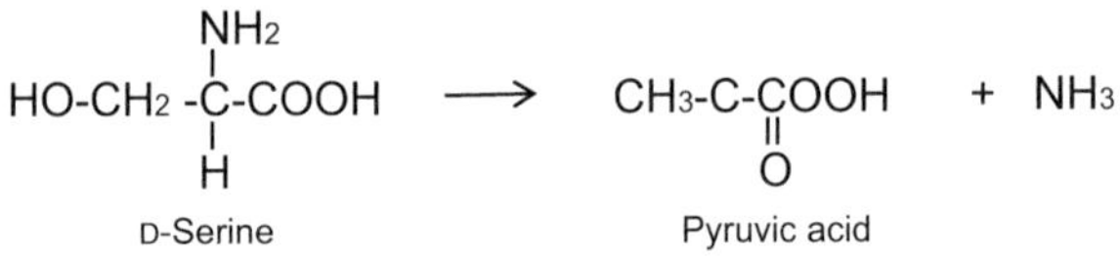

Fig. 16.5 Reaction catalyzed by D-serine dehydratase

eukaryotes and prokaryotes (Ito et al. 2008; Tanaka et al. 2008). Both types of enzymes depend on PLP, but their structures are completely different from each other. As described above, pyridoxal enzymes are classified into seven fold types. The bacterial D-serine dehydratase belongs to the fold-type II, and the other type belongs to the fold-type III. The fold-type III enzyme is unique among the PLP enzymes in its Zn dependency. The fold-type III enzyme of *Saccharomyces cerevisiae* acts predominantly on D-serine, and its strict specificity is exploited in the D-serine assay (Ito et al. 2007). Details about the fold-type III D-serine dehydratase are further described by Tanaka et al. in Chap. 20.

16.1.3 Enzymology of the Fold-Type II Serine Racemase

D-Serine, which is one of the most important D-amino acids in mammals, is synthesized by the fold-type II PLP-dependent serine racemase. Physiological roles and implications of the enzyme in neuronal diseases are described in other sections; here, the focus is on the enzymology of the enzyme. The fold-type II serine racemase was first purified from the rat brain (Wolosker et al. 1999a), and the

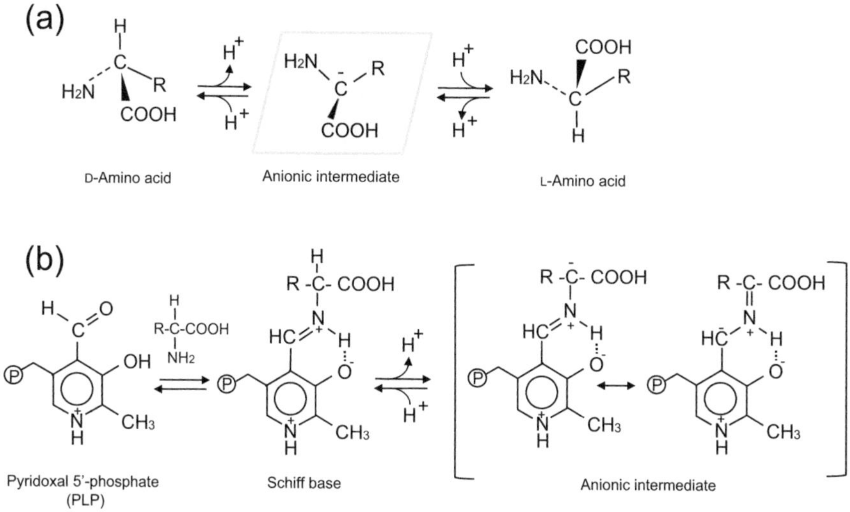

Fig. 16.6 Amino acid racemization catalyzed by PLP-dependent amino acid racemase

gene was cloned from mouse (Wolosker et al. 1999b). The enzyme was found in various organisms and purified from mouse, human (De Miranda et al. 2000), plants such as *A. thaliana* (Fujitani et al. 2006) and *Hordeum vulgare* (Fujitani et al. 2007), fission yeast *Schizosaccharomyces pombe* (Yamauchi et al. 2009), and cellular slime mold *Dictyostelium discoideum* (Ito et al. 2012). Recently, a fold-type II serine racemase was also found in hyperthermophilic archaeon *Pyrobaculum islandicum* (Ohnishi et al. 2008).

The reaction mechanism of amino acid racemases, including serine racemase, is apparently simple: abstraction of the α-hydrogen of a D- or L-amino acid as a proton to form a planar anionic intermediate (Fig. 16.6a). If the proton is added to the α-carbon of the intermediate on the side opposite to the side of proton abstraction, the antipodal amino acid is produced. However, the anionic intermediate is very unstable and hardly formed. In PLP-dependent enzymes, PLP stabilizes the intermediate by serving as an electron sink to delocalize the minus charge on the α-carbon of the intermediate formed by the α-proton abstraction (Fig. 16.6b) (reviewed by Toney 2011). The serine racemase reaction probably proceeds through the scheme shown in Fig. 16.7 (Goto et al. 2009; Ito et al. 2013). PLP binds to the enzyme through the formation of a Schiff base (internal Schiff base) (Fig. 16.7A). In the presence of D- or L-serine, PLP forms a Schiff base (external Schiff base) with the substrate through the transaldimination reaction (Fig. 16.7B, C). Abstraction of α-hydrogen results in the formation of an anionic intermediate (Fig. 16.7D). This is converted to the external Schiff base with an antipodal amino acid (Fig. 16.7C or B) after the proton addition to the anionic intermediate (Fig. 16.7D) on the side opposite to that of the proton abstraction. Transaldimination with the active-site lysine results in the formation of the product

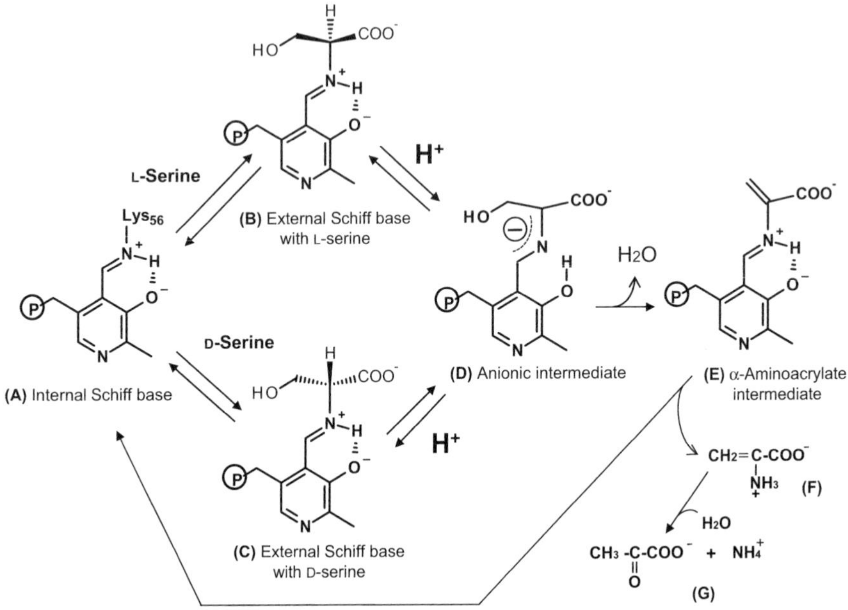

Fig. 16.7 Reaction mechanism of fold-type II serine racemase

amino acid and the internal Schiff base (Fig. 16.7A). The process is common among the fold-type II and III amino acid racemases. The fold-type II serine racemase catalyzes not only racemization but also dehydration (α, β-elimination) of D- and L-serine to produce pyruvate and ammonia (De Miranda et al. 2002; Foltyn et al. 2005). The enzyme differs from the fold-type III racemases in this respect. Dehydration occurs by the elimination of hydroxyl group from the anionic intermediate (Fig. 16.7D) to produce the Schiff base of α-aminoacrylate and PLP (Fig. 16.7E). Because the hydroxyl group is not an efficient leaving group, it is probably eliminated as water. Subsequently, α-aminoacrylate (Fig. 16.7F) is released from the enzyme upon the transaldimination and nonenzymatically hydrolyzed to pyruvate and ammonia (Fig. 16.7G). We studied the structure-function relationship of the fold-type II serine racemases of *S. pombe* and *D. discoideum*.

We carried out the X-ray crystallography of the *S. pombe* serine racemase (PDB ID; 1V71 and 1WTC) (Goto et al. 2009). The enzyme is a homodimer, whose subunit consists of two domains. The active-site structure, shown in Fig. 16.8, resembles that of the rat liver L-serine dehydratase, whose three-dimensional structure bound with *O*-methyl-L-serine was resolved (Yamada et al. 2003). The PLP-binding lysyl residue (Lys57) of serine racemase is situated between the domains of each monomer. In addition to Lys57, PLP is bound to the enzyme through hydrogen bonds formed with the amide group of glycine cluster, ^{183}Gly-Gly-Gly-Gly-Leu, which surrounds the phosphate group of PLP. Ser308 is situated close to the pyridine nitrogen of PLP and probably forms a hydrogen bond. The residue interacting with the pyridine nitrogen is important for the catalysis of PLP

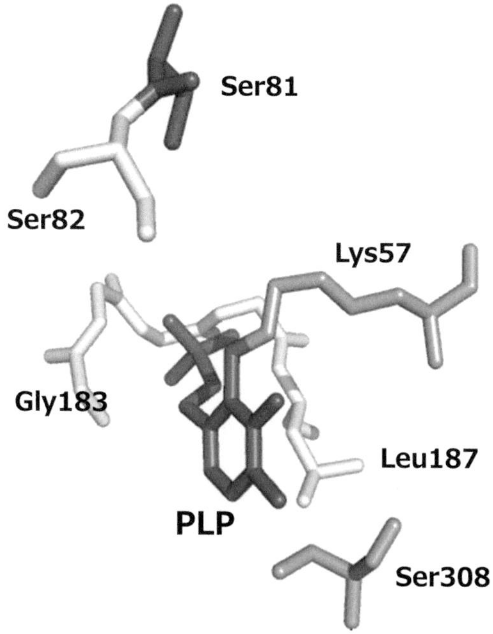

Fig. 16.8 Active-site structure of fold-type II serine racemase of *Schizosaccharomyces pombe*. The figure was written according to the PDB data of 1V71

enzymes, because it regulates electronic conditions of PLP. The active-site structure is common to the mouse and human serine racemases (Smith et al. 2010).

Both racemization and dehydration of D- and L-serine catalyzed by eukaryotic serine racemases are enhanced by divalent cations, such as Ca^{2+} and Mg^{2+} (Cook et al. 2002; De Miranda et al. 2002). The X-ray crystallography of the *S. pombe* enzyme revealed that Mg^{2+} is coordinated with the carboxyl group of Glu208 and Asp214 and the main-chain carbonyl group of Gly212. The fold-type II serine racemase is also enhanced by Mg-ATP, which binds to the domain and the subunit interface (Goto et al. 2009). In addition to the divalent cations and Mg-ATP, the racemase and dehydratase activities of the *D. discoideum* serine racemase are enhanced by Na^+ (Ito et al. 2012). The k_{cat}/K_m values of the L-serine dehydratase activity of the *D. discoideum* enzyme in the presence of 1 mM $MgCl_2$, 20 mM NaCl, or 1 mM Mg-ATP were enhanced 2.6-fold, 10.6-fold, or 42-fold of that without the cofactors, respectively. $MgCl_2$ and NaCl enhanced the k_{cat} value but did not affect the K_m value significantly. In contrast, Mg-ATP increased the k_{cat} value and decreased the K_m value. The Mg^{2+} and Na^+ concentrations required for the half-maximal activation were 1.2 μM and 2.2 mM, respectively. In the *D. discoideum* enzyme, Na^+ and the divalent cations probably bind to the same binding site (Ito et al. 2012).

We studied the catalytic residue(s) abstracting and adding α-hydrogen of D- and L-serine. These residues should be located near PLP and serve as the acid-base catalysts. According to the computer graphics models of the *S. pombe* enzyme with L- and D-serine, PLP-binding Lys57 and Ser82 were proposed as the candidates for the catalytic residues (Goto et al. 2009). Lys57 and Ser82, which are conserved among serine racemases, are situated on the opposite sides of PLP. Ser81 was suggested as the binding site for the α-carboxyl group of both D- and L-serine. We determined the role of these residues using the *D. discoideum* enzyme. We carried out site-directed mutagenesis of Ser80, Ser81, and Lys56 residues, which correspond to Ser81, Ser82, and Lys57 of the *S. pombe* enzyme, respectively (Ito et al. 2013). Mutation of Ser80 of the *D. discoideum* enzyme to an alanyl residue abolished both racemization and dehydration activities against D- and L-serine. In contrast, mutation of Ser80 to a cysteinyl residue retained both activities, suggesting that Ser80 is a binding site of the carboxyl group of D- and L-serine, as expected from the X-ray crystallography. Conversion of Ser81 of the *D. discoideum* enzyme to an alanyl residue abolished the racemase and D-serine dehydratase activities but retained the L-serine dehydratase activity. These results suggested that Ser81 serves as a general acid-base catalyst for the abstraction of the α-hydrogen from D-serine and for the addition of α-hydrogen to the anionic intermediate to produce D-serine. The results also suggested the presence of another catalytic residue for the abstraction and addition of the α-hydrogen of L-serine. We expected that Lys56 would be such catalytic residue, and its mutation to an alanyl residue would abolish the racemase activity and the L-serine dehydrase activity but retain the D-serine dehydrase activity. However, the K56A mutant serine racemase lost all activities including the D-serine dehydrase activity. Lys56 anchors PLP and is responsible for the transaldimination, which is probably why the mutation of Lys56 abolished all of the enzyme's activities. The fold-type II serine racemase reaction probably proceeds through a two-base mechanism, in which two different catalytic residues, i.e., seryl and lysyl residues, abstract and add α-hydrogen of D- and L-serine, respectively. The two-base mechanism is also employed in the fold-type III amino acid racemases, even though their structure differs from that of the fold-type II serine racemases. For example, alanine racemase reaction of the *Geobacillus stearothermophilus* enzyme was reported to proceed through the two-base mechanism, where Lys39 and Tyr265' catalyze the abstraction and addition of α-hydrogen of D- and L-alanine, respectively (Sun and Toney 1999; Watanabe et al. 2002).

As described previously, the eukaryotic serine racemase catalyzes both racemization and dehydration (α,β-elimination) of D- and L-serine to produce pyruvate and ammonia. The dehydration activity of serine racemase was suggested to contribute to the D-serine homeostasis, especially in the brain area lacking D-amino acid oxidase that is principally responsible for the D-serine degradation in the mammalian brain (Foltyn et al. 2005). Efficiencies of the D- and L-serine racemization catalyzed by serine racemase were approximately the same, whereas the dehydration activity of D-serine was lower than that of L-serine. In the case of the mouse enzyme (Strísovský et al. 2005), k_{cat} and K_m values in the racemization of L-serine as a substrate were 45.5 ± 0.5 min^{-1} and 3.8 ± 0.1 mM, and those of D-serine were 113 ± 3 min^{-1} and 14.5 ± 1.1 mM, respectively. The k_{cat} and K_m values for L-serine in the dehydration reaction were 81.3 ± 2.8 min^{-1} and 4.0 ± 0.5 mM, and those for

D-serine were $8.8 \pm 0.2\ \mathrm{min}^{-1}$ and 3.2 ± 0.3 mM, respectively. In the racemization, ratio of the k_{cat}/K_m value for L-serine to that for D-serine was 1.5. In contrast, the ratio in the dehydrase reaction was 7.4. Such imbalance in the dehydration efficiency between D- and L-serine is a common feature of the fold-type II serine racemases. Two hypotheses can possibly explain the imbalance in the dehydration activity. One is that D-serine dehydration does not occur directly from D-serine but after conversion of D-serine to L-serine through racemization. The other is that the efficiency of the OH group elimination differs between D- and L-serine. We verified the first hypothesis by examining the isotope effect upon the D-serine dehydration in ^2H_2O. As described previously, racemization of amino acids proceeds through the formation of anionic intermediate and the following protonation. If the racemization is carried out in ^2H_2O with the D-serine as a substrate, a deuterion is probably incorporated into the Cα of the anionic intermediate instead of a proton to produce L-[α-^{2}H]-serine. Upon the dehydration of L-[α-^{2}H]-serine, deuterion is abstracted from the Cα. If the Cα-hydrogen abstraction and/or addition are rate-limiting step (s), deuterium isotope effect(s) would be expected upon the D-serine dehydration in ^2H_2O via racemization.

We prepared the L-[α-^{2}H]-serine using a mutant alanine racemase and D-serine dehydratase and performed a reaction with the *D. discoideum* serine racemase in H_2O. The ratio of k_{cat}/K_m values of L-[α-^{2}H]-serine dehydration was about three times lower than that of L-serine (L-[α-^{1}H]-serine) dehydration, suggesting that α-hydrogen abstraction is a rate-limiting step in the L-serine dehydration. In contrast, the ratio of k_{cat}/K_m values of the D-serine dehydration in H_2O to that in ^2H_2O was 1.8, and the value was corresponding to the ratio obtained with the L-serine as a substrate. These results disproved our first hypothesis that the D-serine dehydration proceeds through L-serine formed by racemization; instead, the serine racemase catalyzes the D-serine dehydration directly from D-serine. We then verified our second hypothesis that the efficiency of the dehydration depends on that of the OH group elimination by examining the catalytic efficiency of the dehydration of four threonine isomers. The *D. discoideum* serine racemase exhibited the highest dehydrase activity with D-*allo*-threonine (2*R*, 3*R*) followed by L-threonine (2*S*, 3*R*). The catalytic efficiencies of D-*allo*-threonine and L-threonine were about 100 and 20 times of that of D-threonine (2*R*, 3*S*) and L-*allo*-threonine (2*S*, 3*S*), respectively. These results demonstrated that the threonine dehydration catalyzed by serine racemase is affected by the configuration of C3 much more than C2. These results sustained our second hypothesis that orientation of the hydroxyl group of D-serine and L-serine in the active site of the enzyme determines the efficiency of the dehydration reaction.

16.1.4 Aspartate Racemase

Similarly to D-serine, D-aspartate also performs important functions in mammals, as described in Part I. D-Aspartate is synthesized by a cofactor-independent aspartate racemase in some archaea and bacteria (see Chap. 21). Aspartate racemase depending on PLP was also found in the acidothermophilic archaeon *Thermoplasma acidophilum* (Long et al. 2001). The first enzyme responsible for D-aspartate biosynthesis described in eukaryotes was the PLP-dependent aspartate racemase of the bivalve mollusk *Scapharca broughtonii* (Shibata et al. 2003). Its gene was cloned and expressed in the *E. coli* cells. The deduced amino acid sequence of the enzyme showed 43–44 % similarity to that of the mammalian serine racemase (Abe et al. 2006). Similar PLP-dependent aspartate racemase, which is 41 % identical to the mammalian serine racemase, was found in *Aplysia californica* (Wang et al. 2011). In 2010, the mammalian aspartate racemase was identified by Kim et al. (2010). The mammalian enzyme was reported to be encoded by glutamate-oxaloacetate transaminase 1-like 1 (Got1l1) and is only 14 % identical to the enzyme of *A. californica.* Got1l1 is a homologue of the aspartate aminotransferase and was reported to catalyze transamination besides racemization. However, these results could not be reproduced (Tanaka-Hayashi et al. 2015; Matsuda et al. 2015). We showed that the D-aspartate contents in the Got1l1-KO mice were not significantly different from those of the wild-type mice in the testis and hippocampus. The purified Got1l1 expressed in mammalian cells showed only L-aspartate aminotransferase activity but not aspartate racemase activity (Tanaka-Hayashi et al. 2015). According to these findings, we think that Got1l1 is not the main enzyme responsible for D-aspartate synthesis, and there might be another pathway.

References

Abe K, Takahashi S, Muroki Y et al (2006) Cloning and expression of the pyridoxal 5'-phosphate-dependent aspartate racemase gene from the bivalve mollusk *Scapharca broughtonii* and characterization of the recombinant enzyme. J Biochem 139:235–244

Adams E, Frank L (1980) Metabolism of proline and the hydroxyprolines. Annu Rev Biochem 49:1005–1061

Arias CA, Martín-Martinez M, Blundell TL et al (1999) Characterization and modelling of VanT: a novel, membrane-bound, serine racemase from vancomycin-resistant *Enterococcus gallinarum* BM4174. Mol Microbiol 31:1653–1664

Cook SP, Galve-Roperh I, Martínez del Pozo A et al (2002) Direct calcium binding results in activation of brain serine racemase. J Biol Chem 277:27782–27792

De Miranda J, Santoro A, Engelender S et al (2000) Human serine racemase: moleular cloning, genomic organization and functional analysis. Gene 256:183–188

De Miranda J, Panizzutti R, Foltyn VN et al (2002) Cofactors of serine racemase that physiologically stimulate the synthesis of the *N*-methyl-D-aspartate (NMDA) receptor coagonist D-serine. Proc Natl Acad Sci U S A 99:14542–14547

Dowhan W Jr, Snell EE (1970) D-Serine dehydratase from *Escherichia coli*. II. Analytical studies and subunit structure. J Biol Chem 245:4618–4628

Foltyn VN, Bendikov I, De Miranda J et al (2005) Serine racemase modulates intracellular D-serine levels through an α, β-elimination activity. J Biol Chem 280:1754–1763

Fujitani Y, Nakajima N, Ishihara K et al (2006) Molecular and biochemical characterization of a serine racemase from *Arabidopsis thaliana*. Phytochemistry 67:668–674

Fujitani Y, Horiuchi T, Ito K et al (2007) Serine racemases from barley, *Hordeum vulgare* L., and other plant species represent a distinct eukaryotic group: gene cloning and recombinant protein characterization. Phytochemistry 68:1530–1536

Funakoshi M, Sekine M, Katane M et al (2008) Cloning and functional characterization of *Arabidopsis thaliana* D-amino acid aminotransferase - D-aspartate behavior during germination. FEBS J 275:1188–1200

Gallo KA, Knowles JR (1993) Purification, cloning, and cofactor independence of glutamate racemase from Lactobacillus. Biochemistry 32:3981–3990

Goto M, Yamauchi T, Kamiya N et al (2009) Crystal structure of a homolog of mammalian serine racemase from *Schizosaccharomyces pombe*. J Biol Chem 284:25944–25952

Ito T, Takahashi K, Naka T et al (2007) Enzymatic assay of D-serine using D-serine dehydratase from *Saccharomyces cerevisiae*. Anal Biochem 37:167–172

Ito T, Hemmi H, Kataoka K et al (2008) A novel zinc-dependent D-serine dehydratase from *Saccharomyces cerevisiae*. Biochem J 409:399–406

Ito T, Murase H, Maekawa M et al (2012) Metal ion dependency of serine racemase from *Dictyostelium discoideum*. Amino Acids 43:1567–1576

Ito T, Maekawa M, Hayashi S et al (2013) Catalytic mechanism of serine racemase from *Dictyostelium discoideum*. Amino Acids 44:1073–1084

Karakawa S, Miyoshi Y, Konno R et al (2013) Two-dimensional high-performance liquid chromatographic determination of day-night variation of D-alanine in mammals and factors controlling the circadian changes. Anal Bioanal Chem 405:8083–8091

Katane M, Homma H (2010) D-Aspartate oxidase: the sole catabolic enzyme acting on free D-aspartate in mammals. Chem Biodivers 7:1435–1449

Kim PM, Duan X, Huang AS et al (2010) Aspartate racemase, generating neuronal D-aspartate, regulates adult neurogenesis. Proc Natl Acad Sci U S A 107:3175–3179

Long Z, Lee JA, Okamoto T et al (2001) Occurrence of D-amino acids and a pyridoxal 5'-phosphate-dependent aspartate racemase in the acidothermophilic archaeon, *Thermoplasma acidophilum*. Biochem Biophys Res Commun 281:317–321

Matsuda S, Katane M, Maeda K et al (2015) Biosynthesis of D-aspartate in mammals: the rat and human homologs of mouse aspartate racemase are not responsible for the biosynthesis of D-aspartate. Amino Acids 47:975–985

Ohnishi M, Saito M, Wakabayashi S et al (2008) Purification and characterization of serine racemase from a hyperthermophilic archaeon, *Pyrobaculum islandicum*. J Bacteriol 190:1359–1365

Rudnick G, Abeles RH (1975) Reaction mechanism and structure of the active site of proline racemase. Biochemistry 14:4515–4522

Shibata K, Watanabe T, Yoshikawa H et al (2003) Purification and characterization of aspartate racemase from the bivalve mollusk *Scapharca broughtonii*. Comp Biochem Physiol B Biochem Mol Biol 134:307–314

Smith MA, Mack V, Ebneth A et al (2010) The structure of mammalian serine racemase: evidence for conformational changes upon inhibitor binding. J Biol Chem 285:12873–12881

Strísovský K, Jirásková J, Mikulová A et al (2005) Dual substrate and reaction specificity in mouse serine racemase: identification of high-affinity dicarboxylate substrate and inhibitors and analysis of the β-eliminase activity. Biochemistry 44:13091–13100

Sun S, Toney MD (1999) Evidence for a two-base mechanism involving tyrosine-265 from arginine-219 mutants of alanine racemase. Biochemistry 38:4058–4065

Tanaka H, Yamamoto A, Ishida T et al (2008) D-Serine dehydratase from chicken kidney: a vertebral homologue of the cryptic enzyme from *Burkholderia cepacia*. J Biochem 143:49–57

Tanaka-Hayashi A, Hayashi S, Inoue R et al (2015) Is D-aspartate produced by glutamic-oxaloacetic transaminase-1 like 1 (Got1l1): a putative aspartate racemase? Amino Acids 47:79–86

Tanizawa K, Masu Y, Asano S et al (1989a) Thermostable D-amino acid aminotransferase from a thermophilic *Bacillus* species. Purification, characterization, and active site sequence determination. J Biol Chem 264:2445–2449

Tanizawa K, Asano S, Masu Y et al (1989b) The primary structure of thermostable D-amino acid aminotransferase from a thermophilic *Bacillus* species and its correlation with L-amino acid aminotransferases. J Biol Chem 264:2450–2454

Tokuyama S, Hatano K (1996) Overexpression of the gene for *N*-acylamino acid racemase from *Amycolatopsis* sp. TS-1-60 in *Escherichia coli* and continuous produciton of optically active methionine by a bioreactor. Appl Microbiol Biotechnol 44:774–777

Toney MD (2011) Controlling reaction specificity in pyridoxal phosphate enzymes. Biochim Biophys Acta 1814:1407–1418

Urusova DV, Isupov MN, Antonyuk S et al (2012) Crystal structure of D-serine dehydratase from *Escherichia coli*. Biochim Biophys Acta 1824:422–432

Wang L, Ota N, Romanova EV et al (2011) A novel pyridoxal 5'-phosphate-dependent amino acid racemase in the *Aplysia californica* central nervous system. J Biol Chem 286:13765–13774

Watanabe A, Yoshimura T, Mikami B et al (2002) Reaction mechanism of alanine racemase from *Bacillus stearothermophilus*: x-ray crystallographic studies of the enzyme bound with *N*-(5'-phosphopyridoxyl)alanine. J Biol Chem 277:19166–19172

Wolosker H, Mori H (2012) Serine racemase: an unconventional enzyme for an unconventional transmitter. Amino Acids 43:1895–1904

Wolosker H, Sheth KN, Takahashi M et al (1999a) Purification of serine racemase: biosynthesis of the neuromodulator D-serine. Proc Natl Acad Sci U S A 96:721–725

Wolosker H, Blackshaw S, Snyder SH (1999b) Serine racemase: a glial enzyme synthesizing D-serine to regulate glutamate-*N*-methyl-D-aspartate neurotransmission. Proc Natl Acad Sci U S A 96:13409–13414

Yamada T, Komoto J, Takata Y et al (2003) Crystal structure of serine dehydratase from rat liver. Biochemistry 42:12854–12865

Yamauchi T, Goto M, Wu HY et al (2009) Serine racemase with catalytically active lysinoalanyl residue. J Biochem 145:421–424

Yohda M, Okada H, Kumagai H (1991) Molecular cloning and nucleotide sequencing of the aspartate racemase gene from lactic acid bacteria *Streptococcus thermophilus*. Biochim Biophys Acta 1089:234–240

Yoshimura T, Goto M (2008) D-Amino acids in the brain: structure and function of pyridoxal phosphate-dependent amino acid racemases. FEBS J 275:3527–3537

Chapter 17
Alanine Racemase and D-Amino Acid Oxidase in Aquatic Animals

Naoko Yoshikawa, Mohammed Golam Sarower, and Hiroki Abe

Abstract Various aquatic invertebrates including crustaceans and some bivalve mollusks contain a copious amount of free D-alanine in their tissues. In these invertebrates, D-alanine is largely accumulated with the L-form under a high-salinity environment for maintaining cell volume. D-Alanine is a major osmolyte in these invertebrate tissues, together with glycine, L-alanine, L-glutamine, and L-proline, and is responsible for intracellular isosmotic regulation. Alanine racemase, catalyzing the interconversion of D- and L-amino acids, has been isolated to homogeneity from the muscle of black tiger prawn, and its cDNA has been cloned from the muscle and hepatopancreas of kuruma prawn *Marsupenaeus japonicus*. This is the first time cloning was achieved in eukaryotes other than yeast. Common carp *Cyprinus carpio* is an omnivorous fish that often feeds on crustaceans and mollusks containing free D-alanine, but that contains only a trace amount of D-alanine in their tissues. A cDNA of D-amino acid oxidase has been cloned from carp hepatopancreas. Carp D-amino acid oxidase is an inducible enzyme. The activity and mRNA levels of D-amino acid oxidase increase in the intestine and are followed by the hepatopancreas and the kidney. Carp D-amino acid oxidase is structurally similar to the porcine kidney enzyme but is enzymatically similar to the yeast enzyme. D-Amino acid oxidase is thought to be an important enzyme responsible for the efficient utilization of the carbon skeleton of food-derived D-alanine.

Keywords Alanine racemase • Carp • Crustacean • D-alanine • D-amino acid oxidase

---

N. Yoshikawa
Department of Materials and Life Science, Shizuoka Institute of Science and Technology, Fukuroi, Shizuoka 437-8555, Japan

M.G. Sarower
Fisheries and Marine Resource Technology Discipline, Khulna University, Khulna 9208, Bangladesh

H. Abe (✉)
Research Institute of Seafood Biochemistry, Suginami, Tokyo 167-0023, Japan
e-mail: hirokiabe@kxd.biglobe.ne.jp

T. Yoshimura et al. (eds.), *D-Amino Acids*, DOI 10.1007/978-4-431-56077-7_17

17.1 Introduction

In aquatic invertebrates, D-amino acid research dates back to the end of the 1970s. D'Aniello and Giuditta (1977) discovered over 10 μmol/g wet weight of free D-aspartate in the brain of octopus *Octopus vulgaris*. Free D-alanine has also been found in several crustaceans (D'Aniello and Giuditta 1980). Thereafter, these free D-amino acids have been reported to be found in several tissues of crustaceans and bivalve mollusks. In particular, free D-alanine is found in a large number of aquatic invertebrate species. Bivalves belonging to the subclass Heterodonta, such as the small brackish water clam *Corbicula japonica*, short-necked clam *Tapes philippinarum*, and hard clam *Meretrix lamarckii*, contain over 25 μmol/g of D-alanine in the mantle and foot muscles (Matsushima et al. 1984). Otter shell *Tresus keenae* has approximately 50 μmol/g of D-alanine in the siphon, and the ratio of D-alanine to total alanine reaches over 80 % (Okuma et al. 1998). By contrast, bivalves belonging to the subclass Pteriomorphia, such as scallop *Patinopecten yessoensis*, oyster *Crassostrea gigas*, and mussel *Mytilus edulis*, contain almost no D-alanine (Okuma et al. 1998), whereas *M. edulis* has D-aspartate in foot muscle (Felbeck and Wiley 1987). In addition, it has been reported that D-aspartate exists in these molluscan species, including ark shell *Scapharca broughtonii* (Watanabe et al. 1998), and aspartate racemase has been isolated from the foot muscle (Shibata et al. 2003).

In crustaceans, D-alanine has been found in all species and tissues and, depending on the species, reaches 3–16 μmol/g in the muscle and hepatopancreas (Okuma et al. 1995). The percentage of D-alanine to total alanine reaches approximately 30–60 % in crustaceans. When red swamp crayfish *Procambarus clarkii* acclimates from freshwater to 75 % seawater, total free amino acid increased over twofold in the tail muscle (Fig. 17.1). The amino acids most responsible for the increase are D- and L-alanine followed by glycine, L-glutamine, and L-proline. The increase of D- and L-alanine is approximately sevenfold in 75 % seawater, and the percentage of D-alanine to total alanine elevates from 38 % in freshwater to 48 % in 75 % seawater (Okuma and Abe 1994; Fujimori and Abe 2002). The increase of D- and L-alanine is also found in cardiac muscle, hepatopancreas, and nervous tissues with the exception of gill and hemolymph. Further observations have been made in studies of other invertebrate species. One such study has examined an estuary inhabitant kuruma prawn *Marsupenaeus japonicus* transferred from 100 % seawater to 50 and 150 % seawater. After acclimating from 50 to 150 % seawater, the prawn shows a twofold increase of total free amino acids in muscle with the highest increase shown in glycine followed by D- and L-alanine (Abe et al. 2005). The accumulation of D-alanine, along with the hypersalinity acclimation, has also been observed in the hard clam *Meretrix lusoria*, in which only D-alanine and L-alanine increase in the adductor muscle during acclimation to 150 % seawater (Okuma et al. 1998). From these data, one can conclude that D-alanine, together with the L-form, is an effective and compatible osmolyte responsible for intracellular isosmotic and cell volume regulation in invertebrates. The D-form amino acid does not

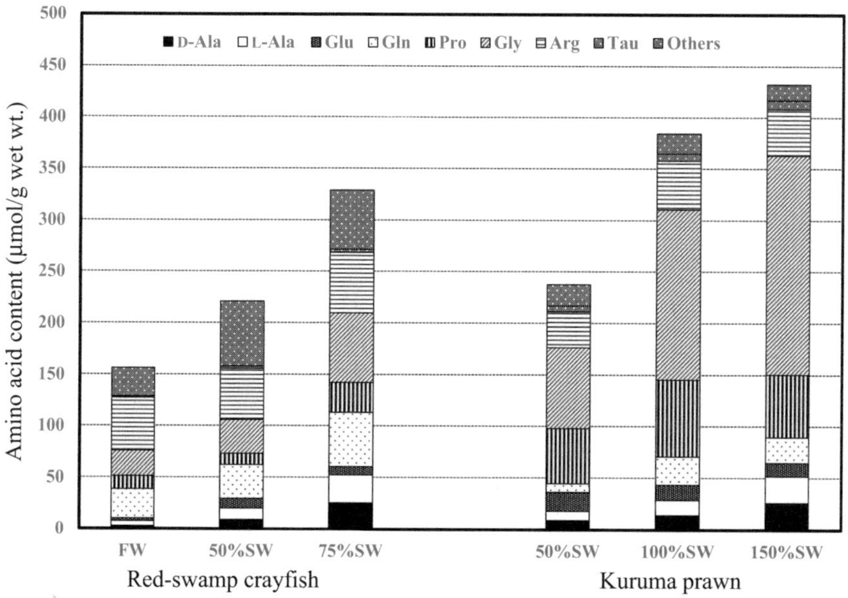

Fig. 17.1 Changes of free amino acid contents in crustacean species, freshwater red swamp crayfish and brackish water kuruma prawn, during the acclimation to different salinities. *FW* freshwater, *SW* seawater

interact with existing proteins that are composed of L-amino acids ("invisible" from proteins). Thus, D-alanine is not harmful to cell environments, even if it accumulates in large amounts in the cell, and may be an ideal osmolyte in cells tightly packed with various proteins and small molecules.

From these data, we are interested in the evolutional significance of the D-alanine synthesizing enzyme, alanine racemase, in these aquatic invertebrates. We are also interested in fish that feed on invertebrates that contain D-alanine, as fish have only trace amounts of D-alanine in all tissues.

17.2 Alanine Racemase in Crustaceans

17.2.1 Distribution of Alanine Racemase in Aquatic Invertebrates

To clarify the physiological roles of free D-amino acids in aquatic invertebrates, it is necessary to investigate the origin of D-amino acids in these animals. Studies on the origin of D-amino acids in bivalve mollusks have put forth hypotheses that bivalves have incorporated and accumulated exogenous D-amino acids from external sources, such as seawater (Bada 1982) or symbiotic bacteria (Felbeck 1985).

However, the activity of alanine racemase has been detected in the muscle of *C. japonica* (Matsushima et al. 1984) as well.

Alanine racemase (ARase, EC 5.1.1.1), a pyridoxal 5′-phosphate (PLP)-dependent enzyme, catalyzes the specific interconversion of D- and L-alanine. Several D-amino acids are essential components of the peptidoglycan layer of the cell wall structure in bacteria, and the racemization of alanine is important for bacteria survival. For this reason, bacterial ARase has been regarded as an attractive target for novel antibacterial drugs and has been studied considerably. Furthermore, racemases have long been considered to exist in eubacteria only. However, in addition to *C. japonica*, ARase has also been found in eukaryotes such as fungus *Tolypocladium niveum* (Hoffmann et al. 1994), fission yeast *Schizosaccharomyces pombe* (Uo et al. 2001a), and several aquatic invertebrates. In animal studies, ARase has been partially purified for the first time from the muscle of black tiger prawn *Penaeus monodon* (Fujita et al. 1997) and purified to homogeneity from the tail muscle of crayfish *Procambarus clarkii* (Shibata et al. 2000), the mantle muscle of *C. japonica* (Nomura et al. 2001), and the hepatopancreas of *P. monodon* (Uo et al. 2001b).

17.2.2 *Properties of ARase Isolated from the Muscle of* P. Monodon

We purified ARase to homogeneity from the tail muscle of *P. monodon* and investigated its structural and enzymatic properties (Yoshikawa et al. 2002). The molecular mass of the purified enzyme is approximately 44 kDa using SDS-PAGE and 90 kDa using gel filtration, indicating that ARase exists as a homodimer in prawn muscle. It has been reported that bacterial ARases can be a 40 kDa monomer or a dimer that has identical subunits with molecular masses of approximately 40 kDa each (Yoshimura and Soda 1994). Thus, the molecular size of the monomer is almost identical for all enzymes isolated from both prokaryotes and eukaryotes, with the exception of a 58 kDa monomeric enzyme from *P. clarkii*.

The apparent K_m values of ARase toward L- and D-alanine are 167 mM and 179 mM, respectively, as depicted on Lineweaver-Burk plots (Table 17.1). Although the V_{max} and k_{cat} of the prawn enzyme are high in both directions, the catalytic efficiency (k_{cat}/K_m) is slightly higher in the L to D directions. Thus, the equilibrium constant (K_{eq}) is calculated to be 0.84 from the Haldane equation (k_{cat}/K_m for D-alanine)/(k_{cat}/K_m for L-alanine), and the value is slightly below the theoretical value for racemization (i.e., 1.0). These data indicate that the prawn ARase catalyzes the equilibrium between D- and L-alanine efficiently, regardless of its low content in muscle and high K_m values, and efficaciously regulates intracellular concentrations of D- and L-alanine in muscle. The optimal pH of the prawn enzyme is approximately 9.5 in the direction of L to D and 10 in the D to L

Table 17.1 Properties of alanine racemase isolated from aquatic crustaceans

Species and tissue	Molecular mass (kDa)	Reaction direction	V_{max} (μmol/min/mg)	K_m (mM)	k_{cat} (s^{-1})	k_{cat}/K_m (s^{-1}/mM)	Reference
Black tiger prawn	46	L→D	3502	167	2568	15.38	Yoshikawa et al. (2002)
(Tail muscle)	(Dimer)	D→L	3155	179	2314	12.93	
Black tiger prawn	41	L→D	460	150			Uo et al. (2001b)
(Hepatopancreas)	(Dimer)	D→L	94	24			
Crayfish	58	L→D	7763	171	7504	43.88	Shibata et al. (2003)
(Tail muscle)	(Monomer)	D→L	3386	73.5	3273	44.53	

direction. Moreover, the enzyme is highly specific to D- and L-alanine and does not catalyze the racemization of other amino acids.

ARases require PLP as a cofactor to form a Schiff base with its substrate (Yoshimura and Soda 1994). Although the purified prawn enzyme is slightly activated by the addition of PLP, the enzyme also exhibits activity even in the absence of PLP. This activity in the absence of PLP has been detected throughout the purification procedure. However, the purified enzyme exhibits absorption at 420 nm, corresponding to the Schiff base complex of PLP with an active site lysine. Moreover, the enzyme is strongly inhibited by hydroxylamine and aminooxyacetic acid, which are known to be strong inhibitors of PLP-dependent enzymes. These data may indicate a strong binding of PLP to the enzyme protein, most likely in the form of a covalent bond via an aldimine linkage to lysine39 (Shaw et al. 1997).

17.2.3 Structural Analysis of Prawn ARase

To clarify the structure of the prawn ARase, we have determined the partial amino acid sequences of the isolated enzyme. Seven peptide fragments have been obtained from the purified enzyme (Yoshikawa et al. 2002). Of these fragments, three peptides are homologous to bacterial ARase, whereas other fragments show no similarity. Based on these homological sequences, we have designed degenerate primers for rapid amplification of cDNA ends (RACE) and have used them to clone the cDNA of prawn ARase. As a result, we have isolated cDNA clones encoding ARase from the muscle and hepatopancreas of kuruma prawn *M. japonicus* for the first time in animals (Yoshikawa et al. 2009). We have expressed its recombinant protein in *Escherichia coli* as well.

The nucleotide sequence of 1798 base pairs (bp) includes a 5′-noncoding region of 459 bp and an open reading frame of 1263 bp that is followed by 62 bp of a 3′-noncoding region and poly(A)$^+$ tail. The amino acid sequence deduced from the nucleotide sequence encoding *M. japonicus* ARase contains 421 residues with a predicted molecular mass of 45,770 Da for the subunit of the dimeric enzyme. The deduced amino acid sequence of the *M. japonicus* enzyme is similar, although not identical, to the six partial amino acid sequences of the *P. monodon* enzyme. Initially, *M. japonicus* had been classified into the same genus as *P. monodon*, and it had been predicted that ARase of these close relatives would show high homology in primary structure.

By contrast, the complete amino acid sequence of *M. japonicus* ARase shows less than 30 % amino acid identity to ARases in yeast or bacteria. Studies of the ARase in *Geobacillus stearothermophilus* using X-ray crystallography (Shaw et al. 1997) and site-directed mutagenesis (Watanabe et al. 1999a, b) reveal that two residues of ARase are essential for catalysis. It is suggested that lysine39, bound to PLP, removes the α-proton from D-alanine and that tyrosine265 removes the α-hydrogen from L-alanine. The lysine and proximal residues are known to be well conserved in all bacterial ARases. In the *M. japonicus* sequence, these active sites and residues that interact with PLP are also conserved. These data indicate that the ARase gene has been conserved from bacteria to *M. japonicus* throughout evolution.

Splice variants derived from the ARase-encoding mRNA have been obtained from the muscle and hepatopancreas of *M. japonicus*. These splice variants in the translational region as well as a partial deletion in the 5′-noncoding region have been detected in the process of cDNA cloning. These products may have arisen to regulate the process of gene transcription and expression. To elucidate the function of these products, further investigation is required.

17.2.4 Additional Physiological Roles of D-Alanine and ARase

In *M. japonicus*, the ARase gene is detected in both muscle and hepatopancreas. The mRNA expression level and the activity of ARase in hepatopancreas are higher than in muscle. By contrast, the amount of D-alanine is lower in hepatopancreas than in muscle (Yoshikawa et al. 2009). Thus, it is possible that the physiological functions of D-alanine and ARase differ in each tissue. In hepatopancreas, the amount of D-alanine and the activity of ARase increase during acclimation from 100 to 150 % seawater. The expression level of mRNA encoding ARase also increases under the same conditions. These data indicate that the increase of ARase activity results from the upregulation of enzyme expression in response to high-salinity stress.

We have determined ARase activity in the muscle and hepatopancreas of *M. japonicus* before and after molting (Fig. 17.2). The activity after molting increases twofold in muscle for both the L to D and D to L directions. By contrast, there was no marked increase in hepatopancreas (Yoshikawa et al. 2011). It is

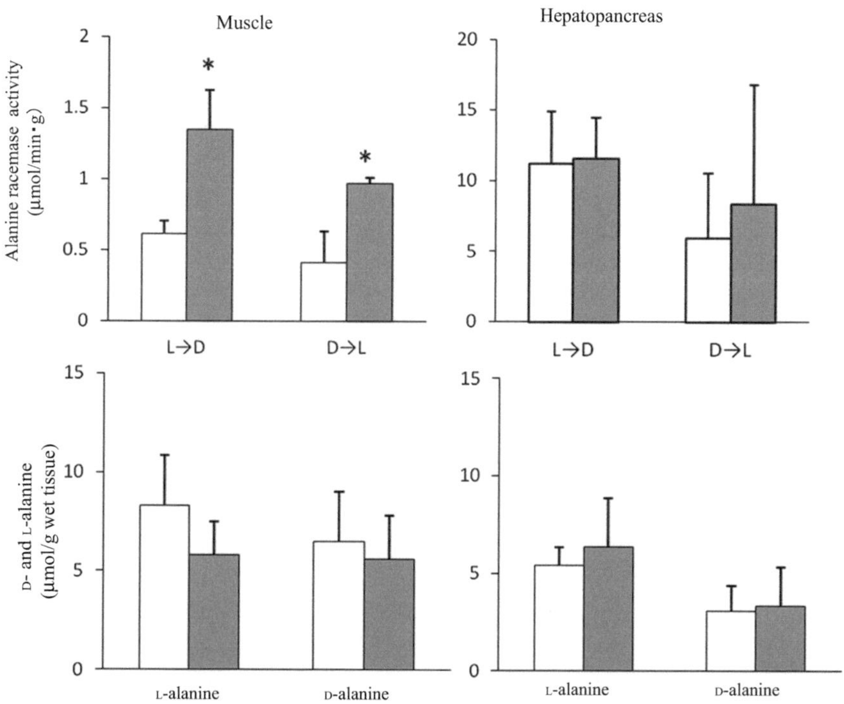

Fig. 17.2 Effects of ecdysis on alanine racemase activities and D- and L-alanine contents in the muscle and hepatopancreas of kuruma prawn. Closed columns show the case for prawns after molting and open columns before molting. Each column and bar represents means and SD, respectively, for five prawns (*$p < 0.05$)

considered that immediately after molting, the soft carapace may be very sensitive to changes in environmental salinity compared with the effects of salinity on the hard carapace during the inter-molting period. In euryhaline macrurans, such as *M. japonicus*, it is possible that a large amount of D-alanine is used for the maintenance of physiological conditions in the muscle during and immediately after molting until the shell hardens. However, the amount of D-alanine in the muscle is nearly identical before and after ecdysis. This contradiction leads us to speculation that the increase of ARase activity is a defense against the environmental salinity stress during the soft and salt permeable carapace stage. Moreover, we postulate that D-alanine produced by ARase in prawn muscle might be utilized to synthesize bound-form components, which are necessary for physiological changes during ecdysis.

From these data, it is clear that D-alanine plays essential roles in aquatic invertebrates and that alanine racemase is an indispensable enzyme conserved from the prokaryote era.

17.3 D-Amino Acid Oxidase in Fish

17.3.1 Distribution of D-Amino Acid Oxidase in Fish

Several fish species feed on juvenile and adult invertebrates, such as crustaceans and mollusks containing a large amount of D-alanine. In contrast to various invertebrates, however, fish contain only a trace amount of D-amino acids in all tissues (Kera et al. 1998). We have found high activities of the D-amino acid-catabolizing enzymes, D-amino acid oxidase (DAO, EC 1.4.3.3) and D-aspartate oxidase (DDO, EC 1.4.3.1), in several tissues of various fish species (Sarower et al. 2003a).

DAO is a typical flavoprotein that has been studied comprehensively in mammals and microbes (refer to Chap. 5.4 of this book). In mammals, DAO is abundant in the kidney, liver, intestine, and brain (Konno and Yasumura 1992). Similarly, in most fish, DAO is found mainly in the kidney, liver, and intestine, and in some fishes, it is also found in the gonads, stomach, pyloric caecum, and brain (Sarower et al. 2003a). In marine fish species, the enzyme activity is high in the kidney and liver and low in the intestine, with some exceptions. DAO activity, however, is much higher in these tissues than is DDO activity, with exceptions in the kidney of several marine fishes. In freshwater fish such as carp, rainbow trout, and landlocked masu salmon, high DAO activity is distributed in the kidney, liver, and intestine. Rainbow trout and masu salmon often eat insects and crustaceans that contain D-amino acids, especially D-alanine (Okuma et al. 1995). It is known that carp is an omnivorous fish that often feeds on crustaceans and bivalves.

Many fish species in the juvenile stage eat zooplankton and small crustaceans, but fishes such as sea bream, mackerel, sea perch, and flounder are fish eaters (piscivorous) in the adult stage and contain only low DAO activity. The other common DAO activity pattern is not found in fish species, and the distribution is highly species specific. However, it is probable that marine fishes feeding on invertebrate have higher DAO activity than piscivorous fish. Thus, for fishes containing high activities, especially in the intestine, DAO may eliminate D-amino acids before they are absorbed by the intestine. By contrast, high DAO activity in the kidney and liver may have a role in the reutilization of D-amino acids as the L-form via their corresponding 2-oxo acids.

An inverse relationship exists between the amount of D-amino acid and the activities of DAO and DDO in the liver and kidneys of rat, mouse, rabbit, and chicken (D'Aniello et al. 1993). Thus, DAO and DDO in fish may function effectively to eliminate D-amino acids from fish tissue.

17.3.2 Physiological Roles of DAO in Fish

The distribution of DAO activity in fish tissue suggests the inducible nature of DAO by fish food. Thus, we have investigated changes in DAO activity after the oral administration of D-amino acids to carp *Cyprinus carpio* (Sarower et al. 2003a). The 30-day oral administration of D-alanine to carp, which often feed on crustaceans and mollusks, increases DAO activity approximately 8-, 3-, and 1.5-fold in the intestine, hepatopancreas, and kidney, respectively. No increase has been found in the brain, an organ in which DAO baseline activity is low. In contrast to DAO, there is no increase in the activity of DDO or DAO in any organs after the ingestion of D-aspartate or D-glutamate. These data suggest that DAO is an inducible enzyme in fish and that D-alanine is a physiological substrate for the enzyme. The inducible nature of fish DAO may explain the species-specific distribution of DAO in fish. Carp DAO and DDO localize subcellularly in peroxisomes (Sarower et al. 2003a), similar to mammalian DAO.

To further confirm the inducibility of DAO in fish, a cDNA of DAO has been cloned from the hepatopancreas of carp. This 1294 bp clone contains an open reading frame of 1041 bp that corresponds to a predicted polypeptide of 347 amino acid residues (Sarower et al. 2003b). Oral administration of D-alanine to carp (5 μmol/g body weight · day) over 14 days has been found to increase both enzyme activity and mRNA levels in the hepatopancreas (Fig. 17.3a, b). The enzyme activity increases rapidly to a plateau after 3 days, and the level of mRNA increases over 7 days. The increases of DAO activity and mRNA level do not coincide, suggesting that the DAO induction may be regulated not only in transcription step but also in a specific step such as transformation step. In addition, tissue-specific mRNA expression and enzyme activity have been examined after 14 days of D-alanine ingestion. As seen in Fig. 17.3c and d, a strong hybridization signal has been detected in the intestine followed by hepatopancreas and kidney. These data clearly show that DAO is inducible in carp tissue with exogenous D-alanine. In mammals, DAO is not an inducible enzyme (Nagata et al. 1991) although DDO is inducible in mouse liver after the oral administration of D-aspartate (Yamada et al. 1989). Thus, it is likely that DAO in carp is not merely a detoxifying agent but an important enzyme responsible for the effective utilization of the carbon skeleton of exogenous D-amino acids in the intestine, hepatopancreas, and kidney. This may reflect the wider and higher distribution of D-alanine in carp food compared to the distribution of D-aspartate or D-glutamate.

17.3.3 Characteristics of Primary Structure

The predicted molecular mass of carp hepatopancreas DAO is 39 kDa, similar to the DAO molecular mass in mammals (38–39 kDa) and larger than the molecular mass in microorganisms (28–34 kDa). The amino acid sequence of carp DAO is highly homologous to DAO sequences in mammals but not to DAO sequences in microorganisms. Carp DAO is 62 %, 61 %, and 60 % identical to DAO in human, mouse, and porcine kidney, respectively, and 29 %, 26 %, 21 %, 21 %, and 19 % identical to DAO in *Streptomyces coelicolor*, *Rhodotorula gracilis*, *Schizosaccharomyces*

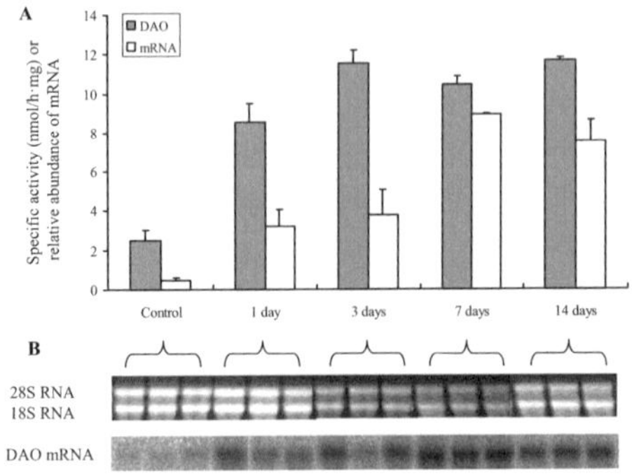

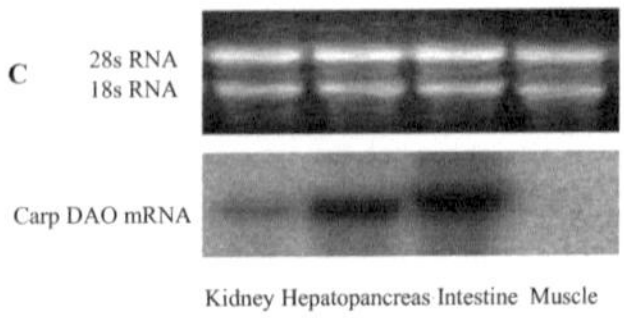

D

Relative abundance of mRNA (%)	5.0	35.5	59.5	0
DAO activity (nmol/h/mg)	3.7	7.7	8.0	ND

Fig. 17.3 Changes of enzyme activities and expression of mRNA of D-amino acid oxidase in carp tissues during D-alanine ingestion (5 μmol/g body weight·day) for 14 days. (**A**) D-Amino acid oxidase activity and relative mRNA expression in hepatopancreas of carp, (**B**) levels of mRNA of D-amino acid oxidase, (**C**) levels of mRNA in several tissues of carp after 14 days, (**D**) relative mRNA expression and activities of D-amino acid oxidase

pombe, *Trigonopsis variabilis*, and *Fusarium solani*, respectively (Sarower et al. 2003b).

The deduced amino acid sequence of carp DAO is very similar to the sequences of many DAOs from other organisms especially with regard to the six highly conserved regions proposed by Faotto et al. (1995). Region I contains the consensus sequence GXGXXG in all DAOs. In particular, the sequence is GAGVIG in DAOs from all animal sources, including carp. Three residues in the active site, Tyr224, Tyr228, and Arg283, have also been conserved in the carp enzyme. Regions II, IV, and V contain these essential residues (Faotto et al. 1995). The last conserved terminal sequence Ser-(Lys/His/Arg)-Leu in region VI is thought to represent the type 1 peroxisomal signal (Subramani 1993). The center residue of the terminal sequence is histidine in mammalian DAO and arginine in the carp enzyme.

The conservation of Arg283, Tyr224, and Tyr228 in carp and mammalian DAO suggests that these amino acids are responsible for the same functions. The high DAO activity of protein expressed in *E. coli* demonstrates that this cDNA encodes a functional DAO and is expressed as a soluble and active holoenzyme. These data indicate the conservation of the DAO gene from microbes to mammals throughout evolution.

17.3.4 Kinetic Properties of Carp DAO

Carp hepatopancreas DAO has been overexpressed in *E. coli* cells, the recombinant DAO has been purified to homogeneity, and the catalytic characteristics have been determined (Sarower et al. 2005). Recombinant carp DAO shows broad substrate specificity. Of all polar and nonpolar amino acids, D-alanine produces the highest V_{max} value. K_m values toward these D-amino acids, however, are different from the order of V_{max}. The lowest K_m value is found for D-tyrosine and D-alanine, followed by D-valine and D-phenylalanine. Carp DAO has a k_{cat} value of approximately 190 s^{-1}, which is higher than that of porcine kidney DAO (10 s^{-1}) and lower than that of yeast DAO (300 s^{-1}). The k_{cat}/K_m values indicate that the enzyme is highly active toward D-alanine, followed by D-valine, D-tyrosine, and D-phenylalanine. The enzyme is inactive toward acidic D-amino acids and all L-amino acids. This substrate specificity is different from that of other species. For mammalian DAO, D-proline and D-methionine are the most specific substrates (Curti et al. 1992), whereas for *R. gracilis* DAO, D-methionine, D-tryptophan, and D-alanine are the most specific substrates (Pollegioni et al. 1992). Thus, carp DAO is thought to have evolved by feeding on invertebrates containing abundant D-alanine.

The optimal temperature of carp DAO, approximately 35 °C, is the same as that of *Rhodosporidium toruloides* DAO, whereas *Candida boidinii* DAO shows a higher optimal temperature (50 °C). Carp DAO, however, maintains high activity over a wide range of temperatures (20–50 °C) and is stable in this temperature range. The pH stability of carp DAO is also high between pH 5.5 and 11.0. Thus, carp DAO is more stable than its yeast and mammalian counterparts over a wide range of temperatures and pH values.

Several DAO inhibitors such as benzoate, *p*-chloromercuribenzoate, and creatinine are inhibitory to carp DAO as well. The inhibitory effects of these compounds are consistent with the effects of these inhibitors on DAO from other sources.

17.3.5 Overall Structure and Topology

The 3D structure obtained from SWISS-MODEL is depicted in Fig. 17.4 (Sarower et al. 2005, 2009). The secondary structure topology consists of 11 α-helices and 17 β-strands. The FAD-binding domain is characterized by 6 α-helices and 10 - β-strands, whereas the interface domain is composed of 5 α-helices and 7 β-strands. This structure topology is close to that of mammalian and yeast DAO. Main topological differences, however, are observed in the active site loop 218–226 (9 residues in carp compared to 13 residues in porcine kidney), and the loop is absent in yeast DAO (Pollegioni et al. 2002). The long active site loop makes DAO a slow turnover enzyme in which product release becomes rate limiting (Wierenga et al. 1986). Thus, the shorter active site loop of carp DAO may be responsible for the low K_m values of D-alanine and for the higher turnover rate than that calculated for the porcine kidney DAO. The other difference is the absence of a long C-terminal loop (302–322) found in yeast DAO (6 residues in carp and 4 residues in porcine kidney compared to 21 residues in yeast). The long C-terminal loop makes a wider interaction area between the two monomers of yeast DAO, resulting in "head-to-tail" dimerization. In contrast to yeast DAO, mammalian DAO shows "head-to-head" dimerization (Mattevi et al. 1996). Carp DAO also forms a "head-to-head" dimer similar to porcine kidney DAO (Sarower et al. 2005, 2009).

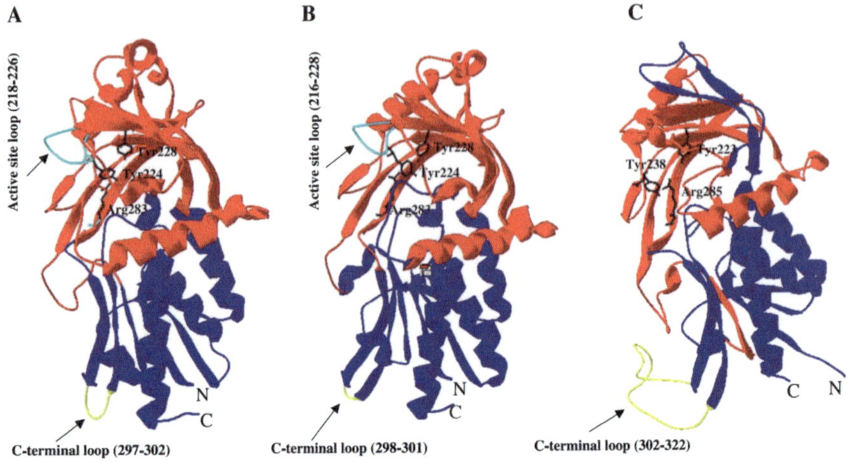

Fig. 17.4 Ribbon representations of monomer structure of D-amino acid oxidase in carp hepatopancreas (**a**), porcine kidney (**b**), and *Rhodotorula gracilis* (**c**) constructed with ProModII Arg283, Tyr224, and Tyr228 in porcine kidney enzyme (Mizutani et al. 1996) and Arg285, Tyr223, and Tyr238 in yeast enzyme (Pollegioni et al. 2002) are thought to be key catalytic residues. The letters N and C indicate the N and C terminus of the protein, respectively

These catalytic and structural characteristics of carp DAO show that although carp DAO is similar to porcine kidney DAO structurally, it is similar to yeast DAO enzymatically, in turnover rate and in pH and temperature stability. This similarity is a reflection of the availability of D-amino acids in food.

17.4 Conclusion

Environmental factors affect cell biochemistry greatly, and aquatic environments differ vastly from terrestrial environments with respect to salinity (0–35 ppt), oxygen concentration (1/30 in the air), temperature, and pressure. The food web structure is more complicated in aquatic environments than in terrestrial environments. Dating back to the Precambrian era, aquatic animals have relied heavily upon each other and have adapted in severe environments. The utilization of D-amino acids may have evolved during the environmental adaptation and survival of aquatic animals. Thus, it is clear that aquatic animals that mainly depend upon L-amino acids also rely on the D-amino acid biosystem for maintaining life-support systems. It is likely that many other D-amino acid subsystems in various organisms, including aquatic animals, will be elucidated in the future.

References

Abe H, Yoshikawa N, Sarower MDG, Okada S (2005) Physiological function and metabolism of free D-alanine in aquatic animals. Biol Pharm Bull 28:1571–1577

Bada JL (1982) Racemization of amino acids in nature. Interdiscip Sci Rev 7:30–46

Curti B, Ronchi S, Pilone MS (1992) D- and L-amino acid oxidase. In: Mller F (ed) Chemistry and biochemistry of flavoenzymes, vol 3. CRC Press, London, pp 69–94

D'Aniello A, Giuditta A (1977) Identification of D-aspartic acid in the brain of *Octopus vulgaris* LAM. J Neurochem 29:1053–1057

D'Aniello A, Giuditta A (1980) Presence of D-alanine in crustacean muscle and hepatopancreas. Comp Biochm Physiol 60B:319–322

D'Aniello A, D'Onofrio G, Pischetola M, D'Aniello G, Vetere A, Petrucelli L, Fisher GH (1993) Biological role of D-amino acid oxidase and D-aspartate oxidase. J Biol Chem 268:26941–26949

Faotto L, Pollegioni L, Ceciliani F, Ronchi S, Pilone MS (1995) The primary structure of D-amino acid oxidase from *Rhodotorula gracilis*. Biotechnol Lett 17:193–198

Felbeck H (1985) Occurrence and metabolism of D-aspartate in the gutless bivalve *Solemya reidi*. J Exp Zool 234:145–149

Felbeck H, Wiley S (1987) Free D-amino acids in the tissues of marine bivalves. Bio Bull 173:252–259

Fujimori T, Abe H (2002) Physiological roles of free D- and L-alanine in the crayfish *Procambarus clarkii* with special reference to osmotic and anoxic stress responses. Comp Biochem Physiol 131A:893–900

Fujita E, Okuma E, Abe H (1997) Partial purification and properties of alanine racemase from the muscle of black tiger prawn *Penaeus monodon*. Fish Sci 63:440–445

Hoffmann K, Schneider-Scherzer E, Kleinkauf H, Zocher R (1994) Purification and characterization of eukaryotic alanine racemase acting as key enzyme in cyclosporine biosynthesis. J Biol Chem 269:12710–12714

Kera Y, Hasegawa S, Watanabe T, Segawa H, Yamada R (1998) D-aspartate oxidase and free acidic D-amino acids in fish tissues. Comp Biochem Physiol 119B:95–100

Konno R, Yasumura Y (1992) D-amino-acid oxidase and its physiological function. Int J Biochem 24:519–524

Matsushima O, Katayama H, Yamada K, Kado Y (1984) Occurrence of free D-alanine and alanine racemase activity in bivalve molluscs with special reference to intracellular osmoregulation. Mar Biol Lett 5:217–225

Mattevi A, Vanoni MA, Todone F, Rizzi M, Teplyakov A, Coda A, Bolognesi M, Cuti B (1996) Crystal structure of D-amino acid oxidase: a case of active site mirror-image convergent evolution with flavocytochrome b_2. Proc Natl Adad Sci USA 93:7496–7501

Mizutani H, Miyahara I, Hirotsu K, Nishina Y, Shiga K, Setoyama C, Miura R (1996) Three-dimensional structure of porcine kidney D-amino acid oxidase at 3.0 Å resolution. J Biochem 120:14–17

Nagata Y, Yamada R, Nagasaki H, Konno R, Yasumura Y (1991) Administration of D-alanine did not cause increase of D-amino acid oxidase activity in mice. Experientia 47:835–838

Nomura T, Yamamoto I, Morishita F, Furukawa Y, Matsushima O (2001) Purification and some properties of alanine racemase from a bivalve mollusc *Corbicula japonica*. J Exp Zool 289:1–9

Okuma E, Abe H (1994) Total D-amino and other free amino acids increase in the muscle of crayfish during seawater acclimation. Comp Biochem Physiol 109A:191–197

Okuma E, Fujita E, Amano H, Noda H, Abe H (1995) Distribution of free D-amino acids in the tissues of crustaceans. Fish Sci 61:157–160

Okuma E, Watanabe K, Abe H (1998) Distribution of free D-amino acids in bivalve mollusks and the effects of physiological conditions on the level of D- and L-alanine in the tissues of the hard clam, *Meretrix lusoria*. Fish Sci 64:606–611

Pollegioni L, Falbo A, Pilone MS (1992) Specificity and kinetics of *Rhodotorula gracilis* D-amino acid oxidase. Biochim Biophys Acta 1120:11–16

Pollegioni L, Diederichs K, Molla G, Umhau S, Welte W, Ghisla S, Pilone MS (2002) Yeast D-amino acid oxidase: structural basis of its catalytic properties. J Mol Biol 324:535–546

Sarower MG, Matsui T, Abe H (2003a) Distribution and characteristics of D-amino acid and D-aspartate oxidases in fish tissues. J Exp Zool 295A:151–159

Sarower MG, Okada S, Abe H (2003b) Molecular characterization of D-amino acid oxidase from common carp *Cyprinus carpio* and its induction with exogenous free D-alanine. Arch Biochem Biophys 420:121–129

Sarower MG, Okada S, Abe H (2005) Catalytic and structural characteristics of carp hepatopancreas D-amino acid oxidase expressed in *Escherichia coli*. Comp Biochem Physiol 140B:417–425

Sarower MG, Okada S, Abe H (2009) Carp D-amino acid oxidase: structural active site basis of its catalytic mechanisms. Science Asia 35:150–155

Shaw JP, Petsko GA, Ringe D (1997) Determination of the structure of alanine racemase from *Bacillus stearothermophilus* at 1.9-Å resolution. Biochemistry 36:1329–1342

Shibata K, Shirasuna K, Motegi K, Kera Y, Abe H, Yamada R (2000) Purification and properties of alanine racemase from crayfish *Procambarus clarkii*. Comp Biochem Physiol 126B:599–608

Shibata K, Watanabe T, Yoshikawa H, Abe K, Takahashi S, Yamada R (2003) Purification and characterization of aspartate racemase from the bivalve mollusk *Scapharca broughtonii*. Comp Biochem Physiol 134B:307–314

Subramani S (1993) Protein import into peroxisomes and biogenesis of the organelle. Ann Rev Cell Biol 9:445–478

Uo T, Yoshimura T, Tanaka N, Takigawa K, Esaki N (2001a) Functional characterization of alanine racemase from *Schizosaccharomyces pombe*: an eukaryotic counterpart to bacterial alanine racemase. J Bacteriol 183:2226–2233

Uo T, Ueda M, Nishiyama T, Yoshimura T, Esaki N (2001b) Purification and characterization of alanine racemase from hepatopancreas of black-tiger prawn, *Penaeus monodon*. J Mol Catal B 12:137–144

Watanabe T, Shibata K, Kera Y, Yamada R (1998) Occurrence of free D-aspartate and aspartate racemase in the blood shell *Scapharca broughtonii*. Amino Acids 14:353–360

Watanabe A, Kurokawa T, Yoshimura T, Kurihara K, Soda K, Esaki N (1999a) Role of lysine 39 of alanine racemase from *Bacillus stearothermophilus* that binds pyridoxal 5′-phosphate. J Biol Chem 274:4189–4194

Watanabe A, Yoshimura T, Mikami B, Esaki N (1999b) Tyrosine 265 of alanine racemase as a base abstracting α-hydrogen from L-alanine: the counterpart residues to lysine 39 specific to D-alanine. J Biochem 126:781–786

Wierenga RK, Terpstra P, Hol WG (1986) Prediction of the occurrence of the ADP-binding bab-fold in proteins, using an amino acid sequence fingerprint. J Mol Biol 187:101–107

Yamada R, Nagasaki H, Nagata Y, Wakabayashi Y, Iwashima A (1989) Administration of D-aspartate increases D-aspartate oxidase activity in mouse liver. Bhichim Biophys Acta 990:325–328

Yoshikawa N, Dohmae N, Takio K, Abe H (2002) Purification, properties, and partial amino acid sequences of alanine racemase from the muscle of black tiger prawn *Penaeus monodon*. Comp Biochem Physiol 133B:445–453

Yoshikawa N, Okada S, Abe H (2009) Molecular characterization of alanine racemase in the kuruma prawn *Marsupenaeus japonicus*. J Biochem 145:249–258

Yoshikawa N, Ashida W, Hamase K, Abe H (2011) HPLC determination of the distribution of D-amino acids and the effects of ecdysis on alanine racemase activity in kuruma prawn *Marsupenaeus japonicas*. J Chromatogr B 879:3283–3288

Yoshimura T, Soda K (1994) Alanine racemase: structure and function. In: Fukui T, Soda K (eds) Molecular aspects of enzyme catalysis. Kodansha, Tokyo, pp 147–163

Chapter 18
Serine Racemase

Darrick T. Balu

Abstract Serine racemase (SR) belongs to the family of fold II pyridoxal 5′-phosphate (PLP) enzymes that catalyzes the racemization of L-serine to D-serine, as well as the α,β-elimination of water from both enantiomers to produce ammonia and pyruvate. This unique property of SR makes it more structurally homologous to bacterial serine/threonine dehydratases, rather than the classical racemases that belong to the fold type III PLP family. The initial evidence for SR and D-serine in eukaryotes came from insects, such as earthworms and silkworms. Following the discovery of D-serine in the rodent brain, mammalian SR was eventually cloned. In humans, SR is primarily found in the brain, heart, liver, kidney, and skeletal muscle. This chapter will review the mechanisms of both the racemization and α,β-elimination reactions catalyzed by SR. Furthermore, the cellular distribution of SR and D-serine, as well as the ways in which SR activity is regulated, will be discussed, with particular emphasis on the mammalian brain.

Keywords Serine racemase • D-serine • *N*-methyl-D-aspartate receptor • Pyridoxal 5′-phosphate

18.1 Enzymatic Activity and Catalytic Mechanisms

Serine racemase (SR; EC 5.1.1.18) is the enzyme responsible for both the conversion of L-serine to D-serine and the α,β-elimination of water from L- or D-serine to yield pyruvate and ammonia. SR is classified as a fold II pyridoxal 5′-phosphate (PLP) enzyme, whereby the cofactor binds to a lysine at the catalytic site of SR (Lys56) to form an internal aldimine (Wolosker et al. 1999). However, SR is more structurally similar to bacterial serine/threonine dehydratases, rather than classical amino acid racemases (De Miranda et al. 2000). Even though the PLP attachment sites are conserved and there is structural similarity between SR and type II PLP

D.T. Balu (✉)
Department of Psychiatry, Harvard Medical School, Boston, MA, USA

Translational Psychiatry Laboratory, McLean Hospital, Belmont, MA, USA
e-mail: dbalu@mclean.harvard.edu

T. Yoshimura et al. (eds.), *D-Amino Acids*, DOI 10.1007/978-4-431-56077-7_18

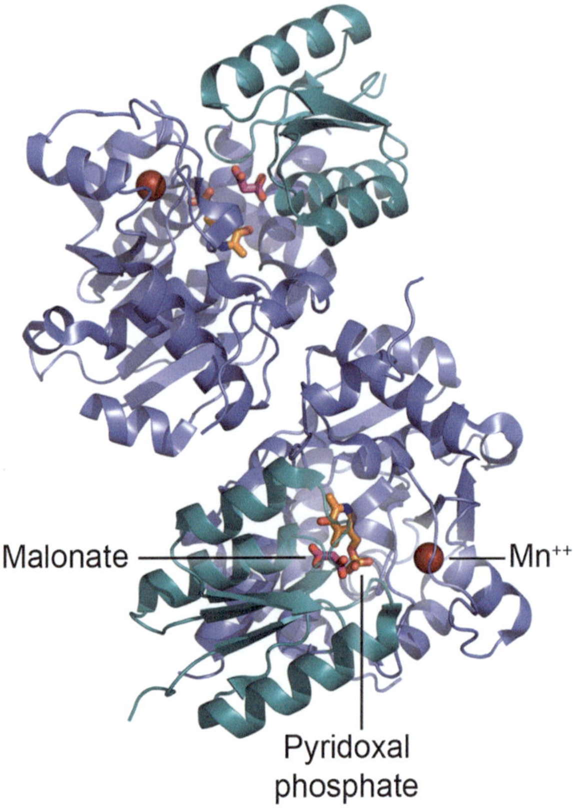

Fig. 18.1 Crystal structure of dimeric human serine racemase showing a detailed active site. PLP = *orange*; malonate = *red*; Mn^{2+} = *red* sphere (Reproduced from Canu et al. 2014 with permission from Frontiers)

family members, there are critical differences regarding allosteric regulation and reaction specificity.

The crystal structure of SR revealed that the enzyme consists of two identical subunits that function as a dimer (Fig. 18.1). The monomers contain two domains that are connected by a flexible loop, with the larger domain responsible for interacting with PLP and dimerization (Goto et al. 2009). In addition to PLP, SR has binding sites for magnesium (Mg^{2+}) and a Mg-ATP complex, both of which lie outside of the catalytic site at the interface of the small and large domains (Table 18.1). Since chelating Mg^{2+} greatly reduces enzymatic activity and the Mg-ATP complex stimulates enzymatic activity, it is likely that both of these factors stabilize the folding of SR. In addition, it is thought that ATP is not an energy requirement for enzyme activity because ADP is also equally effective in SR activation (De Miranda et al. 2002). There are also additional mechanisms involving the active and allosteric sites in human SR that regulate its protein confirmation with respect to its affinity for ATP (Marchetti et al. 2013). For example, the binding of glycine to the active site of SR increases the affinity of the enzyme for ATP.

The reactions catalyzed by SR involve the combination of two reactions through the formation of a common intermediate, the resonance-stabilized carbanion

Table 18.1 Serine racemase cofactors and substrates

Cofactors	Divalent cations, Mg-ATP, ADP
Substrates	L-serine, D-serine, L-threonine

(Foltyn et al. 2005). Racemization is attained by deprotonation of the external aldimine (L-serine-PLP complex), which occurs due to the special orientation of the neutral amino group of Lys56 and the deprotonated hydroxyl group of Ser84 (Goto et al. 2009). At the opposite end of the molecule, D-serine-PLP is generated by the reprotonation of the carbanion. The rate of racemization by SR is about 100-fold lower than the bacterial alanine racemase, with a catalytic constant (K_{cat}) of SR racemization ranging between 3 and 45 min^{-1} that indicates that it takes more than 1 s for SR to release one molecule of D-serine.

In addition to the reversible racemization of L-serine to D-serine, SR also catalyzes the irreversible α,β-elimination of water from both enantiomers to produce ammonia and pyruvate (De Miranda et al. 2002), which is in accordance with its homology to serine dehydratases. For each D-serine generated, about four molecules of pyruvate are produced. The α,β-elimination is attained by protonation of the substrate β-hydroxy group that results in the removal of water. This is associated with the formation of aminoacrylate-PLP, an unstable intermediate that spontaneously hydrolyzes into pyruvate and ammonia.

The exact physiological role SR α,β-elimination activity is unknown. Although pyruvate is an important metabolite, the rate of its formation from other sources, such as glycolysis, is several orders of magnitude faster than the rate of L-serine α,β-elimination, making it unlikely that SR-derived pyruvate plays an important metabolic role (Wolosker and Mori 2012). The production of pyruvate may be the primary role of SR in the liver, since it lacks *N*-methyl-D-aspartate receptors (NMDARs) and where D-serine is believed to not play a functional role (Wolosker et al. 1999). Contrary to the liver, D-serine plays an important role in regulating excitatory neurotransmission in the adult mammalian forebrain. SR expression and D-serine content are highest in the forebrain, due to the low levels of the main degradative, peroxisomal enzyme for D-serine, D-amino acid oxidase (DAO). Furthermore, forebrain D-serine levels are unaltered in mice that have a catalytically inactive form of DAO, demonstrating the lack of importance for this enzyme in regulating D-serine concentration (Hashimoto et al. 1993). Thus, the SR α,-β-elimination reaction might provide a mechanism by which to control intracellular D-serine levels in certain brain regions (Wolosker 2011).

18.2 Cellular Localization of Serine Racemase and D-Serine

SR and D-serine were first observed in eukaryotic insects, such as silkworms and earthworms (Corrigan and Srinivasan 1966). D-Amino acids, including D-serine, are now well-established modulators of neuronal activity in mammals (Boehning and Snyder 2003; Wolosker et al. 2008). Activation of the NMDAR requires the binding of either glycine or D-serine at the glycine modulatory site (GMS) on the NR1 subunit (Kishi and Macklis 2004). D-Serine is enriched in corticolimbic regions of the brain and is localized to the same areas as NMDARs (Schell et al. 1995). Thus, D-serine is believed to be the primary NMDAR co-agonist in the forebrain because of its high concentration (Papouin et al. 2012; Rosenberg et al. 2013), while genetic deletion of SR, which nearly eliminates brain D-serine levels, impairs NMDAR-mediated currents (Basu et al. 2009; Balu et al. 2013).

Initial in vitro studies suggested that SR was an astrocytic enzyme, and therefore astrocytes were the major source of brain D-serine (Mothet et al. 2005; Schell et al. 1995; Wolosker et al. 1999). However, recent immunohistochemical studies have suggested that SR is more prominently expressed in neurons, rather than in astrocytes (Ding et al. 2011; Miya et al. 2008; Ehmsen et al. 2013). Furthermore, using mice with conditional deletions of SR either in excitatory forebrain neurons (nSR-/-) or astrocytes (aSR-/-), it was shown that the majority of SR was expressed in forebrain excitatory neurons, particularly in the neocortex and hippocampus, and that approximately 15 % was expressed in astrocytes (Benneyworth et al. 2012). This pattern of neuronal expression was also observed in human postmortem neocortex (Fig. 18.2), as SR was found in both excitatory and inhibitory neurons, but not in astrocytes (Balu et al. 2014). It should be noted that L-serine derived from astrocytes is the precursor for D-serine in the rodent brain, as mice that lack the astrocyte-enriched enzyme phosphoglycerate 3-dehydrogenase, which catalyzes the first step in L-serine biosynthesis, display marked reductions in brain D-serine (Yang et al. 2010).

As was the case with SR, D-serine was originally believed to be stored in and released from astrocytes (Fossat et al. 2012; Schell et al. 1995; Martineau et al. 2013). More recent studies demonstrated that D-serine was located primarily in neurons (Kartvelishvily et al. 2006; Curcio et al. 2013). However, up until that point, the specificity of D-serine immunostaining in brain tissue was not yet established. Utilizing SR−/− mice as a negative control to optimize staining conditions (Balu et al. 2014; Ehmsen et al. 2013), D-serine was shown to be almost exclusively stored in neurons, particularly GABAergic neurons in the neocortex and hippocampus (Balu et al. 2014). Interestingly, SR and D-serine were only co-localized in approximately ~50 % of neurons, depending on the brain region (Balu et al. 2014). This separation of SR and D-serine might be related to the enzymatic profile of SR, as the α,β-elimination reaction that metabolizes D-serine is more efficient than the racemization of L-serine to D-serine.

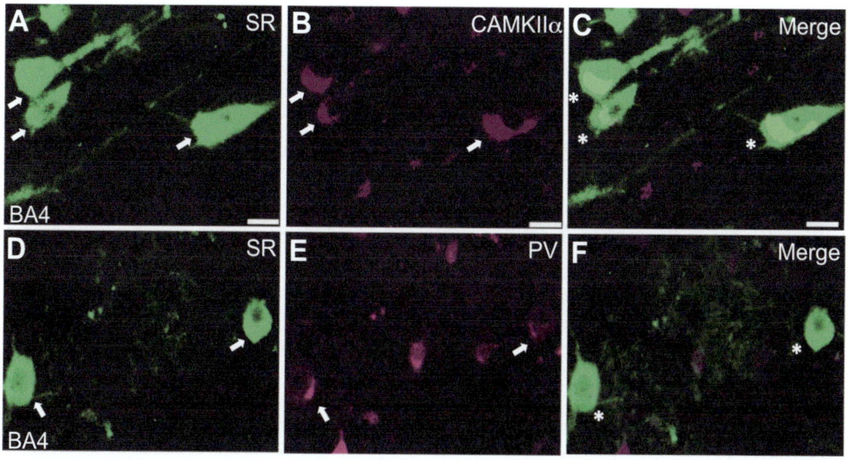

Fig. 18.2 Serine racemase is expressed in neurons in human postmortem cortex. Representative brain sections of human primary motor cortex (BA4; 40x). SR (**a**, **d**; *arrows*) was expressed mainly by excitatory neurons (**b**; CAMKIIα *arrows*) and a subset of GABAergic neurons (**e**; PV; *arrows*). *Merged images* are shown in *panels* (**c**, **f**; SR, *green*; CAMKIIα and PV, magenta). CAMKIIα, calcium-calmodulin kinase II alpha; PV, parvalbumin; SR, serine racemase. *Scale bars* represent 25 μm (This figure has been modified from Balu et al. 2014)

18.3 Regulation of Serine Racemase Activity

In the mammalian brain, there are still many unresolved questions to how SR activity is regulated in vivo. Many of the early studies focused on how SR was regulated in cultured astrocytes. More recent studies have expanded this line of research to include neuronal cultures and rodent brain tissue (Table 18.2).

In cultured astrocytes, SR activity is enhanced by a C-terminal interaction with glutamate receptor-interacting protein (GRIP1), which also binds to α-amino-3-hydroxy-5-methyl-4-isoxazolepropionic acid (AMPA) glutamate-type receptors (Kim et al. 2005). On the other hand, astrocytic SR is inhibited by association with phosphatidylinositol (4, 5)-bisphosphate (PIP2) in cell membranes, which inhibits the enzyme by competing for ATP at the nucleotide-binding site (Mustafa et al. 2009). This inhibition of SR by PIP2 can be relieved though stimulation of metabotropic glutamate receptor 5 (mGluR5) that degrades PIP2 through phospholipase 2 activation (Mustafa et al. 2009). However, it is now known that the majority of SR in vivo is found in neurons. Furthermore, the conditions used to culture astrocytes during the time of these initial studies are more a model of reactive astrocytes, rather than astrocytes found in the brain under physiological conditions (Foo et al. 2011). Thus, it is not known if the aforementioned regulatory mechanisms hold true in vivo.

There have also been studies investigating how SR activity is regulated in cell and neuronal cultures, as well as SR isolated from brain tissue. SR interacts with the Golgin subfamily A member 3 (Golga3) protein, whereby the binding of SR to

Table 18.2 Mechanisms of serine racemase regulation

	Effect	Mechanism	References
Cultured astrocytes			
GRIP1	Activation	Interacting partner	Kim et al. (2005)
PIP2	Inhibition	Cell membrane association	Mustafa et al. (2009)
Cell culture/brain tissue			
GRIP1	Activation	Interacting partner	Kim et al. (2005)
Golga3	Activation	Ubiquitin-proteasome interaction	Dumin et al. (2006)
NMDAR stimulation	Inhibition	Palmitoylation and plasma membrane association	Balan et al. (2009)
		Nitrosylation at Cys113	Mustafa et al. (2007)
FBXO22	Activation	Cytosolic targeting	Dikopoltsev et al. (2014)
AMPAR stimulation	Activation	Internalization and disruption of stargazin interaction	Ma et al. (2014)
SR phosphorylation	Activation	Unknown (cytosolic SR)	Foltyn et al. (2010)

Astrocytic and neuronal cultures, as well as cell lines and brain tissue extracts, have been used to elucidate the mechanisms by which serine racemase is regulated in cells
Abbreviations: *AMPAR* α-Amino-3-hydroxy-5-methyl-4-isoxazolepropionic acid receptor; *FBXO22* F-box only protein 22, *GRIP1* glutamate receptor-interacting protein, *Golga3* Golgin subfamily A member 3, *NMDAR* N-methyl-D-aspartate receptor, *SR* serine racemase

Golga3 prevents its ubiquitylation and subsequent breakdown by the ubiquitin-proteasome system (Dumin et al. 2006). In addition to degradation, the subcellular localization of SR also plays a role in regulating its activity. There are different intracellular pools of SR, with the cytosolic fraction having high racemase activity and the plasma membrane SR fraction having low activity. In primary neuronal cultures, NMDAR activation inhibits D-serine production by promoting the translocation of SR from the cytosol to the plasma membrane, a process that is dependent on the palmitoylation of serine/threonine residues of SR (Balan et al. 2009). It is thought that the irreversible NMDAR-mediated translocation of SR to the plasma membrane is a mechanism used by neurons committed to dying to reduce D-serine synthesis and curb the spread of excitotoxicity (Wolosker and Mori 2012). F-box only protein 22 (FBXO22), a protein traditionally associated with the ubiquitin-proteasome system, enhances SR activity and D-serine production by preventing the enzyme from associating with intracellular membranes (Dikopoltsev et al. 2014). Interestingly, the effects of FBXO22 on D-serine synthesis are not related to the targeting of SR to the proteasome for degradation (Dikopoltsev et al. 2014).

Ionotropic glutamatergic receptors are another mechanism by which cells regulate SR and the production of D-serine. NMDAR activation inhibits SR activity by nitrosylating SR at amino acid Cys113, which displaces the cofactor ATP at its binding site (Mustafa et al. 2007). In addition to NMDAR activity, recent in vitro evidence suggests that AMPARs also regulate SR activity. SR interacts with and forms a ternary complex with two synaptic proteins, stargazin and postsynaptic density protein 95 (PSD-95) (Ma et al. 2014). Stargazin acts to sequester SR at the membrane, thereby inhibiting its activity. Upon AMPAR stimulation, SR is internalized and the interaction with stargazin is disrupted, which ultimately leads to increased D-serine production. These findings suggest that AMPAR activity can regulate NMDAR activity via the regulation of SR and subsequent D-serine production (Ma et al. 2014).

Phosphorylation of the murine SR enzyme is another form of posttranslational modification that regulates its activity (Foltyn et al. 2010). There are two threonine residues in SR that are phosphorylated. Phosphorylation of SR at Thr71 was confirmed in vivo by phosphoproteome analysis of mouse brain (Huttlin et al. 2010). Thr71 is the predominant phosphorylation site and has a motif for proline-directed kinases, although the specific kinase has not been identified. In vitro evidence suggests that Thr71 phosphorylation enhances D-serine synthesis. However, it should be noted that this phosphorylation site is not conserved in the human SR.

References

Balan L, Foltyn VN, Zehl M, Dumin E, Dikopoltsev E, Knoh D, Ohno Y, Kihara A, Jensen ON, Radzishevsky IS, Wolosker H (2009) Feedback inactivation of D-serine synthesis by NMDA receptor-elicited translocation of serine racemase to the membrane. Proc Natl Acad Sci U S A 106:7589–7594. doi:0809442106 [pii] 10.1073/pnas.0809442106

Balu DT, Li Y, Puhl MD, Benneyworth MA, Basu AC, Takagi S, Bolshakov VY, Coyle JT (2013) Multiple risk pathways for schizophrenia converge in serine racemase knockout mice, a mouse model of NMDA receptor hypofunction. Proc Natl Acad Sci U S A 110(26):E2400–E2409. doi:10.1073/pnas.1304308110

Balu DT, Takagi S, Puhl MD, Benneyworth MA, Coyle JT (2014) D-serine and serine racemase are localized to neurons in the adult mouse and human forebrain. Cell Mol Neurobiol 34 (3):419–435. doi:10.1007/s10571-014-0027-z

Basu AC, Tsai GE, Ma CL, Ehmsen JT, Mustafa AK, Han L, Jiang ZI, Benneyworth MA, Froimowitz MP, Lange N, Snyder SH, Bergeron R, Coyle JT (2009) Targeted disruption of serine racemase affects glutamatergic neurotransmission and behavior. Mol Psychiatry 14 (7):719–727. doi:mp2008130 [pii] 10.1038/mp.2008.130

Benneyworth MA, Li Y, Basu AC, Bolshakov VY, Coyle JT (2012) Cell selective conditional null mutations of serine racemase demonstrate a predominate localization in cortical glutamatergic neurons. Cell Mol Neurobiol 32:613–624. doi:10.1007/s10571-012-9808-4

Boehning D, Snyder SH (2003) Novel neural modulators. Annu Rev Neurosci 26:105–131. doi:10.1146/annurev.neuro.26.041002.131047

Canu N, Ciotti MT, Pollegioni L (2014) Serine racemase: a key player in apoptosis and necrosis. Front synaptic neurosci 6:9. doi:10.3389/fnsyn.2014.00009

Corrigan JJ, Srinivasan NG (1966) The occurrence of certain D-amino acids in insects. Biochemistry 5(4):1185–1190

Curcio L, Podda MV, Leone L, Piacentini R, Mastrodonato A, Cappelletti P, Sacchi S, Pollegioni L, Grassi C, D'Ascenzo M (2013) Reduced D-serine levels in the nucleus accumbens of cocaine-treated rats hinder the induction of NMDA receptor-dependent synaptic plasticity. Brain 136:1216–1230. doi:10.1093/brain/awt036

De Miranda J, Panizzutti R, Foltyn VN, Wolosker H (2002) Cofactors of serine racemase that physiologically stimulate the synthesis of the N-methyl-D-aspartate (NMDA) receptor coagonist D-serine. Proc Natl Acad Sci U S A 99(22):14542–14547. doi:10.1073/pnas.222421299

De Miranda J, Santoro A, Engelender S, Wolosker H (2000) Human serine racemase: molecular cloning, genomic organization and functional analysis. Gene 256(1–2):183–188

Dikopoltsev E, Foltyn VN, Zehl M, Jensen ON, Mori H, Radzishevsky I, Wolosker H (2014) FBXO22 is required for optimal synthesis of the NMDA receptor Co-agonist D-serine. J Biol Chem 289:33904–33915. doi:10.1074/jbc.M114.618405

Ding X, Ma N, Nagahama M, Yamada K, Semba R (2011) Localization of D-serine and serine racemase in neurons and neuroglias in mouse brain. Neurol Sci 32(2):263–267. doi:10.1007/s10072-010-0422-2

Dumin E, Bendikov I, Foltyn VN, Misumi Y, Ikehara Y, Kartvelishvily E, Wolosker H (2006) Modulation of D-serine levels via ubiquitin-dependent proteasomal degradation of serine racemase. J Biol Chem 281(29):20291–20302. doi:10.1074/jbc.M601971200

Ehmsen JT, Ma TM, Sason H, Rosenberg D, Ogo T, Furuya S, Snyder SH, Wolosker H (2013) D-serine in glia and neurons derives from 3-phosphoglycerate dehydrogenase. J Neurosci 33 (30):12464–12469. doi:10.1523/JNEUROSCI.4914-12.2013

Foltyn VN, Bendikov I, De Miranda J, Panizzutti R, Dumin E, Shleper M, Li P, Toney MD, Kartvelishvily E, Wolosker H (2005) Serine racemase modulates intracellular D-serine levels through an alpha, beta-elimination activity. J Biol Chem 280(3):1754–1763. doi:10.1074/jbc.M405726200

Foltyn VN, Zehl M, Dikopoltsev E, Jensen ON, Wolosker H (2010) Phosphorylation of mouse serine racemase regulates D-serine synthesis. FEBS Lett 584(13):2937–2941. doi:10.1016/j.febslet.2010.05.022

Foo LC, Allen NJ, Bushong EA, Ventura PB, Chung WS, Zhou L, Cahoy JD, Daneman R, Zong H, Ellisman MH, Barres BA (2011) Development of a method for the purification and culture of rodent astrocytes. Neuron 71(5):799–811. doi:10.1016/j.neuron.2011.07.022

Fossat P, Turpin FR, Sacchi S, Dulong J, Shi T, Rivet JM, Sweedler JV, Pollegioni L, Millan MJ, Oliet SH, Mothet JP (2012) Glial D-serine gates NMDA receptors at excitatory synapses in prefrontal cortex. Cereb Cortex 22(3):595–606. doi:10.1093/cercor/bhr130

Goto M, Yamauchi T, Kamiya N, Miyahara I, Yoshimura T, Mihara H, Kurihara T, Hirotsu K, Esaki N (2009) Crystal structure of a homolog of mammalian serine racemase from schizosaccharomyces pombe. J Biol Chem 284(38):25944–25952. doi:10.1074/jbc.M109.010470

Hashimoto A, Nishikawa T, Konno R, Niwa A, Yasumura Y, Oka T, Takahashi K (1993) Free D-serine, D-aspartate and D-alanine in central nervous system and serum in mutant mice lacking D-amino acid oxidase. Neurosci Lett 152(1–2):33–36

Huttlin EL, Jedrychowski MP, Elias JE, Goswami T, Rad R, Beausoleil SA, Villen J, Haas W, Sowa ME, Gygi SP (2010) A tissue-specific atlas of mouse protein phosphorylation and expression. Cell 143(7):1174–1189. doi:10.1016/j.cell.2010.12.001

Kartvelishvily E, Shleper M, Balan L, Dumin E, Wolosker H (2006) Neuron-derived D-serine release provides a novel means to activate N-methyl-D-aspartate receptors. J Biol Chem 281 (20):14151–14162. doi:10.1074/jbc.M512927200

Kim PM, Aizawa H, Kim PS, Huang AS, Wickramasinghe SR, Kashani AH, Barrow RK, Huganir RL, Ghosh A, Snyder SH (2005) Serine racemase: activation by glutamate neurotransmission via glutamate receptor interacting protein and mediation of neuronal migration. Proc Natl Acad Sci U S A 102(6):2105–2110. doi:10.1073/pnas.0409723102

Kishi N, Macklis JD (2004) MECP2 is progressively expressed in post-migratory neurons and is involved in neuronal maturation rather than cell fate decisions. Mol Cell Neurosci 27 (3):306–321. doi:S1044-7431(04)00170-8 [pii] 10.1016/j.mcn.2004.07.006

Ma TM, Paul BD, Fu C, Hu S, Zhu H, Blackshaw S, Wolosker H, Snyder SH (2014) Serine racemase regulated by binding to stargazin and PSD-95: POTENTIAL N-METHYL-d-ASPARTATE-alpha-AMINO-3-HYDROXY-5-METHYL-4-ISOXAZOLEPROPIONIC ACID (NMDA-AMPA) GLUTAMATE NEUROTRANSMISSION CROSS-TALK. J Biol Chem 289(43):29631–29641. doi:10.1074/jbc.M114.571604

Marchetti M, Bruno S, Campanini B, Peracchi A, Mai N, Mozzarelli A (2013) ATP binding to human serine racemase is cooperative and modulated by glycine. FEBS J 280(22):5853–5863. doi:10.1111/febs.12510

Martineau M, Shi T, Puyal J, Knolhoff AM, Dulong J, Gasnier B, Klingauf J, Sweedler JV, Jahn R, Mothet JP (2013) Storage and uptake of D-serine into astrocytic synaptic-like vesicles specify gliotransmission. J Neurosci 33(8):3413–3423. doi:10.1523/JNEUROSCI.3497-12.2013

Miya K, Inoue R, Takata Y, Abe M, Natsume R, Sakimura K, Hongou K, Miyawaki T, Mori H (2008) Serine racemase is predominantly localized in neurons in mouse brain. J Comp Neurol 510(6):641–654. doi:10.1002/cne.21822

Mothet JP, Pollegioni L, Ouanounou G, Martineau M, Fossier P, Baux G (2005) Glutamate receptor activation triggers a calcium-dependent and SNARE protein-dependent release of the gliotransmitter D-serine. Proc Natl Acad Sci U S A 102(15):5606–5611. doi:10.1073/pnas.0408483102

Mustafa AK, Kumar M, Selvakumar B, Ho GP, Ehmsen JT, Barrow RK, Amzel LM, Snyder SH (2007) Nitric oxide S-nitrosylates serine racemase, mediating feedback inhibition of D-serine formation. Proc Natl Acad Sci U S A 104(8):2950–2955. doi:10.1073/pnas.0611620104

Mustafa AK, van Rossum DB, Patterson RL, Maag D, Ehmsen JT, Gazi SK, Chakraborty A, Barrow RK, Amzel LM, Snyder SH (2009) Glutamatergic regulation of serine racemase via reversal of PIP2 inhibition. Proc Natl Acad Sci U S A 106(8):2921–2926. doi:10.1073/pnas.0813105106

Papouin T, Ladepeche L, Ruel J, Sacchi S, Labasque M, Hanini M, Groc L, Pollegioni L, Mothet JP, Oliet SH (2012) Synaptic and extrasynaptic NMDA receptors are gated by different endogenous coagonists. Cell 150(3):633–646. doi:10.1016/j.cell.2012.06.029

Rosenberg D, Artoul S, Segal AC, Kolodney G, Radzishevsky I, Dikopoltsev E, Foltyn VN, Inoue R, Mori H, Billard JM, Wolosker H (2013) Neuronal D-serine and glycine release via the Asc-1 transporter regulates NMDA receptor-dependent synaptic activity. J Neurosci 33 (8):3533–3544. doi:10.1523/JNEUROSCI.3836-12.2013

Schell MJ, Molliver ME, Snyder SH (1995) D-serine, an endogenous synaptic modulator: localization to astrocytes and glutamate-stimulated release. Proc Natl Acad Sci U S A 92 (9):3948–3952

Wolosker H (2011) Serine racemase and the serine shuttle between neurons and astrocytes. Biochim Biophys Acta 1814(11):1558–1566. doi:10.1016/j.bbapap.2011.01.001

Wolosker H, Blackshaw S, Snyder SH (1999) Serine racemase: a glial enzyme synthesizing D-serine to regulate glutamate-N-methyl-D-aspartate neurotransmission. Proc Natl Acad Sci U S A 96(23):13409–13414

Wolosker H, Dumin E, Balan L, Foltyn VN (2008) D-amino acids in the brain: D-serine in neurotransmission and neurodegeneration. FEBS J 275(14):3514–3526. doi:10.1111/j.1742-4658.2008.06515.x

Wolosker H, Mori H (2012) Serine racemase: an unconventional enzyme for an unconventional transmitter. Amino Acids 43(5):1895–1904. doi:10.1007/s00726-012-1370-3

Yang JH, Wada A, Yoshida K, Miyoshi Y, Sayano T, Esaki K, Kinoshita MO, Tomonaga S, Azuma N, Watanabe M, Hamase K, Zaitsu K, Machida T, Messing A, Itohara S, Hirabayashi Y, Furuya S (2010) Brain-specific Phgdh deletion reveals a pivotal role for L-serine biosynthesis in controlling the level of D-serine, an N-methyl-D-aspartate receptor co-agonist, in adult brain. J Biol Chem 285(53):41380–41390. doi:10.1074/jbc.M110.187443

Chapter 19
D-Amino Acid Oxidase and D-Aspartate Oxidase

Yusuke Kato, Diem Hong Tran, Huong Thi Thanh Trinh, and Kiyoshi Fukui

Abstract D-Amino acids are the enantiomers of L-amino acids used as building blocks of proteins and were previously unknown as to pathophysiological roles. However, increasing numbers of recent studies have suggested importance of D-amino acids. D-Serine (D-Ser) works as a co-agonist of the NMDA receptor, an ion channel for neurotransmission. A decrease in the amount of D-Ser in brain has been observed in patients of schizophrenia. It has been suggested that D-aspartate (D-Asp) is related with the regulation of motor neurons, memory, and mental disorders. Cognate flavoenzymes such as D-amino acid oxidase (DAO) and D-aspartate oxidase (DDO) regulate the amount of D-Ser and D-Asp in our body, respectively, to modulate such biological events. We therefore have proposed that the "D-amino acid biosystems" play important roles in our bodies. Chemicals that modulate the enzymatic activity of DAO and DDO are expected as potent therapeutic drugs for schizophrenia and other mental disorders.

Keywords D-Amino acid oxidase • Glutamatergic neurotransmission • Schizophrenia • Glial cells • Choroid plexus • D-Aspartate oxidase • Pituitary gland

19.1 Introduction

Naturally occurring proteins are composed only of L-amino acids, although their enantiomers, D-amino acids, are never used as the building blocks of proteins. L- and D-amino acids are mirror images of each other and thereby never superposable, but otherwise their structures are the identical. So their relationship is similar to that between our left and right hands. All living organisms including viruses, bacteria, plants, animals, and humans exclusively use L-amino acids as the building blocks of proteins, which exhibits specificity of the nature of living organisms. D- and L-amino acids possess a chiral carbon, Cα, which causes the chirality. D-amino

Y. Kato • D.H. Tran • H.T.T. Trinh • K. Fukui (✉)
Division of Enzyme Pathophysiology, The Institute for Enzyme Research (KOSOKEN), Tokushima University, 3-18-15 Kuramoto, Tokushima 770-8503, Japan
e-mail: kiyo.fukui@tokushima-u.ac.jp

T. Yoshimura et al. (eds.), *D-Amino Acids*, DOI 10.1007/978-4-431-56077-7_19

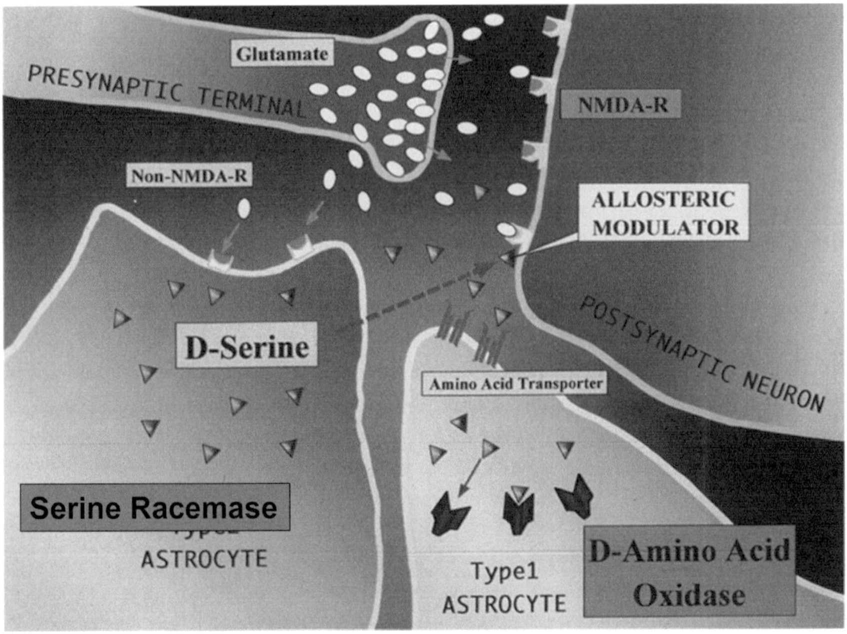

Fig. 19.1 The metabolic system of D-Ser around a glutamate synapse and regulation of the NMDA receptors (From Fukui 2012, reproduced with permission of the copyright owner)

acid oxidase (DAO) as well as D-aspartate oxidase (DDO) takes only D-amino acids as its substrate to catalyze the oxidative removal of amino group. This reaction is important not only because this is a regulatory mechanism of asymmetric synthesis but also because this reaction profoundly contributes to the function of the central nervous systems of mammals. Recent studies have shown and proposed that "D-amino acid biosystems" that depend on D-amino acids deeply relate with the regulation of our bodies (Fig. 19.1).

19.2 D-Amino Acid Oxidase

DAO is one of flavoproteins, which possesses flavin adenine dinucleotide (FAD) as a coenzyme, and localizes in peroxisomes in the cells of kidney, liver and brain. Since DAO was found by a Nobel Prize winner, Hans Krebs, in 1935 (Krebs 1935), physiological importance of DAO had been unclear for a long time. However, recent studies have revealed that one of D-amino acids, D-serine (D-Ser), modulates the activity of the NMDA receptor to have an impact on the mechanism of glutamate neurotransmission in the central nervous system of human (Hashimoto et al. 1993; Mothet et al. 2000; Schell et al. 1995). We have therefore proposed that DAO is an important regulator of the concentration of D-Ser and that an increase in the enzymatic activity of DAO causes a decrease in the concentration of D-Ser, leading to the onset of schizophrenia by the abnormal regulation of the NMDA receptors in the central nervous system.

19.2.1 D-Serine

19.2.1.1 Distribution, Synthesis, and Physiological Importance of D-Ser

D-Serine (D-Ser) is present in the central nervous system, and the distribution of D-Ser is almost consistent with that of a subtype of glutamate receptors, the N-methyl-D-aspartate (NMDA) receptor (Hashimoto et al. 1993). In the brain, D-Ser is localized in type 2 astrocytes and released from the cells upon activation of non-NMDA receptors (Schell et al. 1995). The NMDA receptor is one of the ion channels and plays a pivotal role in fast neurotransmission to contribute to the excited neurotransmission, plasticity of synapse, learning, and memory. D-Ser is a co-agonist of the NMDA receptor and influences the "glycine" site of the NMDA receptor to enhance the excitation by glutamate (Matsui et al. 1995; Mothet et al. 2000). It is presumed that endogenous D-Ser is synthesized from cellular L-Ser by Ser racemase in mammalian brain (Wolosker et al. 1999).

19.2.1.2 D-Ser Metabolism and Pathology

D-Ser is a neural regulatory factor that works as a co-agonist of the NMDA receptor in excitatory neurotransmission. D-Ser plays important roles in various diseases caused by dysfunction in the regulation of the NMDA receptor. It was suggested that a decrease in activity of the NMDA receptor is related to schizophrenia because both of the positive and negative symptoms and cognitive impairment in schizophrenia are improved by medication of D-Ser to patients of schizophrenia (Tsai et al. 1998). Mice with the mutated NR1 subunit of the NMDA receptor showed abnormal behavior relevant to schizophrenia (Mohn et al. 1999).

Schizophrenia is presumed to be a multifactorial disease caused by both genetic predisposition and environmental factors, although the mechanism of the onset and pathology remains obscure. Several working hypotheses, such as dopamine, serotonin, and glutamate theories, have been proposed as to the mechanism of schizophrenia. It is expected that abnormality not only in a single neural transmission system but also in multiple transmission systems leads to various symptoms and pathologies of schizophrenia. Among these transmission systems, the glutamate system mediated by the NMDA receptor is that to which D-Ser primarily contributes.

It has been well recognized that schizophrenia-like symptom is induced by medication of the antagonists of the NMDA receptor. This is one of the reasons that support the glutamate theory, so it is likely that activating the NMDA receptor leads to relieving the symptoms of schizophrenia.

Abnormal behavior of animals by drugs that induce the schizophrenia-like symptoms such as antagonists to the NMDA receptor is suppressed by administration of D-Ser. Agonists to the NMDA receptor, Gly and D-Ser, and a partial agonist to the NMDA receptor, D-cycloserine, which is also known as an antituberculosis drug, are effective for the treatment of schizophrenia. So, it is likely that abnormality in the amount of D-Ser is linked to that in the neurotransmission related to the NMDA receptor.

Based on these lines of evidence, we propose a working hypothesis that an increase in the DAO activity causes a decrease in the amount of D-Ser, a neuromodulator, in the synaptic cleft leading to the onset and pathology of schizophrenia caused by degeneracy in the regulation of the NMDA receptor.

19.2.2 D-Amino Acid Oxidase (DAO, EC 1.4.3.3)

19.2.2.1 Reaction Mechanism of DAO

DAO was first found in kidneys of pigs by Krebs in 1935 (Krebs 1935) and contains noncovalently bound flavin adenine dinucleotide (FAD) as its prosthetic group. The enzymatic reaction by DAO contains two steps as shown in Fig. 19.2. In the first step, FAD at the active site of DAO catalyzes dehydrogenation from the amino group of a D-amino acid to produce an α-imino acid and concomitantly converts to $FADH_2$. In the second step, the $FADH_2$ produced in the first step reduces O_2 to produce H_2O_2. The α-imino acid produced in the first step spontaneously (nonenzymatically) converts to an α-keto acid and an ammonia in aqueous solution. DAO is highly specific to D-amino acids and inactive to L-amino acids but shows broad specificity among various kinds of neutral D-amino acids and marginal preference to basic D-amino acids.

19.2.2.2 Crystal Structure of Human DAO (hDAO)

It is expected that novel drugs that inhibit the activity of DAO may contribute to the medication of schizophrenia, because the inhibition of DAO by chlorpromazine (CPZ) was notably observed in the studies using the cultured cell line. In order to design novel drugs based on protein structure, we first aimed at determining the structure of hDAO. Using an *E. coli* expression system, we successfully produced large amount of the hDAO protein (Kawazoe et al. 2007). K_m, k_{cat}, and K_i values for a competitive inhibitor, benzoate, were almost the same between human and porcine DAOs.

By contrast, the affinity of hDAO to FAD is low (Raibekas et al. 2000), and the reaction rate of the FAD reduction is far slower than DAO from porcine kidney

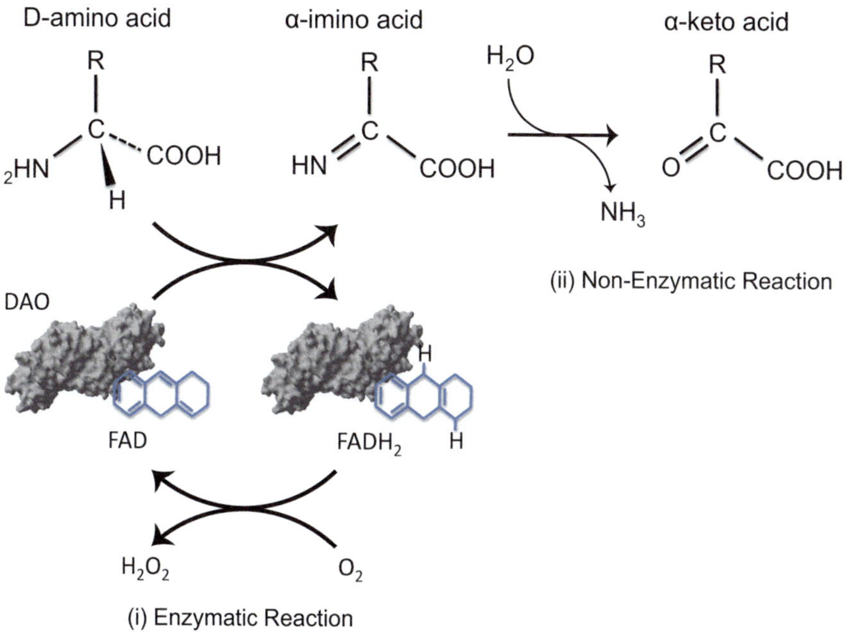

Fig. 19.2 Reaction of D-amino acid by DAO

(Molla et al. 2006). However, the molecular mechanism to explain the kinetic difference of these DAOs had been unclear, so we have determined the crystal structure of hDAO at 2.5 Å resolution (Kawazoe et al. 2006). hDAO forms a dimer composed of 39 kDa polypeptides akin to porcine DAO. Important residues and their structure for the DAO activity are completely conserved on the *re* side of the flavin group of FAD. However, the structure of the VAAGL hydrophobic stretch of hDAO on the *si* side of the flavin group differs from that of porcine DAO (Fig. 19.3). This is considered to be a case of structurally ambivalent peptide (SAP), which is presumably caused by the structural diversity of the Gly residue in the hydrophobic stretch. The structural diversity in the hydrophobic stretch of DAOs may confer the difference in the catalytic nature of these DAOs.

19.2.3 Cloning of cDNAs of DAOs

Gene encoding DAO is located in chromosome 12 in human genome. We cloned the cDNAs of DAO from cDNA libraries of kidneys from pig (Fukui et al. 1987), human (Momoi et al. 1988), mouse (Tada et al. 1990), and even rabbit in which DAO activity was undetected (Momoi et al. 1990) and determined the sequences of cDNAs of DAOs from those mammals. The cDNA sizes are 3.2 kilobases (kb) for pig, 1.6 kb for human, 1.6 kb for mouse, and 2.1 kb for rabbit. The corresponding

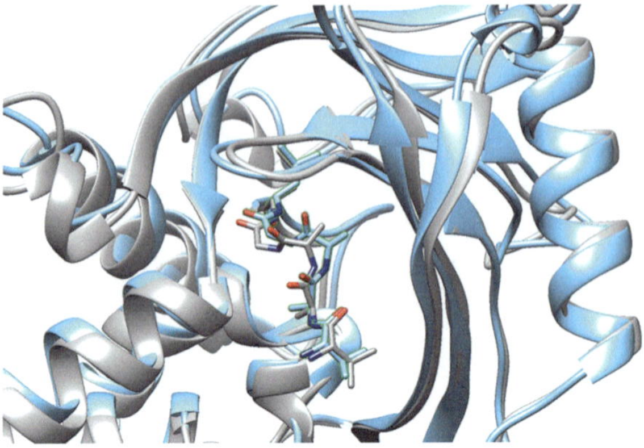

Fig. 19.3 Comparison of crystal structures of DAO from human (*gray*) and pig (*blue*). The VAAGL stretch is indicated as stick models (The illustration was drawn by UCSF Chimera Pettersen et al. 2004)

proteins were revealed to be composed of 347 amino acid residues except for mouse, whose DAO is composed of 345 amino acid residues.

The primary structures of DAOs from various species are quite similar to each other. The amino acid sequence identities to porcine DAO are 85 % for human, 77 % for mouse, and 80 % for rabbit DAOs. Cys and Pro residues are highly conserved among species. In addition, a hydrophobic region composed of 17 amino acid residues including the Gly-Xxx-Gly-Xxx-Xxx-Gly sequence in the N-terminal region and the Ser-His-Leu sequence of peroxisomal targeting signal in the C-terminal region are conserved among DAOs of the four species.

19.2.4 Expression of DAO

19.2.4.1 Localization of DAO

DAO is expressed exclusively in the kidney, brain and liver (Fukui et al. 1988). In the kidney, the expression of DAO is highly detected in proximal tubule. It is presumed that DAO localizes in peroxisome after the synthesis as a mature protein in cytosol because the amino acid sequence of DAO has peroxisomal targeting signal 1, PTS1, at the C-terminus. DAO might be involved in immune system as this enzyme produces H_2O_2, an antimicrobial reagent. Excess H_2O_2 produced by either hyperactivity of DAO or excess supply of substrates could also induce cell death because H_2O_2 is one of sources of oxidative stress. By taking advantage of the nature as a supplier of oxidative stress, polyethylene-glycol-conjugated DAO has been developed for the treatment of cancer (Fang et al. 2008).

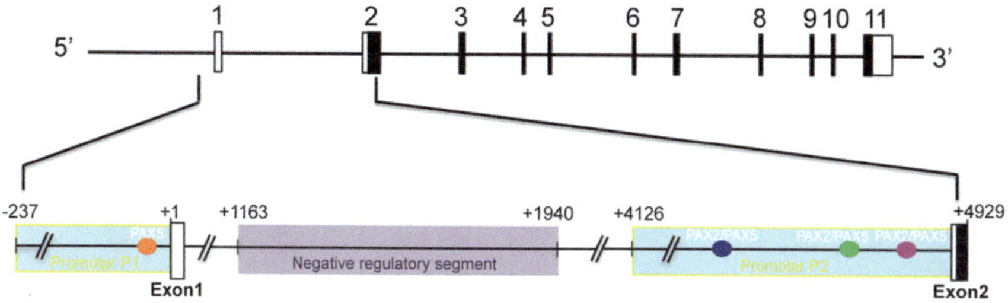

Fig. 19.4 Regulation of the hDAO gene. The hDAO gene consists of 11 exons and 10 introns. The exons are numbered and indicated as boxes. Closed and open boxes indicate coding and untranslated regions, respectively. Enlargement shows regions of the promoter P1 and P2 (*blue bands*) and the negative regulatory segment (*purple band*) as well as the PAX2/PAX5 binding sites

19.2.4.2 Regulatory Mechanism of Human DAO Expression

We have recently studied the expression mechanism of hDAO using the luciferase reporter system and revealed that the proximal upstream regions of exon1 (−237/+1) and exon2 (+4,126/+4,929), called P1 and P2, are promoter regions, respectively (Tran et al. 2015) (Fig. 19.4). Although the PAX5 transcription factor can bind to the P1 and P2 sites, PAX2, one of the orthologous of PAX5, can bind only to the P2 site. We also found that there is a negative regulatory segment (+1,163/+1,940) of the hDAO gene in intron1.

19.2.4.3 DAO Expression and D-Ser Metabolism in Astrocytes

In order to elucidate the physiological importance of DAO, we have established the primary cell culture of glial cells and examined expression of DAO using RT-PCR. The expression of DAO was observed in glial cells not only from the cerebellum but also those from the cerebrum. We also demonstrated that the expression level is high in type 1 astrocytes but low in type 2 astrocytes. This is an interesting contrast because type 2 astrocytes produce D-Ser (Schell et al. 1995). These together indicate that astrocytes contribute not only to supplying neurotrophic factors but also to the metabolism of D-Ser. Because the distribution of DAO and D-Ser in animal bodies inversely correlates, we propose that DAO is one of the negative regulators of the concentration of D-Ser as a physiological component.

We next examined whether extracellular D-Ser is metabolized by intracellular DAO with the use of primary astrocyte culture incubated with D-Ser. Survival rate of the primary astrocytes from both cerebellum and cerebrum decreased with high concentration of D-Ser in media (Park et al. 2006). Both the original C6 that is glioma cell line from rat and the stable transformant of C6 with forced expression of DAO (C6/DAO) showed increased cell death depending on the concentration of D-Ser in media. In addition, C6/DAO was killed with lower concentration of D-Ser.

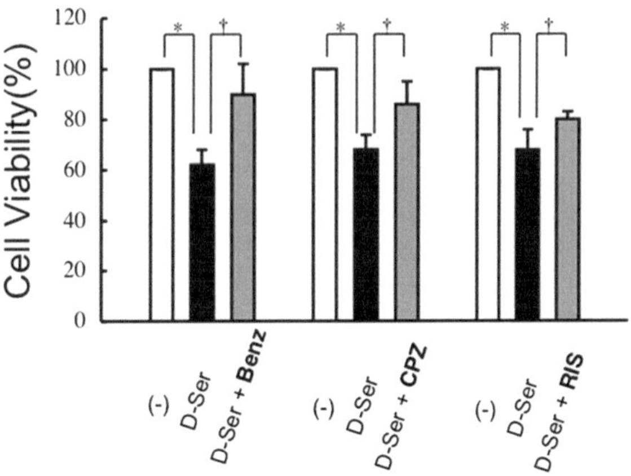

Fig. 19.5 Rescue of D-Ser-induced cell death of C6/DAO cells by DAO inhibitors. The concentrations of the chemicals were 10 mM for D-Ser, 20 mM for benzoate (Benz), 1 μM for CPZ, and 400 nM for risperidone (RIS) (From Fukui 2012, reproduced with permission of the copyright owner)

These suggest that H_2O_2 produced through the catalysis of DAO as the result of deamination of D-Ser induced strong cytotoxic effect.

By contrast, such cell death was rescued by the addition of DAO inhibitors such as benzoate and CPZ (Park et al. 2006) (Fig. 19.5). CPZ has been used as a classical antipsychotic and is one of competitive inhibitors of DAO competing with FAD, a coenzyme of DAO (Yagi et al. 1956). These support the idea that the death of C6 cells is due to the metabolic reaction of D-Ser by DAO. Taken together, it was suggested that DAO in astroglial cells may catalyze the metabolic reaction of D-Ser in the brain to control the concentration of D-Ser that works as a neuromodulator.

19.2.5 Pathophysiology of Schizophrenia and the Novel Strategy for the Treatment

19.2.5.1 Link Between Schizophrenia and DAO

The prevalence rate for schizophrenia is approximately 1 % of the population over the age of 18. Schizophrenia is presumably caused by both genetic predisposition and environmental factors. There have been several proposed systems connected to the pathology of schizophrenia, including the dopamine and glutamate systems. The symptoms of schizophrenia also have broad spectrum: positive, negative, and cognitive symptoms. Although various drugs with great curative effect on the positive symptoms have been developed, the variety and curative effect of drugs on the negative and cognitive symptoms have been fewer and lower, respectively. While the dopamine system is related mainly to the positive symptoms, the abnormality in the glutamate system could cause all of the positive, negative, and cognitive symptoms. So the development of new drugs for the treatment of the abnormalities in the glutamate system is important. In the glutamate system, abnormal level of D-Ser and abnormal activity of DAO in brain are presumably

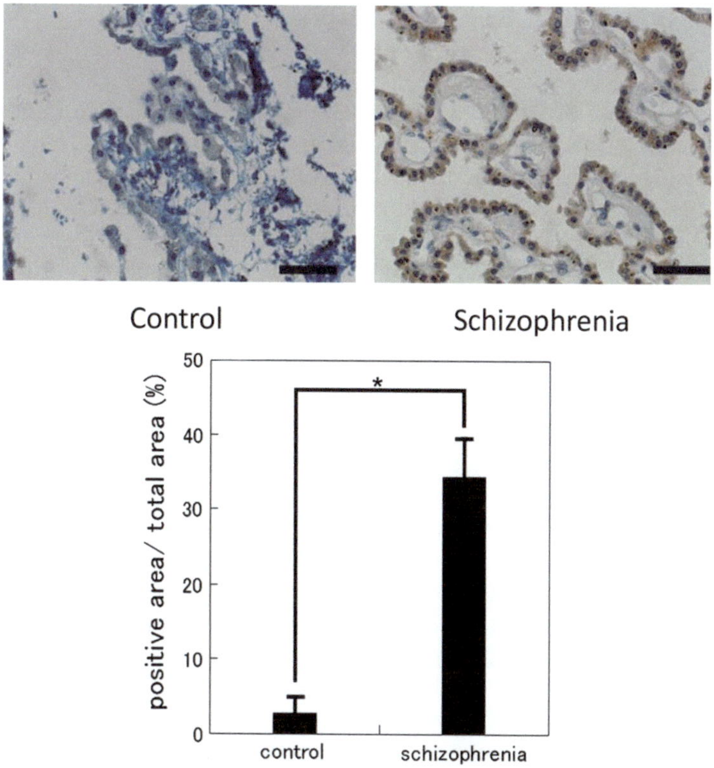

Fig. 19.6 Expression of hDAO in human tissue. *Upper panels* display the results of immunohistochemical analysis of the human choroid plexus with antibody against hDAO. *Lower panel* shows the quantitative analysis of hDAO-expressing area in choroid plexus. These data indicate an increase in the expression of hDAO in the choroid plexus of schizophrenia patients (From Ono et al. 2009, modified and reproduced with permission of the copyright owner)

linked to the pathology. In order to test the link among D-Ser, DAO, and the glutamate system, we examined the expression level of DAO in the brain of both postmortem rat and human tissues and observed that DAO is expressed in the cerebellum, pons, medulla oblongata, and choroid plexus (Ono et al. 2009). Especially, epithelial cells showed high-level expression of DAO in the choroid plexus (Fig. 19.6). In addition, the expression level of hDAO is higher in the choroid plexus of the patients than that of the controls. These suggest that hDAO expressed in the choroid plexus may regulate the D-amino acid concentration in cerebrospinal fluid. We propose that a decrease in the concentration of D-Ser caused by an increase in the activity of hDAO could be linked to the pathology of schizophrenia caused by the misregulation of the NMDA receptor, because D-Ser is one of the co-agonists of the NMDA receptor to enhance the excitatory neurotransmission by glutamate.

19.2.5.2 DAO Inhibition by Antipsychotics

Although it has been well known that the curative effect of CPZ, one of the antipsychotics, is displayed through inhibiting dopamine receptors, CPZ also inhibits enzymatic activity of DAO (Yagi et al. 1960). We have demonstrated that the inhibitory mechanism of CPZ to recombinant hDAO is competitive inhibition (Iwana et al. 2008). Although both porcine and human DAOs are inhibited competitively by CPZ, the inhibitory constant of CPZ to hDAO is higher than that to porcine DAO. CPZ is photochemically reactive and has clinical side effects such as phototoxicity and photodermatosis. In addition, it has been reported that CPZ is converted in vitro to radicals by irradiation of UV and enzymes such as peroxidases and others. So, we speculate that CPZ may be converted to compounds with higher inhibitory activity in our body by exposure of the skin to sunlight and drug metabolism. To test this speculation, we irradiated white light to CPZ to test the change in inhibitory effects on the catalytic activity of DAO and thereby found out that CPZ trimer exhibited stronger inhibition than CPZ monomer (Iwana et al. 2008). We have demonstrated that risperidone also inhibits the enzymatic activity of DAO. Risperidone is an antipsychotic drug used to treat schizophrenia (Abou El-Magd et al. 2010). These results suggest that pharmacological activities of those antipsychotic drugs may be in part through inhibition of DAO.

It is therefore expected that inhibitors of DAO can be effective antipsychotic drugs to treat schizophrenia. Such kinds of chemicals are considered to be important candidates for treatment of mental disorders caused by the abnormal regulation of the NMDA receptors.

19.2.6 D-Amino Acid Biosystems

We have reviewed the evidences to indicate the importance of D-Ser and DAO in the pathophysiology of schizophrenia and proposed the strategy for the development of the new treatment. Misregulation and dysfunction of the NMDA receptor are related to various mental diseases. Based on the expression of DAO in the astroglial cells and choroid plexus, we have proposed that the D-Ser-DAO system as one of the “D-amino acid biosystems” plays a pivotal role in regulating the function of the NMDA receptor in the glutamate neuron (Fig. 19.1). We believe that the evidences and views discussed in the present review will be useful for the development of new-generation drugs for schizophrenia.

19.3 D-Aspartate and D-Aspartate Oxidase

19.3.1 Introduction

In addition to D-Ser, D-aspartate (D-Asp) has also been reported to be involved in important biological events such as an increase in the production of testosterone in testis (D'Aniello et al. 1996) and inhibition of the production of melatonin in the pineal glands (Ishio et al. 1998). D-Asp is the enantiomer of L-Asp (Fig. 19.7). Recent studies have also suggested that D-Asp is related with schizophrenia in human and cognitive ability in mouse (Errico et al. 2011, 2013). D-Asp is an intermediate product for the biosynthesis of N-methyl-D-aspartate (NMDA) (D'Aniello et al. 2000) and important for the maintenance of long-term potentiation (LTP) working as a co-agonist at the NMDA receptor (Fagg and Matus 1984). D-Aspartate oxidase (DDO) is found in the kidney, liver, brain, and the pituitary gland and involved in the metabolism of D-Asp, thereby presumably regulating the concentration of D-Asp in tissues.

19.3.2 Studies on D-Asp Oxidase (EC 1.4.3.1)

An enzyme that catalyzes degradation of D-Asp is first reported in 1949 by Still et al. (Still et al. 1949). This enzyme, D-Asp oxidase (DDO), stereospecifically cleaves acidic D-amino acids such as D-Asp, D-Glu, and NMDA. On the other hand, biosynthetic pathway of D-Asp in mammalian had long been unknown until mammalian aspartate racemase (DR) was found in 2010 (Kim et al. 2010). Localization of DR is in the brain, testis, and heart.

19.3.2.1 Enzymatic Reaction of DDO

The holoenzyme of DDO contains flavin adenine dinucleotide (FAD) as coenzyme akin to DAO. The overall reaction scheme of DDO is almost the same as that of DAO except substrates (Katane and Homma 2010) (Fig. 19.2).

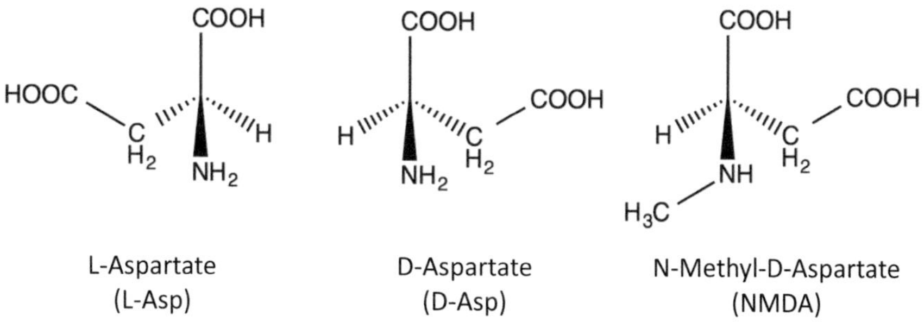

Fig. 19.7 Molecular structure of L-aspartate, D-aspartate, and N-methyl-D-aspartate

19.3.2.2 Distribution of DDO and D-Asp

DDO activity in a mammalian body is high in the kidney, liver, and brain (Van Veldhoven et al. 1991). DDO is localized in peroxisomes in a cell (Beard 1990). DDO content in the cerebrum, liver, and kidney of a newborn rat is relatively low and rapidly increases after birth, while the content of D-Asp in those organs decreases (Dunlop et al. 1986; Van Veldhoven et al. 1991). Histochemical studies revealed that DDO is present in the hippocampus, cerebral cortex, olfactory epithelium, choroid plexus, and ependymal of rat brains (Schell et al. 1997). In the pituitary gland of an adult rat, the content of D-Asp is high in the anterior and posterior lobes, but low in the intermediate lobe, whereas DDO activity is present only in the intermediate lobe (Lee et al. 1999; Schell et al. 1997) (Fig. 19.8). In the pineal gland, DDO activity is absent, while the content of D-Asp is high. Thus, the cells with high content of DDO generally show low D-Asp concentration, suggesting that D-Asp in a mammalian cell is catalyzed by DDO.

19.3.2.3 Molecular Studies of DDO

Several kinds of DDOs have been purified from various sources including octopus, pig, and bovine (D'Aniello and Rocca 1972; Negri et al. 1987; Yamamoto et al. 2007). DDOs from bovine and octopus are monomeric, whereas that from pig is composed of a homotetramer. cDNA cloning of mammalian DDOs was carried out from sources such as bovine, human, mouse, and pig (Katane et al. 2007; Setoyama and Miura 1997; Simonic et al. 1997; Yamamoto et al. 2007). Each of all the cloned DDOs encodes 341 amino acid residues and has 37–38 kDa of estimated molecular weight for a subunit. Similarity in the primary structure is high among the mammalian DDOs with 75–89 % identity. The FAD binding consensus sequences, Gly-Xaa-Gly-Xaa-Xaa-Gly, are present close to the N-termini of the DDOs. Type 1 peroxisome targeting signal (PTS1) sequences are also present in those of the DDOs. Human DDO has a noncanonical PTS1 sequence, although human DDO is targeted to a peroxisome like the other mammalian DDOs (Amery et al. 1998).

Crystal structure of DDO is yet to be determined, although model structures were built for mouse and bovine DDOs based on the homology modeling method (Katane et al. 2007; Sacchi et al. 2002). The models suggested that Arg216, Tyr223, Arg237, Arg278, and Ser308 are presumably important for the activity of DDOs (Fig. 19.9).

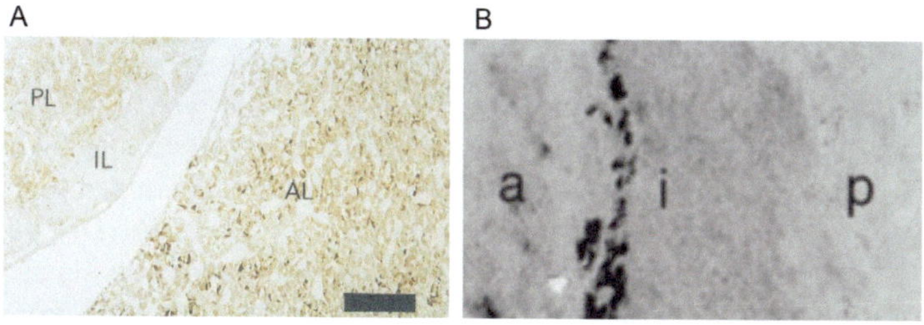

Fig. 19.8 Distribution of D-Asp and DDO in the pituitary gland. (**a**) Immunohistochemical staining of D-Asp is observed in the *anterior* (AL) and *posterior* (PL) lobes of the pituitary gland of a 6-week-old rat, whereas no staining is observed in the intermediate lobe (IL). (**b**) Staining of DDO is seen in the intermediate lobe (i), whereas it not seen in the *anterior* (a) and *posterior* (p) lobes. Intense staining in the intermediate lobe is seen in the region adjacent to the anterior lobe. From Lee et al. (1999) and Schell et al. (1997), modified and reproduced with permission of the copyright owner

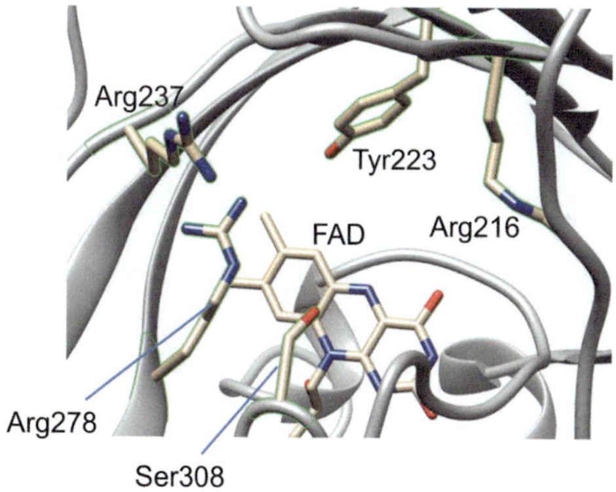

Fig. 19.9 The catalytic site of DDO in a structure model made by the Swiss model (Guex et al. 2009). Residues mentioned in the text and a part of FAD are indicated as stick models, whereas the other residues are shown as a *gray* ribbon model. *Khaki*, *blue*, and *red* atoms are carbon, nitrogen, and oxygen atoms, respectively (This figure is produced by UCSF Chimera Pettersen et al. 2004)

19.3.3 Studies for DDO-Deficient Mice

A couple of research groups have reported establishment of DDO-deficient mice (Errico et al. 2006; Huang et al. 2006). Those mice showed increases in the D-Asp concentration in several tissues and difference in several phenotypes from those of the wild type.

19.3.3.1 Phenotypes Related With α-Melanocyte-Stimulating Hormone

The intermediate lobe of the pituitary gland of a wild-type mouse produces proopiomelanocortin (PONC), the precursor of α-melanocyte-stimulating hormone (α-MSH), whereas expression level of POMC and α-MSH decreased in the intermediate lobe of the DDO-deficient mouse (Huang et al. 2006). The DDO-deficient mouse also showed α-MSH-related symptoms such as sexual deficits, an increase in body mass and decrease in the frequency of autogrooming. These together suggest that D-Asp suppresses the α-MSH level in the intermediate lobe.

19.3.3.2 Impacts on Brain Research

Bergmann glia of the cerebellum of the DDO-deficient mouse showed D-Asp immunoreactivity undetectable in the wild type. The cerebellum controls motor neurons. Results from the rotor rod test showed shorter latency of the DDO-deficient mouse, suggesting the possibility that D-Asp is involved in the control of motor neurons (Weil et al. 2006).

The DDO-deficient mice also showed a selective enhancement of hippocampus-mediated spatial memory in the tests of learning and memory such as Morris water maze and contextual fear conditioning (Errico et al. 2008a). The amount of D-Asp in the hippocampus of the DDO-deficient mouse was 13-fold greater than that of the wild type. Electrophysiological experiments using hippocampal slices of the DDO-deficient mouse showed delayed decay of LTP (Errico et al. 2008b). These suggest that D-Asp contributes to the maintenance of LTP.

19.3.3.3 Implications for Mental Diseases

Dysfunction of the NMDA receptor is observed in patients of depression and schizophrenia. Thus, studies to make use of D-Asp for the treatment of those mental diseases are carried out recently. The DDO-deficient mouse showed shortened immobility time in the Porsolt forced swim test, suggesting that the DDO-deficient mouse is in a less negative mood than the wild type (Weil et al. 2006). Prepulse inhibition undermined by amphetamine or MK-801 restores in the DDO-deficient mouse (Errico et al. 2008a). These suggest that D-Asp exhibits resistance against the sensorimotor-gating deficits, which are seen in patients of schizophrenia (Braff et al. 1992). It should be noted that there has been, however, a report that the DDO-deficient mouse exhibited the prepulse inhibition (Weil et al. 2006). Collectively inhibitors that increase the concentration of D-Asp in the brain may be possible antipsychotic drugs against mental diseases related with the NMDA receptor.

Although roles of D-Asp in our bodies had long been unclear, recent studies using molecular and DDO knockout approaches have revealed the importance of

D-Asp and DDO. We believe that studies to produce inhibitors of DDO will be important for the biological and medical fields.

References

Abou El-Magd RM, Park HK, Kawazoe T et al (2010) The effect of risperidone on D-amino acid oxidase activity as a hypothesis for a novel mechanism of action in the treatment of schizophrenia. J Psychopharmacol 24:1055–1067

Amery L, Brees C, Baes M et al (1998) C-terminal tripeptide Ser-Asn-Leu (SNL) of human D-aspartate oxidase is a functional peroxisome-targeting signal. Biochem J 336(Pt 2):367–371

Beard ME (1990) D-aspartate oxidation by rat and bovine renal peroxisomes: an electron microscopic cytochemical study. J Histochem Cytochem 38:1377–1381

Braff DL, Grillon C, Geyer MA (1992) Gating and habituation of the startle reflex in schizophrenic patients. Arch Gen Psychiatry 49:206–215

D'Aniello A, Rocca E (1972) D-Aspartate oxidase from the hepatopancreas of Octopus vulgaris Lam. Comp Biochem Physiol B 41:625–633

D'Aniello A, Di Cosmo A, Di Cristo C et al (1996) Involvement of D-aspartic acid in the synthesis of testosterone in rat testes. Life Sci 59:97–104

D'Aniello A, Di Fiore MM, Fisher GH et al (2000) Occurrence of D-aspartic acid and N-methyl-D-aspartic acid in rat neuroendocrine tissues and their role in the modulation of luteinizing hormone and growth hormone release. Faseb J 14:699–714

Dunlop DS, Neidle A, McHale D et al (1986) The presence of free D-aspartic acid in rodents and man. Biochem Biophys Res Commun 141:27–32

Errico F, Pirro MT, Affuso A et al (2006) A physiological mechanism to regulate D-aspartic acid and NMDA levels in mammals revealed by D-aspartate oxidase deficient mice. Gene 374:50–57

Errico F, Rossi S, Napolitano F et al (2008a) D-aspartate prevents corticostriatal long-term depression and attenuates schizophrenia-like symptoms induced by amphetamine and MK-801. J Neurosci 28:10404–10414

Errico F, Nistico R, Palma G et al (2008b) Increased levels of D-aspartate in the hippocampus enhance LTP but do not facilitate cognitive flexibility. Mol Cell Neurosci 37:236–246

Errico F, Nistico R, Napolitano F et al (2011) Persistent increase of D-aspartate in D-aspartate oxidase mutant mice induces a precocious hippocampal age-dependent synaptic plasticity and spatial memory decay. Neurobiol Aging 32:2061–2074

Errico F, Napolitano F, Squillace M et al (2013) Decreased levels of D-aspartate and NMDA in the prefrontal cortex and striatum of patients with schizophrenia. J Psychiatr Res 47:1432–1437

Fagg GE, Matus A (1984) Selective association of N-methyl aspartate and quisqualate types of L-glutamate receptor with brain postsynaptic densities. Proc Natl Acad Sci U S A 81:6876–6880

Fang J, Deng D, Nakamura H et al (2008) Oxystress inducing antitumor therapeutics via tumor-targeted delivery of PEG-conjugated D-amino acid oxidase. Int J Cancer 122:1135–1144

Fukui K (2012) Disease-targeted enzymology on the function and structure of D-amino acid metabolic system. Article in Japanese Vitamin 86:63–73

Fukui K, Watanabe F, Shibata T et al (1987) Molecular cloning and sequence analysis of cDNAs encoding porcine kidney D-amino acid oxidase. Biochemistry 26:3612–3618

Fukui K, Momoi K, Watanabe F et al (1988) In vivo and *in vitro* expression of porcine D-amino acid oxidase: *in vitro* system for the synthesis of a functional enzyme. Biochemistry 27:6693–6697

Guex N, Peitsch MC, Schwede T (2009) Automated comparative protein structure modeling with SWISS-MODEL and Swiss-PdbViewer: a historical perspective. Electrophoresis 30(Suppl 1): S162–173

Hashimoto A, Nishikawa T, Oka T et al (1993) Endogenous D-serine in rat brain: N-methyl-D-aspartate receptor-related distribution and aging. J Neurochem 60:783–786

Huang AS, Beigneux A, Weil ZM et al (2006) D-aspartate regulates melanocortin formation and function: behavioral alterations in D-aspartate oxidase-deficient mice. J Neurosci 26:2814–2819

Ishio S, Yamada H, Hayashi M et al (1998) D-aspartate modulates melatonin synthesis in rat pinealocytes. Neurosci Lett 249:143–146

Iwana S, Kawazoe T, Park HK et al (2008) Chlorpromazine oligomer is a potentially active substance that inhibits human D-amino acid oxidase, product of a susceptibility gene for schizophrenia. J Enzyme Inhib Med Chem 23:901–911

Katane M, Homma H (2010) D-aspartate oxidase: the sole catabolic enzyme acting on free D-aspartate in mammals. Chem Biodivers 7:1435–1449

Katane M, Furuchi T, Sekine M et al (2007) Molecular cloning of a cDNA encoding mouse D-aspartate oxidase and functional characterization of its recombinant proteins by site-directed mutagenesis. Amino Acids 32:69–78

Kawazoe T, Tsuge H, Pilone MS et al (2006) Crystal structure of human D-amino acid oxidase: context-dependent variability of the backbone conformation of the VAAGL hydrophobic stretch located at the si-face of the flavin ring. Protein Sci 15:2708–2717

Kawazoe T, Park HK, Iwana S et al (2007) Human D-amino acid oxidase: an update and review. Chem Rec 7:305–315

Kim PM, Duan X, Huang AS et al (2010) Aspartate racemase, generating neuronal D-aspartate, regulates adult neurogenesis. Proc Natl Acad Sci U S A 107:3175–3179

Krebs HA (1935) Metabolism of amino-acids: deamination of amino-acids. Biochem J 29:1620–1644

Lee JA, Homma H, Tashiro K et al (1999) D-aspartate localization in the rat pituitary gland and retina. Brain Res 838:193–199

Matsui T, Sekiguchi M, Hashimoto A et al (1995) Functional comparison of D-serine and glycine in rodents: the effect on cloned NMDA receptors and the extracellular concentration. J Neurochem 65:454–458

Mohn AR, Gainetdinov RR, Caron MG et al (1999) Mice with reduced NMDA receptor expression display behaviors related to schizophrenia. Cell 98:427–436

Molla G, Sacchi S, Bernasconi M et al (2006) Characterization of human D-amino acid oxidase. FEBS Lett 580:2358–2364

Momoi K, Fukui K, Watanabe F et al (1988) Molecular cloning and sequence analysis of cDNA encoding human kidney D-amino acid oxidase. FEBS Lett 238:180–184

Momoi K, Fukui K, Tada M et al (1990) Gene expression of D-amino acid oxidase in rabbit kidney. J Biochem 108:406–413

Mothet JP, Parent AT, Wolosker H et al (2000) D-serine is an endogenous ligand for the glycine site of the N-methyl-D-aspartate receptor. Proc Natl Acad Sci U S A 97:4926–4931

Negri A, Massey V, Williams CH Jr (1987) D-aspartate oxidase from beef kidney. Purification and properties. J Biol Chem 262:10026–10034

Ono K, Shishido Y, Park HK et al (2009) Potential pathophysiological role of D-amino acid oxidase in schizophrenia: immunohistochemical and in situ hybridization study of the expression in human and rat brain. J Neural Transm 116:1335–1347

Park HK, Shishido Y, Ichise-Shishido S et al (2006) Potential role for astroglial D-amino acid oxidase in extracellular D-serine metabolism and cytotoxicity. J Biochem 139:295–304

Pettersen EF, Goddard TD, Huang CC et al (2004) UCSF Chimera--a visualization system for exploratory research and analysis. J Comput Chem 25:1605–1612

Raibekas AA, Fukui K, Massey V (2000) Design and properties of human D-amino acid oxidase with covalently attached flavin. Proc Natl Acad Sci U S A 97:3089–3093

Sacchi S, Lorenzi S, Molla G et al (2002) Engineering the substrate specificity of D-amino-acid oxidase. J Biol Chem 277:27510–27516

Schell MJ, Molliver ME, Snyder SH (1995) D-serine, an endogenous synaptic modulator: localization to astrocytes and glutamate-stimulated release. Proc Natl Acad Sci U S A 92:3948–3952

Schell MJ, Cooper OB, Snyder SH (1997) D-aspartate localizations imply neuronal and neuroendocrine roles. Proc Natl Acad Sci U S A 94:2013–2018

Setoyama C, Miura R (1997) Structural and functional characterization of the human brain D-aspartate oxidase. J Biochem 121:798–803

Simonic T, Duga S, Negri A et al (1997) cDNA cloning and expression of the flavoprotein D-aspartate oxidase from bovine kidney cortex. Biochem J 322(Pt 3):729–735

Still JL, Buell MV, Knox WE (1949) Studies on the cyclophorase system: VII. D-aspartic oxidase. J Biol Chem 179:831–837

Tada M, Fukui K, Momoi K et al (1990) Cloning and expression of a cDNA encoding mouse kidney D-amino acid oxidase. Gene 90:293–297

Tran DH, Shishido Y, Chung SP et al (2015) Identification of two promoters for human D-amino acid oxidase gene: implication for the differential promoter regulation mediated by PAX5/PAX2. J Biochem 157:377–387

Tsai G, Yang P, Chung LC et al (1998) D-serine added to antipsychotics for the treatment of schizophrenia. Biol Psychiatry 44:1081–1089

Van Veldhoven PP, Brees C, Mannaerts GP (1991) D-aspartate oxidase, a peroxisomal enzyme in liver of rat and man. Biochim Biophys Acta 1073:203–208

Weil ZM, Huang AS, Beigneux A et al (2006) Behavioural alterations in male mice lacking the gene for D-aspartate oxidase. Behav Brain Res 171:295–302

Wolosker H, Blackshaw S, Snyder SH (1999) Serine racemase: a glial enzyme synthesizing D-serine to regulate glutamate-N-methyl-D-aspartate neurotransmission. Proc Natl Acad Sci U S A 96:13409–13414

Yagi K, Nagatsu T, Ozawa T (1956) Inhibitory action of chlorpromazine on the oxidation of D-amino-acid in the diencephalon part of the brain. Nature 177:891–892

Yagi K, Ozawa T, Nagatsu T (1960) Mechanism of inhibition of D-amino acid oxidase. IV. Inhibitory action of chlorpromazine. Biochim Biophys Acta 43:310–317

Yamamoto A, Tanaka H, Ishida T et al (2007) Functional and structural characterization of D-aspartate oxidase from porcine kidney: non-Michaelis kinetics due to substrate activation. J Biochem 141:363–376

Chapter 20
Eukaryotic D-Serine Dehydratase

Tomokazu Ito, Tohru Yoshimura, Tetsuo Ishida, and Hiroyuki Tanaka

Abstract In mammals, D-serine is endogenously synthesized by serine racemase (SR), and it plays important roles in memory and learning by binding to the *N*-methyl-D-aspartate (NMDA) receptor and δ2 glutamate receptor. D-Amino acid oxidase (DAO) is the only mammalian enzyme that degrades D-serine. Many eukaryotes, excluding mammals, have another D-serine-degrading enzyme, D-serine dehydratase (DSD). DSD is a pyridoxal 5′-phosphate (PLP)- and Zn^{2+}-dependent enzyme that catalyzes the deamination of D-serine to produce pyruvate and ammonia, and it is distinct from the classical D-serine dehydratases such as DsdA of *Escherichia coli*. In chicken, DSD is expressed in the kidney, liver, and brain, and it plays a primary role in removing D-serine from the circulating blood. The cellular slime mold *Dictyostelium discoideum* contains three D-serine-metabolizing enzymes, SR, DAO, and DSD. Among these enzymes, DSD is responsible for D-serine degradation. Although the physiological significance of DSD is still unclear, it is thought to contribute to the maintenance of D-serine at low levels in these organisms. This article describes the biological distribution, physiological role, enzyme properties, structure, reaction mechanism, and application of DSD based on recent studies of the enzyme in chicken and yeast.

Keywords D-Serine • D-Serine dehydratase • Zn^{2+}

T. Ito (✉) • T. Yoshimura
Department of Applied Molecular Biosciences, Graduate School of Bioagricultural Sciences, Nagoya University, Nagoya, Aichi, Japan
e-mail: ito-t@agr.nagoya-u.ac.jp

T. Ishida
Department of Chemistry, Biology and Marine Science, University of the Ryukyus, Naha, Okinawa, Japan

H. Tanaka (✉)
Department of Biochemistry and Molecular Biology, Shiga University of Medical Science, Otsu, Shiga, Japan
e-mail: tanakah@belle.shiga-med.ac.jp

T. Yoshimura et al. (eds.), *D-Amino Acids*, DOI 10.1007/978-4-431-56077-7_20

20.1 Biological Distribution and Physiological Function of DSD

D-Serine dehydratase (DSD) is a pyridoxal 5′-phosphate (PLP)- and Zn^{2+}-dependent enzyme that catalyzes the deamination of D-serine to produce pyruvate and ammonia, and it is distinct from the classical D-serine dehydratases such as DsdA of *Escherichia coli*. The biological distribution of DSD is represented on a phylogenetic tree (Fig. 20.1). A database search with DSD gene sequences from *Saccharomyces cerevisiae* (scDSD) and chicken (chDSD) revealed that many eukaryotes, including fungi, cellular slime molds, invertebrates, and vertebrates, possess DSD homologs. In higher plants and arthropods, no DSD homolog was found. Among vertebrates, only mammals did not possess a DSD homolog (Fig. 20.1). DSD homologs were also found in the genomes of *Actinobacteria*, *Bacteroidetes*, *Planctomycetes*, and α-, β-, γ-, and δ-*Proteobacteria*. DSD homologs were not detected in Archaea. This pattern of DSD distribution among taxa was completely different from that of DsdA, the classical D-serine dehydratase. The classical enzymes were found only in bacteria such as *Firmicutes* and β- and γ-*Proteobacteria*, and homologs were not detected in either Archaea or Eukaryota, demonstrating a more restricted distribution than that of DSD.

The expression of DSD homologs in vertebrates has been examined enzymatically using a highly sensitive high-performance liquid chromatography (HPLC) method (Tanaka et al. 2007). Tissue homogenates of various vertebrates (such as the kidney, liver, and brain of birds, reptiles, amphibians, and fishes) were incubated with D-serine and the resultant pyruvate was quantified after derivatization with 3-methyl-2-benzothiazolinone hydrazone (MBTH) (Tanaka et al. 2009). All of the non-mammalian tissues examined exhibit DSD activity (Fig. 20.1). In contrast, no DSD activity is detected for mammalian tissue homogenates including the kidney, liver, and brain homogenates of rats, mice, rabbits, and pigs.

Based on the observed biological distribution of DSD (Fig. 20.1), there are three possible scenarios for the evolution of this enzyme. (1) An ancestral DSD homolog has been transferred through horizontal gene transfer from bacteria to cellular slime molds, fungi, and urochordates. (2) DSD has been vertically transferred and then repeatedly lost in many lineages, including archaea, higher plants, arthropods, and mammals. (3) Ancestral DSD homologs are present in the genomes of many species, but their sequences are highly diverged from the ancestral sequence owing to strong selective pressures on DSD.

What is the physiological role of DSD? In mammals, D-serine is one of the major physiological co-agonists of NMDA receptors, and the activity of the receptors is finely tuned by D-serine in various brain regions. NMDA receptors play critical roles in excitatory glutamatergic synaptic transmission, and both hyper- and hypofunctions of these receptors are believed to cause serious dysfunction of mammalian brains (Scolari and Acosta 2007 and Nishikawa 2011). Therefore, the synthesis, transport, and degradation of D-serine need to be balanced in a concerted manner to finely control the local concentration of D-serine. In mammals, D-serine is

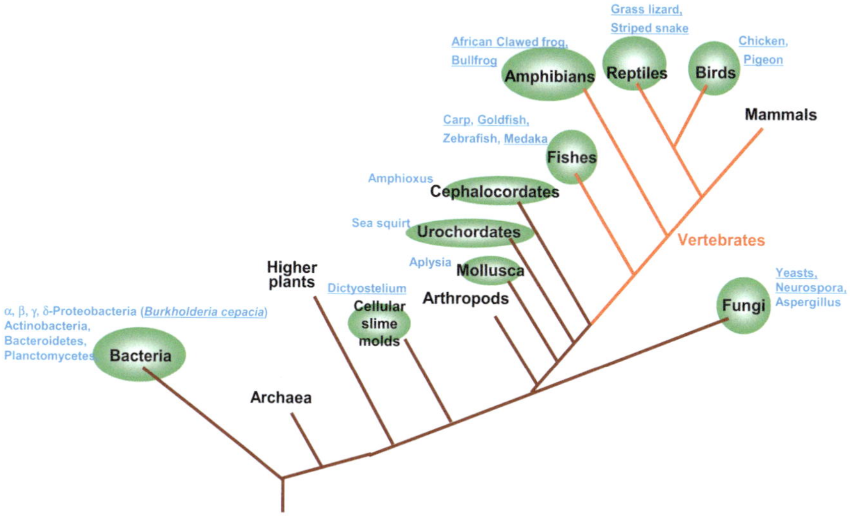

Fig. 20.1 Biological distribution of Zn^{2+}-dependent D-serine dehydratase (DSD). Presence or absence of DSD is indicated on a schematic phylogenetic tree showing major phyla. The branching part of the vertebrate (*orange*) is enlarged. Zn^{2+}-dependent DSD homologs are found for animals belonging to the *circled* phyla (*light green*). Species which possess DSD homologs are shown in *blue* text and among them are *underlined* to which the enzymatic activities have been actually confirmed. Note that among the vertebrates, only mammals lack DSD homologs

produced by serine racemase (SR) and is mainly degraded by DAO (Schell 2004). As described before, DSD is not presented in mammals. In rat brains, the localization of DAO is reciprocal to that of D-serine (Horiike et al. 1987), suggesting that, in mammalian brains, DAO determines the basal levels of D-serine.

As described above, many eukaryotes excluding mammals have two D-serine-degrading enzymes, DSD and DAO. Currently, there are few reports concerning the physiological significance of DSD, and therefore, it is not clear yet why these organisms need two D-serine-degrading enzymes.

Recently, the tissue and cellular distribution of DSD has been examined in chicken using a specific anti-chicken DSD antibody (Nishimura et al. 2014). In the chicken renal cortex, only epithelial cells of the proximal tubules are strongly positive for the DSD antibody. In the chicken liver, only hepatocytes exhibit intense staining for DSD, with the intensity levels even throughout the liver lobule. The cellular localization of DSD in these tissues was identical to that of DAO in mammals (Horiike et al. 1985), suggesting that DSD plays a role in removing D-serine from the circulating blood. In the chicken cerebellar cortex, DSD localizes exclusively in Bergmann glial cells and astrocytes. This DSD localization is identical to that of DAO in the mammalian cerebellum (Horiike et al. 1987). In contrast to the mammalian cerebrum, which does not contain DAO-expressing cells, astrocytes in the chicken cerebrum show intense staining for DSD in the cytoplasm of the cell body and processes. In accordance, chicken brain homogenates exhibit only DSD activity. These results indicate that, in the chicken brain, DSD is the only enzyme responsible for the degradation of D-serine and may regulate brain function through the control of D-serine level (Tanaka et al. 2007).

A cellular slime mold *D. discoideum* has three D-serine metabolic enzymes, SR, DAO, and DSD, but does not have NMDA and δ2 glutamate receptors. The enzymological studies of these three enzymes showed that the D-serine-degrading activity of DAO and SR was 1700- and 460-fold lower, respectively, than that of DSD. In wild-type *D. discoideum* cells, intracellular D-serine is maintained at considerably low levels. However, the DSD-knockout cells accumulate intracellular D-serine. The cell-free extract of this mutant strain does not show D-serine degradation activity, and the mutant cells are highly sensitive to exogenous D-serine. These observations suggest that, in *D. discoideum*, DSD is responsible for the degradation of both endogenous and exogenous D-serine.

These observations demonstrate that, in some organisms and/or tissues, DSD plays a primary role in D-serine regulation. However, further analysis about spatial and temporal control of DSD and DAO will be required for elucidating the respective contribution of these two enzymes in D-serine regulation.

20.2 Enzymatic Properties of DSD

The enzymatic properties of Zn^{2+}-dependent D-serine dehydratase (DSD) have been studied extensively using chDSD and scDSD, which exhibit 31 % sequence identity with each other. DSD exists as a homodimer in solution. As described in detail in the next section, DSD consists of an N-terminal $(\alpha/\beta)_8$ TIM-barrel domain and a C-terminal β-barrel domain. Therefore, DSD is a fold-type III PLP-dependent enzyme. The fold and subunit composition of DSD are completely different from those of the classical D-serine dehydratases such as DsdA of *E. coli*. DsdA is a monomeric protein and is a fold-type II PLP-dependent enzyme. Importantly, DsdA does not require Zn^{2+} for catalysis.

The substrate specificity and kinetic parameters of the two Zn^{2+}-dependent DSDs are summarized in Table 20.1. As described below, DSD enzymes are highly specific to D-serine. The k_{cat} and K_m values of chDSD for D-serine were 0.13 mM and 0.81 s^{-1}, respectively (Tanaka et al. 2008), and those for scDSD were 0.19 mM and 17 s^{-1}, respectively (Ito et al. 2012). Although D-threonine serves as a substrate for both chDSD and scDSD, the k_{cat} value of chDSD and scDSD for D-threonine is about 15- and 43-fold lesser than that for D-serine. D-*allo*-Threonine is also a substrate for scDSD, with a k_{cat} value about one-third that of D-threonine, indicating that DSD prefers the *3S* configuration for the hydroxyl group. In addition to these amino acids, β-Cl-D-alanine, an analogue of D-serine in which the hydroxyl group is replaced with a chlorine atom, is a substrate for scDSD, with a k_{cat}/K_m value about one-fiftieth that of D-serine (Ito et al. 2008). On the other hand, chDSD shows slight activity for L-serine, but not for L-threonine. The k_{cat}/K_m value of chDSD for L-serine is 16 $M^{-1}\cdot s^{-1}$, 390-fold lower than that of D-serine (6.2×10^3 $M^{-1}\cdot s^{-1}$). In

Table 20.1 Kinetic parameters of chDSD and scDSD

	K_m (mM)	k_{cat} (s^{-1})	k_{cat}/K_m ($M^{-1} \cdot s^{-1}$)
(chDSD)[a]			
D-Serine	0.13	0.81	6.2×10^3
D-Threonine	0.31	0.051	1.6×10^2
L-Serine	12	0.20	16
L-Threonine	n. d.[c]	n. d.	n. d.
(scDSD)[b]			
D-Serine	0.19	17	8.9×10^4
D-Threonine	0.039	0.39	1.0×10^4
D-*allo*-Threonine	0.033	0.13	3.9×10^3
β-Cl-D-alanine	0.88	0.19	2.2×10^2
L-Serine	n. d.	n. d.	n. d.
L-Threonine	n. d.	n. d.	n. d.

[a]Values were from Tanaka et al. (2008).
[b]Values were from Ito et al. (2012)
[c]Not detected

contrast, scDSD does not show detectable levels of L-serine dehydratase activity. Other proteinogenic L-amino acids and their corresponding enantiomers (D-amino acids) do not function as substrates for these DSDs. The primary structures of the DSDs are similar to those of metal-activated D-threonine aldolases (Liu et al. 1998) and bacterial alanine racemases (Inagaki et al. 1986), but neither chDSD nor scDSD have detectable levels of racemase activity or aldolase activity for D-serine and/or D-threonine. In addition, transamination and decarboxylation activity is not found for these DSDs (Tanaka et al. 2011 and Ito et al. 2014). DSD is highly sensitive to dehydration.

It is well known that both scDSD and chDSD require not only PLP but also a Zn^{2+} ion for catalytic reactions (Ito et al. 2008, 2012, and Tanaka et al. 2011). This Zn^{2+} requirement is one of the main characteristics of this DSD family, because DSD orthologs of *Schizosaccharomyces pombe*, *D. discoideum*, and *Bacillus licheniformis* with 13.3–28.1 % sequence identity with scDSD were all found to be Zn^{2+}-dependent enzymes (Ito et al. 2014), whereas many other PLP enzymes including the classical D-serine dehydratase do not require Zn^{2+} for their activity. As confirmed using atomic absorption spectrometry, scDSD binds one mole of Zn^{2+} per monomer (Ito et al. 2008). Crystallographic analysis of chDSD has revealed that one Zn^{2+} is coordinated by His347 and Cys349 at the active site (Tanaka et al. 2011; see Fig. 20.2c).

Removal of Zn^{2+} from scDSD completely abolishes the D-serine dehydratase activity of the enzyme, and the activity of scDSD samples is correlated with their Zn^{2+} content (Ito et al. 2008). An intensive EDTA treatment of chDSD reduces activity 20-fold (Tanaka et al. 2011). In scDSD and chDSD, the decreased activities of the samples with low concentrations of Zn^{2+} are fully recovered by adding ~5 μM and ~50 μM Zn^{2+}, respectively. Higher concentrations of Zn^{2+} inhibit the activity of the enzyme. Other metal ions such as K^+, Na^+, Mg^{2+}, Ca^{2+}, Cu^{2+}, and Ni^{2+}

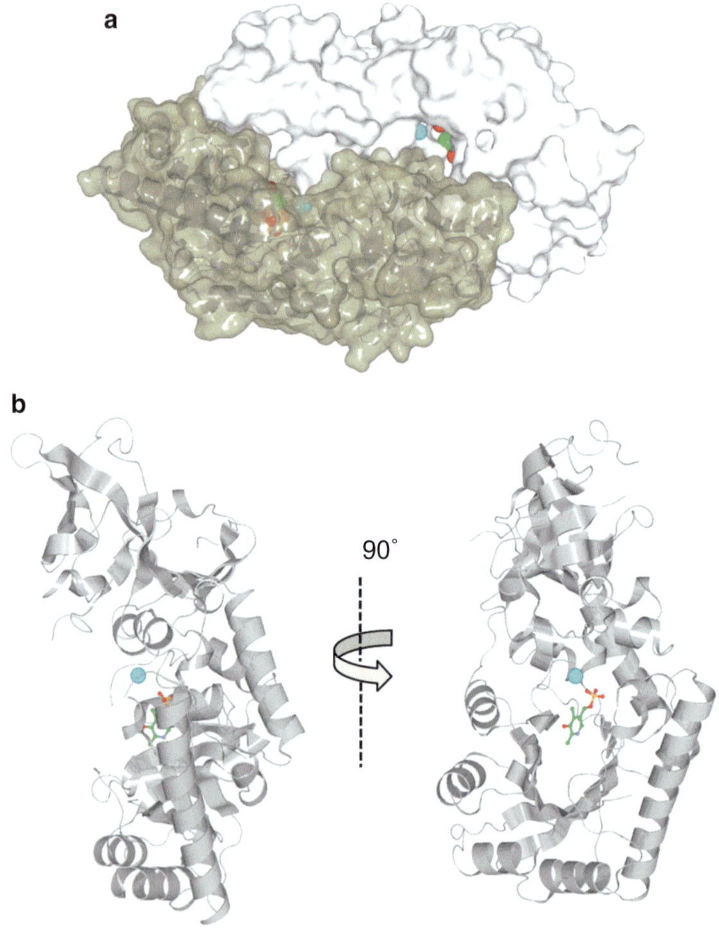

Fig. 20.2 Structure of chDSD. (**a**) chDSD dimer (PDB code 3ANU). One of the subunits is colored *white*. Two active site pockets are formed between the subunits. The C4a atom of PLP and Zn^{2+} are shown as *green* and *cyan spheres*, respectively. (**b**) Subunit structure of chDSD. The PLP is shown as a stick model. The Zn^{2+} is shown as a cyan sphere. (**c**) Stereoview of the active site of chDSD. The Zn^{2+} is shown as a cyan *sphere*. The carbon atoms of PLP are colored *light green*. PLP and Lys45 form internal Schiff base. (**d**) A structural model of the chDSD-D-serine complex (external aldimine intermediate). The corresponding residues of scDSD are shown in parentheses. The ε-amino group of Lys45 is located at a distance of 3.2 Å from the C_α atom and the hydroxyl O atom of D-serine. The Tyr174 interacts with the pyridine nitrogen of PLP

do not restore enzyme activity. The addition of Mn^{2+} partially restores the activity of chDSD (up to ~20 % of the original activity) but does not recover scDSD activity. For scDSD, the enzymatic activities for D-threonine and D-*allo*-threonine are also completely abolished by EDTA treatment. Interestingly, the activity of the EDTA-treated scDSD for β-Cl-D-alanine is fully retained. In accordance with these observations, the enzymes with mutations at Zn^{2+}-binding residues

(His398Ala and/or Cys400Ala for scDSD) have no D-serine activity but retain β-Cl-D-alanine activity (Ito et al. 2012). The difference of Zn^{2+} dependency is probably due to the difference in leaving group efficiency between D-serine and β-Cl-D-alanine. The dehydration of D-serine and the α,β-elimination of β-Cl-D-alanine probably proceed through common steps, α-proton abstraction and elimination of the leaving group on the β-carbon. As a leaving group, chloride is more efficient than a hydroxyl group, and therefore, it can be eliminated spontaneously after the deprotonation at Cα. Zn^{2+} seems to be indispensable for the elimination of the hydroxyl group of substrates.

20.3 X-Ray Structure of DSD

The crystal structures of chDSD have been determined (Tanaka et al. 2011). The overall and subunit structures of chDSD are shown in Figs. 20.2a, b. chDSD forms a head-to-tail dimer in the crystal, similar to that observed in the solution. chDSD consists of an N-terminal $(\alpha/\beta)_8$ TIM-barrel domain and a C-terminal β-barrel domain. This overall structure is very similar to that of prokaryotic alanine racemases (Couñago et al. 2009), which are representative members of the fold-type III PLP-dependent enzyme group.

The active site structure of free chDSD is shown in Fig. 20.2c. The dimeric DSD has two active sites in the dimer. Each active site is located at the dimer interface and contains one PLP molecule and one Zn^{2+} ion, and many water molecules are found in the active site cleft (w1–w9, in Fig. 20.2b, c). The PLP forms a typical Schiff base with the ε-amino group of Lys45, and the Schiff base bond is nearly coplanar with the pyridine ring of PLP (Fig. 20.2c). The pyridine ring is sandwiched by the two histidine residues His43 and His176 from the *re*- and *si*-face sides, respectively (Fig. 20.2c). A Zn^{2+} is coordinated with His347 and Cys349 at distances of 2.2 Å and 2.4 Å, respectively. In addition to these two residues, the three water molecules coordinate in Zn^{2+} binding (w4, w5, and w6 in Fig. 20.2c). The phosphate group of PLP interacts with the main-chain N atoms of Thr222, Gly241, Asn242, and the hydroxyl groups of Tyr181 and Thr222. The OP3 atom of the phosphate group forms hydrogen bonds with two water molecules (w5 and w9), and one of these water molecules (w5) coordinates Zn^{2+}. Unlike other PLP-dependent enzymes, the N1 atom of PLP seems to form a hydrogen bond with the phenolic hydroxyl group of Tyr174 (Fig. 20.2c). Since the N1 atom is likely to act as an acceptor of this hydrogen bond, the N1 atom is probably in an unprotonated state in chDSD.

Recently, the crystal structure of DsdA from *E. coli*, the most frequently investigated classical D-serine dehydratase, was reported by Urusova et al. (2012). DsdA is a monomeric protein with two domains, one of which is much larger than the other. The large domain is composed of 15 α-helices, seven-stranded β-sheets, and one PLP molecule. No Zn^{2+} was found in the crystal structure of DsdA. This DsdA structure is completely different from the chDSD structure described above, and instead resembles that of mammalian serine racemase (Goto et al. 2009) and/or L-serine/L-threonine dehydratase (Yamada et al. 2003). DsdA therefore belongs to the fold-type II PLP-dependent enzymes. These observations indicate that classical D-serine dehydratase and Zn^{2+}-dependent D-serine dehydratase have evolved in a convergent manner.

20.4 The Role of Zn^{2+} in DSD

How is the active site Zn^{2+} involved in enzymatic activity? Some PLP-dependent enzymes inevitably require a metal ion for their enzyme activity. For example, the classical D-serine dehydratase and the L-serine/L-threonine dehydratase require K^+ (Urusova et al. 2012 and Yamada et al. 2003), and the mammalian serine racemase is activated by Mg^{2+} (De Miranda et al. 2002). However, in these enzymes the activity-relevant metal ion locates outside the active site. Therefore, it is suggested that metal ions play a structural role in these enzymes (Goto et al. 2009 and Yamada et al. 2003).

Crystal structures of native- and EDTA-treated chDSD demonstrated that the active site structure is not significantly changed by Zn^{2+} removal. The circular dichroism (CD) and the tryptophan fluorescence analyses of the EDTA-treated wild-type scDSD and the Zn^{2+}-binding site-mutated scDSD (Cys400Ala) demonstrate the absence of a gross conformational change. These structural and biochemical results suggest that Zn^{2+} does not play a structural role in the maintenance of a catalytically competent conformation. In fact, even after the removal of Zn^{2+}, both chDSD and scDSD retain the ability to form D-serine-PLP external aldimine intermediates. This indicates that Zn^{2+} is required for the reaction step(s) after the formation of the external aldimine intermediate. Based on the crystal structure of the EDTA-treated chDSD-D-serine complex, a structural model of the chDSD-D-serine complex was built (Fig. 20.2d). In this model, Zn^{2+} is able to interact directly with the hydroxyl group of D-serine (the distance between the Zn^{2+} and the hydroxyl group of D-serine is 2.4 Å). These observations support the notion that Zn^{2+} activates the hydroxyl group toward elimination via direct interaction.

20.5 The Reaction Mechanism of DSD

Based on the results described above, the following reaction mechanism is proposed for DSD (see Fig. 20.3). The resting enzyme forms the internal aldimine between PLP and the active site Lys residue (Lys45 in chDSD and Lys57 in scDSD) ([I] in Fig. 20.3). The substrate, D-serine, enters the active site from the *si*-face side of the internal aldimine and forms the D-serine-PLP external aldimine by replacing

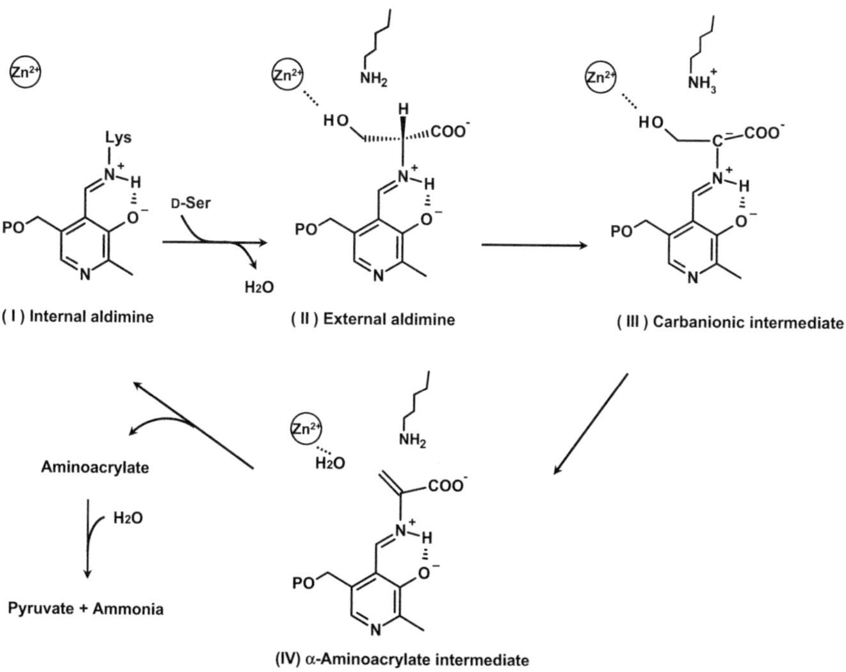

Fig. 20.3 Reaction mechanism of Zn^{2+}-dependent DSD. The resting enzyme forms internal aldimine with the active site Lys residue (Lys45 in chDSD and Lys57 in ScDSD) (I). First, D-serine forms the D-serine-PLP external aldimine (II). Second, the liberated Lys residue abstracts the C_α proton to form a carbanionic intermediate (III). Third, a proton is donated to the β-hydroxyl group by the protonated Lys residue, and the hydroxyl group is eliminated (IV). In this step, Zn^{2+} probably activates the hydroxyl group toward elimination by direct interaction with the β-hydroxyl group. Finally, the resulting α-aminoacrylate intermediate undergoes transaldimination to regenerate the enzyme, and α-aminoacrylate is released. The released α-aminoacrylate is non-enzymatically hydrolyzed to produce pyruvate and ammonia

the Lys residue that initially forms the internal aldimine with PLP ([II] in Fig. 20.3). The liberated ε-amino group of the Lys residue approaches from the *re*-face of the intermediate to the Cα proton of D-serine and abstracts the proton to form a carbanion intermediate ([III] in Fig. 20.3). This α-deprotonation is probably part of the rate-limiting steps in the reaction of scDSD, judging by the UV-visible spectra analyses and the primary deuterium isotope effect observed for D-[α-^{2}H]-serine. Then, a proton is donated to the hydroxyl group of the substrate, and the hydroxyl group is eliminated to form an α-aminoacrylate intermediate ([IV] in Fig. 20.3). In the structural model of the chDSD-D-serine external aldimine complex (Fig. 20.2d), the ε-amino group of Lys45 is located at a distance of 3.2 Å from both the Cα atom and the β-hydroxyl O atom of D-serine. It is expected that this Lys residue is involved in both α-deprotonation and hydroxyl group protonation. Finally, the resulting α-aminoacrylate intermediate undergoes transaldimination to regenerate the resting internal aldimine, and α-aminoacrylate is released. The α-aminoacrylate is non-enzymatically hydrolyzed to produce pyruvate and ammonia.

This reaction mechanism is different from that of DsdA. In DsdA, it is suggested that the hydroxyl group of D-serine is coordinated with the carboxyl group of Asp238 and the phosphate group of PLP via water molecule. It is proposed that the phosphate group of PLP donates a proton to the hydroxyl group and the α-proton is abstracted by the hydroxyl group of Thr168, and they occur in a concerted fashion (Urusova et al. 2012).

The structure of chDSD shows that the pyridine nitrogen of PLP (N1) is within the distance for hydrogen bonding with the highly conserved Tyr residue (Tyr174) (Fig. 20.2d). Unlike in chDSD, in many PLP enzymes catalyzing transamination, decarboxylation, and aldol cleavage, N1 interacts with the acidic side chain of the Asp or Glu residue, which can protonate N1. The protonation is advantageous for PLP to function as a resonance-stabilized electron sink. However, in the case of DSD, based on the phenolic p*Ka* value of Tyr, it is difficult to form a stabilized carbanionic intermediate such as a quinonoid during the DSD reaction. Indeed, quinonoid intermediates with absorption bands around 500 nm are not observed during the reaction of DSDs. The mutants of scDSD, in which Tyr203 is changed to nonpolar or basic residues, retain substantial activity levels. In contrast, the mutation of Tyr203 to Asp and Glu decreases the k_{cat}/K_m value by 10^5-fold compared with that of the wild-type enzyme. These results suggest that the Tyr residue does not participate in the protonation of the N1 of PLP, and that the pyridine nitrogen of PLP is unprotonated in DSD. The protonation of the PLP N1 atom seems to be disadvantageous for the catalysis of DSD.

In a reaction of scDSD with D-serine (D-[α-^{1}H]-serine) in buffered D_2O, the rate constants for the α-proton abstraction and the hydroxyl group elimination were separately determined by ^{1}H-NMR. The rate of α-deprotonation (9.7 s^{-1}) was comparable to that of the rate of hydroxyl group elimination (8.5 s^{-1}) and a small but significant amount of D-[α-^{2}H]-serine is formed during the incubation. This shows that scDSD yields a carbanionic intermediate before eliminating the hydroxyl group via an E1cB-type mechanism.

Interestingly, the ^{1}H-NMR measurement demonstrates that the EDTA-treated scDSD abolished both α-proton abstraction and hydroxyl group elimination activities. Mutation of the C400, one of the Zn^{2+}-binding residues, to alanine results in exiguous α-deprotonation activity and the complete loss of hydroxyl group elimination activity for D-serine; the rate constant for α-deprotonation is 0.065 s^{-1}, a value 150-fold lower than that of the wild-type enzyme. Moreover, scDSD never catalyzes α-deprotonation of D-alanine regardless of whether Zn^{2+} is present, although D-alanine forms an external aldimine with scDSD as confirmed by a UV-visible spectra analysis. These results indicate that the α-proton is abstracted from the amino acids bearing a hydroxyl group only in the presence of Zn^{2+}. It is known that PLP with unprotonated N1 is less efficient for increasing C_α-H acidity compared to PLP with protonated N1. The above results suggest that Zn^{2+} contributes to the increased C_α-H acidity of the external aldimine intermediate of scDSD. The carbanionic intermediate formed after the α-deprotonation is a common intermediate for many PLP-dependent enzyme reactions, including racemization and transamination. If a carbanionic intermediate is protonated at a C4′ atom instead of a C_α atom, pyridoxamine 5′-phosphate (PMP) and α-keto acid is formed (i.e., abortive transamination), possibly leading to enzyme inactivation. The unprotonated N1 and the Zn^{2+} ion seem to avoid the formation of the stable carbanionic intermediate to minimize unwanted transamination as a side reaction.

20.6 Application of DSD

Several lines of evidence suggest that abnormality in the D-serine system is associated with the pathophysiology of neuropsychiatric disorders including schizophrenia, ischemic neuronal cell death, Alzheimer's disease, and amyotrophic lateral sclerosis (ALS). Some studies have reported that in schizophrenics, the D-serine content and the relative amount of D-serine (D-serine/total serine) in the serum and cerebrospinal fluid are decreased. Transient ischemia elevates the extracellular concentrations of D-serine in the rat cerebra and rabbit piriform cortex. In ALS patients, the levels of D-serine and serine racemase in the spinal cord are elevated according to the disease progress.

Conventionally, D-serine is assayed by HPLC after the amino acid is derivatized to a fluorescent diastereomer. The HPLC method is accurate and sensitive, but time-consuming. Therefore, utilizing the high specificity of DSDs for D-serine, we have developed three kinds of enzymatic D-serine assay methods (Ito et al. 2007; Naka et al. 2010, and Suzuki et al. 2011). With these assay methods, the D-serine contents of several hundred samples can be analyzed within a few hours. This is potentially useful for studying the physiological significance and the clinical relevance of D-serine. For all methods, D-serine in biological samples is specifically degraded by DSD, and the resultant pyruvate is determined. The micromolar range of D-serine can be assayed by quantifying the pyruvate using NADH and lactate dehydrogenase (LDH) or 2,4-dinitrophenylhydrazine (2,4-DNP) (Ito et al. 2007 and Suzuki et al. 2011). The lower detection limit of the DSD/LDH coupling assay and the DSD/2,4-DNP assay is 20 μM and 5 μM, respectively. The DSD/LDH method is applicable for the quantitative determination of urinary D-serine. Submicromolar D-serine can be assayed in the presence of pyruvate oxidase, horseradish peroxidase (POD), and *N*-acetyl-3,7-dihydroxyphenoxazine (Amplex Red) (Naka et al. 2010). Amplex Red is oxidized by HRP in the presence of H_2O_2 (generated by the oxidation of pyruvate by POX) to produce highly fluorescent resorufin. With this method, a linear calibration curve for D-serine is obtained that ranges between 0.1 μM and 5.0 μM. This assay method can be used to determine the content of D-serine in serum. These assay methods can also be used for the quantitative determination of the total (D- + L-) serine content by coupling an amino acid racemase to convert L-serine into D-serine. The Tyr354Asn mutant of *Geobacillus stearothermophilus* alanine racemase, catalyzing serine racemization, can be used for this assay method (Ito et al. 2007 and Naka et al. 2010).

References

Couñago RM et al (2009) Biochemical and structural characterization of alanine racemase from *Bacillus anthracis* (Ames). BMC Struct Biol 9:53

De Miranda J et al (2002) Cofactors of serine racemase that physiologically stimulate the synthesis of the *N*-methyl-D-aspartate (NMDA) receptor coagonist D-serine. Proc Natl Acad Sci U S A 99 (22):14542–14547

Goto M et al (2009) Crystal structure of a homolog of mammalian serine racemase from *Schizosaccharomyces pombe*. J Biol Chem 284:25944–25952

Horiike K et al (1985) Histochemical staining of cells containing flavoenzyme D-amino-acid oxidase based on its enzymatic activity: application of a coupled peroxidation method. Acta Histochem Cytochem 18:539–550

Horiike K et al (1987) Localization of D-amino acid oxidase in Bergmann glial cells and astrocytes of rat cerebellum. Brain Res Bull 19:587–596

Inagaki K et al (1986) Thermostable alanine racemase from *Bacillus stearothermophilus*: molecular cloning of the gene, enzyme purification, and characterization. Biochemistry 25 (11):3268–3274

Ito T et al (2007) Enzymatic assay of D-serine using D-serine dehydratase from *Saccharomyces cerevisiae*. Anal Biochem 371(2):167–172

Ito T et al (2008) A novel zinc-dependent D-serine dehydratase from *Saccharomyces cerevisiae*. Biochem J 409(2):399–406

Ito T et al (2012) Role of zinc ion for catalytic activity in D-serine dehydratase from *Saccharomyces cerevisiae*. FEBS J 279(4):612–624

Ito T et al (2014) Reaction mechanism of Zn^{2+}-dependent D-serine dehydratase: role of a conserved tyrosine residue interacting with pyridine ring nitrogen of pyridoxal 5′-phosphate. J Biochem 156(3):173–180

Liu JQ et al (1998) A novel metal-activated pyridoxal enzyme with a unique primary structure, low specificity D-threonine aldolase from *Arthrobacter sp*. Strain DK-38. Molecular cloning and cofactor characterization. J Biol Chem 273(27):16678–16685

Naka T et al (2010) A highly sensitive enzymatic assay for D- and total serine detection using D-serine dehydratase from *Saccharomyces cerevisiae*. J Mol Catal B Enzym 67:150–154

Nishikawa T (2011) Analysis of free D-serine in mammals and its biological relevance. J Chromatogr B Anal Technol Biomed Life Sci 879(29):3169–3183

Nishimura Y et al (2014) Immunohistochemical localization of D-serine dehydratase in chicken tissues. Acta Histochem 116:702–707

Schell MJ (2004) The N-methyl D-aspartate receptor glycine site and D-serine metabolism: an evolutionary perspective. Philos Trans R Soc Lond B 359:943–964

Scolari MJ, Acosta GB (2007) D-Serine: a new word in the glutamatergic neuro-glial language. Amino Acids 33:563–574

Suzuki C et al (2011) Rapid determination of free D-serine with chicken D-serine dehydratase. J Chromatogr B Anal Technol Biomed Life Sci 879(29):3326–3330

Tanaka H et al (2007) Simultaneous measurement of D-serine dehydratase and D-amino acid oxidase activities by the detection of 2-oxo acid formation with reverse-phase high-performance liquid chromatography. Anal Biochem 362:83–88

Tanaka H et al (2008) D-Serine dehydratase from chicken kidney: a vertebral homologue of the cryptic enzyme from *Burkholderia cepacia*. J Biochem 143:49–57

Tanaka H et al (2009) Discovery of D-serine dehydratase in vertebrate and its deficiency in mammals. Tanpakushitsu Kakusan Koso 54:1190–1196

Tanaka H et al (2011) Crystal structure of a zinc-dependent D-serine dehydratase from chicken kidney. J Biol Chem 286(31):27548–27558

Urusova DV et al (2012) Crystal structure of D-serine dehydratase from Escherichia coli. Biochim Biophys Acta 1824(3):422–432

Yamada T et al (2003) Crystal structure of serine dehydratase from rat liver. Biochemistry 42 (44):12854–12865

Chapter 21
Aspartate Racemase: Function, Structure, and Reaction Mechanism

Masafumi Yohda

Abstract Aspartate racemases distribute and function to produce D-aspartate in eubacteria, archaea, invertebrates, and vertebrates. The aspartate racemases of eubacteria and hyperthermophilic archaea are pyridoxal 5′-phosphate (PLP) independent, and two conserved cysteine residues constitute the catalytic center. The crystal structure of the aspartate racemase of hyperthermophilic archaeon was determined. Based on this structure, the detailed reaction mechanism of the pyridoxal 5′-phosphate-independent aspartate racemase was studied by characterizing mutants and molecular dynamics simulations. However, it is still unclear how the catalytic cysteine residue can abstract a proton from the α-carbon. The aspartate in hyperthermophilic archaea is highly racemized, but the physiological role of aspartate racemase and D-aspartate in hyperthermophilic archaea is unknown. The aspartate racemases in invertebrates and vertebrates are PLP dependent. The aspartate racemases from invertebrates, bivalves, and *Aplysia californica* are homologous to serine racemases, but it has taken many years to identify the aspartate racemase responsible for the synthesis of D-Asp in mammals due to the lack of other amino acid racemases. The gene for the mammalian aspartate racemase was obtained *via* its homology with glutamate-oxaloacetate transaminase. Further studies on aspartate racemase will promote research on the mysterious functions of D-Asp in various organisms.

Keywords Aspartate racemase • PLP independent • D-Aspartate

21.1 Introduction

Aspartate racemase activity was first reported in *Lactobacillus fermenti* (Johnston and Diven 1969). In 1972, the incorporation of D-aspartate into a peptidoglycan in lactic acid bacteria (Staudenbauer and Strominger 1972) and the partial purification of aspartate racemase from *Enterococcus faecalis* (formerly *Streptococcus faecalis*)

M. Yohda (✉)
Department of Biotechnology and Life Science, Tokyo University of Agriculture and Technology, Naka, Koganei, Tokyo 184-8588, Japan
e-mail: yohda@cc.tuat.ac.jp

T. Yoshimura et al. (eds.), *D-Amino Acids*, DOI 10.1007/978-4-431-56077-7_21

were reported (Lamont et al. 1972). However, a detailed characterization of aspartate racemase, including the kinetics and cofactor dependency, was not described, most likely due to the low purification yield. I became involved in the study for aspartate racemase when I was working for the Central Research Center of Asahi Glass Co. Ltd. Our group has purified aspartate racemase and cloned the gene for it from *Streptococcus thermophilus* (Okada et al. 1991; Yohda et al. 1991). This was the first purified and cloned aspartate racemase. The subsequent amino acid sequencing and biochemical analysis showed that it is pyridoxal 5′-phosphate (PLP) independent. Since the project on the application of aspartate racemase was terminated and I have retired from Asahi Glass Co. Ltd., no further study has been performed on the enzyme.

I encountered aspartate racemase again while working for RIKEN. It is a very curious story. When I was attending the annual meeting of the Japanese Society for Molecular Biology in Kobe, my old friend, Professor Toshiko Ohta, showed me a slide for the sequence of a gene from a hyperthermophilic archaeon. I was very much astonished to see that it was highly homologous to *Streptococcus* aspartate racemase. Since then, I have performed various studies to reveal the reaction mechanism of PLP-independent aspartate racemases (Yohda et al. 1996; Matsumoto et al. 1999; Liu et al. 2002a, b; Yoshida et al. 2006; Ohtaki et al. 2008).

D-Aspartates also exist in invertebrates and vertebrates as well. Aspartate racemases that are involved in the synthesis of D-aspartate have recently been identified and were found to be PLP dependent. Professors Yamada and Kera have performed excellent studies on D-aspartate and aspartate racemase in the blood shell *Scapharca broughtonii* (Shibata et al. 2003a, b; Abe et al. 2006; Yamada et al. 2006). Although D-aspartate exists in various mammalian organs and is especially abundant in the developing brain, the search for the enzyme responsible for the formation of D-aspartate has been *in vain*. The difficulty is due to the small quantity in the cell and also the lack of sequence homology with known amino acid racemases. Finally, Professor Snyder's group succeeded in cloning aspartate racemase *via* an ingenious speculation (Kim et al. 2010). In this article, I will review the studies on both types of aspartate racemases.

21.2 PLP-Independent Aspartate Racemases

21.2.1 Aspartate Racemases from Lactic Acid Bacteria

Okada et al. examined the distribution of aspartate racemases in lactic acid bacteria (Okada et al. 1991). They purified aspartate racemase from *Streptococcus thermophilus* IAM1198 (StAspR) to homogeneity with a 3400-fold purification. The purified StAspR is likely to be a homodimer with a molecular mass of 28 kDa. StAspR is specific to aspartate and does not catalyze the racemization of alanine and glutamate, and the addition of pyridoxal 5′-phosphate has no effect on its activity. On the contrary, the addition of SH-protecting reagents, 2-mercaptoethanol, or dithiothreitol activates StAspR, suggesting the involvement of cysteine residues in the catalysis. The gene encoding StAspR was cloned from the *S. thermophilus* genome using a probe designed from the N-terminal amino acid sequence (Yohda et al. 1991). The gene contains an open reading frame of 729 nucleotides coding for 243 amino acids. The calculated molecular mass of 27,945 Da agrees well with the apparent molecular mass. The recombinant StAspR was expressed in *Escherichia coli* and was used for the study of the detailed reaction mechanism (Yamauchi et al. 1992). StAspR does not contain PLP or other cofactors such as FAD, NAD^+, and metal ions. Neither carbonyl reagents, such as hydroxylamine, nor sodium borohydride affects StAspR. On the contrary, StAspR is strongly inhibited by iodoacetamide and other thiol reagents. In addition to aspartate, cysteate and cysteine sulfinate are good substrates for StAspR. The K_m values for L- and D-aspartate are 35 and 8.7 mM, respectively. StAspR catalyzes the exchange of the α-hydrogen from the substrate with the solvent's hydrogen. The racemization of L-aspartate in D_2O shows an overshoot in the optical rotation of aspartate before the substrate is fully racemized. This shows that the removal of the α-hydrogen from the substrate is at least partially rate limiting. When L- or D-aspartate is incubated with aspartate racemase in tritiated water, the tritium is preferentially incorporated into the product enantiomer. The results strongly suggest that aspartate racemase functions *via* a two-base mechanism, similar to other PLP-independent amino acid racemases, glutamate racemases and proline racemases. In this mechanism, an α-hydrogen from the substrate amino acid is abstracted on one face as a proton, while a proton is incorporated on the other face. Thus, two conserved cysteine residues are thought to be involved in the reaction of these amino acid racemases.

21.2.2 *Aspartate Racemases from Hyperthermophilic Archaea*

As murein (peptidoglycan) is the only cell wall polymer that forms rigid cell walls in eubacteria, almost all eubacteria have D-amino acids and amino acid racemases. Archaea have a variety of cell wall and cell envelope polymers. Many archaea (including all crenarchaeotes, euryarchaeotes, halophiles, and methanogens) have an outer envelope (or S-layer) composed of hexagonally or tetragonally arranged proteins or glycoproteins that are easily disintegrated by mechanical shearing or detergents. In methanogens, the structure of this polymer is similar to that of murein, and it is referred to as "pseudomurein." Pseudomurein differs from murein in several respects. In particular, no D-amino acids are present in pseudomurein. Thus, there have been only a few reports on D-amino acids or amino acid racemases in archaea (Nagata et al. 1998).

The existence of aspartate racemase and D-aspartate in hyperthermophilic archaea was revealed by accident. While examining the roles of heat shock proteins in the thermotolerance of hyperthermophiles, Professor Toshiko Ohta (Emeritus Professor of Tsukuba University) intended to clone the genes encoding heat shock proteins from hyperthermophilic archaea. She attempted to obtain a DNA fragment from the *hsp70* by PCR amplification from the genomic DNAs of several hyperthermophilic archaea using a set of primers designed from the consensus sequences of *hsp70s*. Among those tested, a fragment of approximately 400 bp was amplified from the genomic DNA of *Desulfurococcus* strain SY. However, the sequence of the amplified DNA was not homologous to other *hsp70s*, but rather, it was similar to *StAspR*. Next, we examined whether aspartate racemase really exists in hyperthermophilic archaea (Yohda et al. 1996). The full-length gene was cloned from the genome of *D.* SY. It contains a 705-bp open reading frame, encoding a 235-residue polypeptide with a molecular mass of 25,977 Da. It shares considerable homology (35.2 % identity and 63.1 % similarity in the amino acid sequence) with StAspR. The amino acid sequence is also homologous to those of glutamate racemases from *Bacillus sphaericus*, *E. coli*, *Lactobacillus brevis*, *Lactobacillus fermenti*, and *Pediococcus pentosaceus*. The homology scores are 28.2, 24.0, 21.6, 27.6, and 27.7 %, respectively.

The putative hyperthermophilic aspartate racemase was expressed in *E. coli*. The recombinant protein exhibited amino acid racemase activity, which was highly specific for aspartate and increased proportionally with the temperature from 37 to 90 °C. Therefore, the protein was identified as the first hyperthermophilic archaeal amino acid racemase. Aspartate racemase activity was also detected in the crude extract of *D.* SY.

Matsumoto et al. investigated the occurrence of free D-amino acids and aspartate racemases in several hyperthermophilic archaea (Matsumoto et al. 1999). The aspartate in all hyperthermophilic archaea is highly racemized. The ratio of D-aspartate to total aspartate is in the range of 43.0–49.1 %. The crude extracts of the hyperthermophiles exhibit aspartate racemase activity at 70 °C, and their homologous aspartate racemase genes have been identified by PCR. The D-enantiomers of other amino acids (alanine, leucine, phenylalanine, and lysine) in *Thermococcus* strains have also been detected. Although the presence of D-aspartate in some hyperthermophilic archaea has been proven, their function is unclear.

21.2.3 The Structure and Reaction Mechanism of PLP-Independent Aspartate Racemase

21.2.3.1 The Structure of Aspartate Racemase from *Pyrococcus horikoshii* OT3

Liu et al. determined the three-dimensional structure of aspartate racemase from the hyperthermophilic archaeon *P. horikoshii* OT3 (PhAspR) at a 1.9-Å resolution

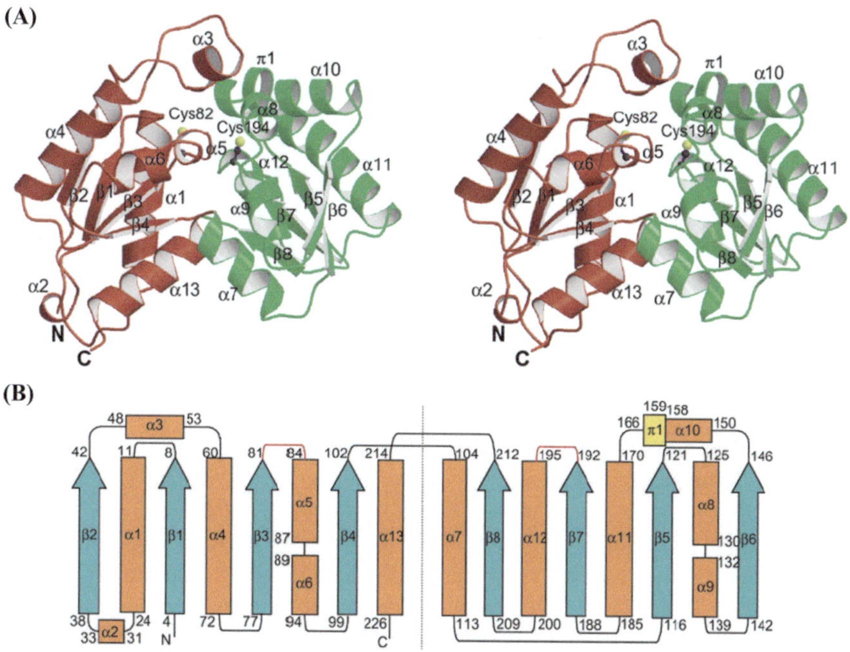

Fig. 21.1 (**a**) A stereoview of the ribbon diagram of the overall monomer structure of *P. horikoshii* OT3 AspR. The N- and C-terminal domains are colored *red* and *green*, respectively. All of the helices and strands, represented as coils and arrows, are labeled sequentially. The positions of Cys82 and Cys194 are labeled, and their side chains are depicted with ball-and-stick models. (**b**) A topology diagram of the secondary structural elements of *P. horikoshii* OT3 AspR. The α-helices are represented by *orange* bars, the π-helix by a fat *yellow* bar, and the β-strands by *blue arrows*. The two loops containing the two active-site cysteine residues are colored *red* (Reproduced from Figure 1 of Liu et al. 2002a)

using X-ray crystallography and refined it to a crystallographic *R* factor of 19.4 % (R_{free} of 22.2 %) (Liu et al. 2002a) (Fig. 21.1). PhAspR forms a stable dimeric structure *via* a strong three-layered inter-subunit interaction. The subunit consists of two structurally homologous α/β-domains, each containing a four-stranded parallel β-sheet flanked by six α-helices. Two strictly conserved cysteine residues (Cys82 and Cys194) are located on both sides of a cleft between the two domains. The spatial arrangement of these two cysteine residues supports the "two-base" mechanism. The previous hypothesis that the active site of aspartate racemase is located at the dimeric interface is clearly incorrect. The structure revealed a unique pseudo-mirror symmetry in the spatial arrangement of the residues around the active site, which may explain the molecular recognition mechanism of the mirror-symmetric aspartate enantiomers by the non-mirror-symmetric aspartate racemase.

The structural homology and functional similarity between the two domains suggest that this enzyme evolved from an ancestral domain *via* gene duplication and gene fusion. Liu et al. expressed only the C-terminal domain of PhAspR (SC-domain) and determined its three-dimensional structure by X-ray crystallography (Liu et al. 2002b). The SC-domain structure has the same folding pattern as that of the C-domain in the intact PhAspR, containing a four-stranded parallel β-sheet flanked by six α-helices from two sides. In both cases, all of the secondary structural elements, including the π1 helix as well as the C- and N-termini, showed no significant structural differences. A superimposition of the SC-domain and C-domain revealed an root-mean-square deviation of the Cα atoms of 0.50 Å, implying that the domain structure is highly stable. The high structural stability of this domain supports the existence of the ancestral domain. When compared with other amino acid racemases, we suggest that gene duplication and gene fusion are conventional ways in the evolution of PLP-independent amino acid racemases.

21.2.3.2 The Catalytic Center of PLP-Independent Aspartate Racemase

The mutation of either Cys82 or Cys194 to alanine (C82A, C194A) markedly reduced the reaction rate but only marginally affected the K_m constant (Yoshida et al. 2006) (Table 21.1). These mutants do not show any specificity in either reaction direction, suggesting cooperation between the two cysteine residues. As the transitional state of proton transfer between two atoms is essentially a hydrogen bond interaction, the proton transfer in the process of aspartate racemization ($-S^-$ · H–Cα · ·H–S–) will proceed through the approximate distance of two hydrogen bonds. Thus, the "cooperation distance" between the γ-sulfur atoms of the two active cysteine residues could be assigned as approximately 8.0 Å, which corresponds to twice the sulfur-involved hydrogen bond distance. However, the distance between the two γ-sulfur atoms of Cys82 and Cys194 is 9.6 Å, which is beyond this cooperation distance. At such a large distance, cooperation between the two cysteine residues is likely difficult, although the two-base mechanism could be supported by the pseudosymmetrical arrangement of these residues at the well-conserved active site.

This contradiction appears to be dissolved by the conformational change. One possibility is the so-called hinge motion between the two domains. As PhAspR is hyperthermophilic and functions at high temperatures, the distance between the side chains of the two cysteine residues may be reduced by the motion. To test this hypothesis, molecular dynamics simulations of PhAspR were performed in a wide temperature range from 300 to 425 K, and the temperature dependence of the intramolecular motion was investigated. The time evolution of the distance between the two catalytic γ-sulfur atoms at 300, 375, 400, and 425 K was calculated. Although the average distance remained over 10 Å up to 375 K, a noticeable change in the time evolution of the distance was observed in the simulations over 400 K. During the simulation time range from 500 to 1000 ps at 425 K, the distance between the two γ-sulfur atoms decreased from 10 to 6 Å and then increased back up to 10 Å. At 680 ps, the two γ-sulfur atoms approached each other at a distance of

Table 21.1 Kinetic parameters of mutant PhAspRs

	L → D direction			D → L direction		
	K_m (mM)	k_{cat} (min^{-1})	k_{cat}/K_m	K_m (mM)	k_{cat} (min^{-1})	k_{cat}/K_m
WT	1.3×10^{-1}	1.9×10^4	1.5×10^5	3.1×10^{-1}	2.6×10^4	8.4×10^4
C82A	1.8	4.0×10^{-1}	2.2×10^{-1}	3.6	1.6×10^{-1}	4.5×10^{-2}
C194A	1.3	9.3×10^{-1}	7.3×10^{-1}	4.5×10^{-1}	1.4×10^{-1}	3.0×10^{-1}
R48I	3.0×10^2	1.0×10^2	3.5×10^{-1}	4.0×10	6.8	1.7×10^{-1}
R48K	2.3×10	5.5×10^2	2.4×10	3.3	5.8×10	1.8×10
R48A	N.D.	N.D.	N.D.	1.5×10^2	1.1×10	7.7×10^{-2}
K164L	4.3	2.8×10	6.6	5.4	2.6×10	4.9
K164R	1.0×10^2	4.1×10	4.1×10^{-1}	3.1×10	1.1×10	3.4×10^{-1}
K164A	2.3×10^2	1.0×10	4.6×10^{-2}	8.7×10	2.9	3.3×10^{-2}
N83G	3.9	2.4×10^4	6.1×10^3	1.8	7.8×10^3	4.3×10^3
T84A	8.2×10^{-2}	9.0×10^2	1.1×10^4	2.6×10^{-1}	1.6×10^3	6.0×10^3
T124A	1.8×10^{-1}	1.8×10^3	9.7×10^3	3.2×10^{-1}	1.5×10^3	4.7×10^3
T127A	1.4×10^{-1}	3.5×10^3	2.4×10^4	1.6×10^{-1}	3.5×10^3	2.2×10^4

Reproduced from Table 21.1 of Yoshida et al. 2006
k_{cat} and K_m values of mutant PhAspRs for L- and D-aspartate at 70 °C are shown
N.D. not determined

5.8 Å. This distance is much smaller than the predicted cooperation distance of 8.0 Å.

Another possibility is the substrate-induced conformational change. Ohtaki et al. determined the crystal structure of an inactive mutant PhAspR complexed with citric acid (Cit) at a resolution of 2.0 Å (Ohtaki et al. 2008). Cit contains the substrate analogue moieties of both L- and D-aspartate and exhibits a low competitive inhibition activity against PhAspR. In the structure, Cit binds to the catalytic site of PhAspR, which induces a conformational change to close the active site. The distance between the thiolates was estimated to be 7.4 Å, suggesting a conformational change in PhAspR following the substrate binding.

However, it is still unclear how the catalytic cysteine residues can abstract protons from an α-carbon.

21.2.3.3 The Study of the Reaction Mechanism by the Characterization of Mutants and Molecular Dynamics Simulations

Yoshida et al. examined the molecular mechanism of PhAspR by mutational analyses and molecular dynamics simulations (Yoshida et al. 2006) (Fig. 21.2, Table 21.1). The two putative catalytic cysteine residues and the surrounding amino acid sequences (Cys-Asn-Thr and Gly-Cys-Thr) are highly conserved among all PLP-independent aspartate and glutamate racemases. In addition to these conserved fragments, Arg48 and Lys164 are strictly conserved in all AspRs. In addition, Thr124 and Thr127 are also conserved in both AspRs and GluRs. Arg48 and

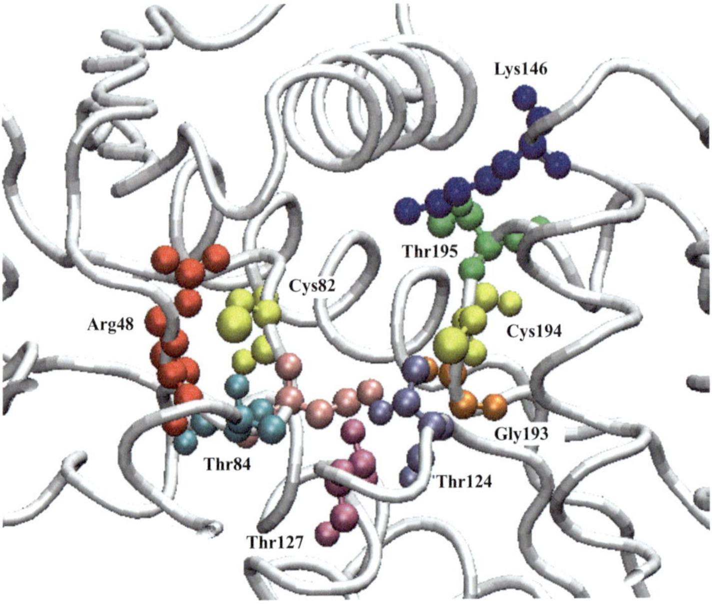

Fig. 21.2 The structure of the catalytic site of PhAspR. The backbone is drawn as a tube model, and the important putative catalytic amino acid residues (Arg48, Cys82, Asn83, Thr84, Thr124, Thr127, Lys164, Gly193, Cys194, and Thr195) are shown as ball-and-stick models. The *arrow* designates the distance between the two catalytic cysteine residues (Reproduced from Figure 1 of Yoshida et al. 2006)

Thr84 are located near Cys82, and Lys164 and Thr124 are located near Cys194. The spatial arrangement of these residues is quasi-mirror symmetrical. The residues of Arg48/Lys164, with a positively charged long side chain, can ionically interact with the side-chain carboxylate of L-/D-Asp via its guanidinium/ammonium group, and Thr84/Thr124 can form a hydrogen bond with the amino acid. Arg48 was replaced with alanine, isoleucine, or lysine, and Lys164 was replaced with alanine, leucine, or arginine. All of the mutants showed a significant decrease in the k_{cat} compared with the wild-type enzyme, but the effects were relatively small when compared with the mutations of the catalytic cysteine residues. Therefore, both Arg48 and Lys164 are important but not indispensable for catalysis. In comparison with other mutants, R48K exhibits relatively strong catalytic activity as well as substrate affinity, suggesting that the positively charged residue of R48 contributes to both the substrate binding and the reaction rate of PhAspR. K164A has the weakest activity among the K164 mutants, and K164L shows a significantly lower K_m than the other mutants of R48 and K164. This result suggests that not only the charge but also the size of the side chain at the 164th position is related to the binding of both aspartate enantiomers to PhAspR. Other mutations (N83G, T84A, T124A, and T127A) also affect the activity. However, the effects are marginal compared to the mutations at Cys82, Cys194, Arg48, and Lys164. Thus, Asn83, Thr84, Thr124, and Thr127 are not essential for catalysis.

Molecular dynamics simulations of PhAspR also provided important insights into the roles of the amino acid residues at the catalytic site and also the activation mechanism of a hyperthermophilic aspartate racemase at high temperatures. As described above, at high temperatures, the γ-sulfur atoms of the cysteine residues oscillate to periodically become in closer proximity than the predicted cooperative distance. The conformation of Tyr160, which is located at the entrance of the cleft and inhibits the entry of a substrate, changes periodically to open the entrance at 375 K. The opening of the gate is likely to be induced by the motion of the adjacent amino acid, Lys164. The entrance of an aspartate molecule was also observed in the molecular dynamics simulations, driven by the force of the electrostatic interaction with Arg48, Lys164, and Asp47 (Fig. 21.3).

21.3 PLP-Dependent Aspartate Racemases

21.3.1 PLP-Dependent Aspartate Racemases in Invertebrates and Archaea

A high concentration of D-aspartate exists in the tissues of the blood shell *Scapharca broughtonii*, and aspartate racemase activity was detected in the extracts of its foot muscle and mantle. Shibata et al. purified aspartate racemase (SbAspR) from the foot muscle to homogeneity (Shibata et al. 2003b). The apparent molecular mass shown by SDS-PAGE is 39 kDa, and it appears at 51–63 kDa in a gel filtration. The absorption spectrum and the effects of aminooxyacetate or $NaBH_4$ show that SbAspR is dependent on pyridoxal 5′-phosphate. SbAspR is highly specific to aspartate and does not racemize alanine, serine, and glutamate. Curiously, its activity is modulated by nucleotides. The activity increases with purine nucleoside monophosphates, while purine nucleoside triphosphates decrease the activity. AMP increases the activity as high as sevenfold, while ATP at saturating concentrations decreases the activity to 7 %. These results suggest the possibility that SbAspR may be involved in energy metabolism.

The cDNA clone for SbAspR was cloned and sequenced (Abe et al. 2006). It contains an open reading frame of 1017 bp encoding a protein of 338 amino acids with a molecular mass of 37.1 kDa, which is in good agreement with the apparent molecular mass of the native protein, at 39 kDa. The deduced amino acid sequence contains a highly conserved PLP-binding motif. Thus, this confirms that this enzyme is the first eukaryotic and PLP-dependent aspartate racemase. SbAspR shares significant amino acid sequence homologies with mammalian serine racemase, which is also PLP dependent, and the highest identity is 44–43 %. SbAspR is also homologous to other PLP-dependent enzymes, such as threonine dehydratases from various sources, with a 39–33 % identity (Fig. 21.4). On the contrary, no microbial aspartates or glutamate racemases show any significant identity because they are PLP-independent enzymes.

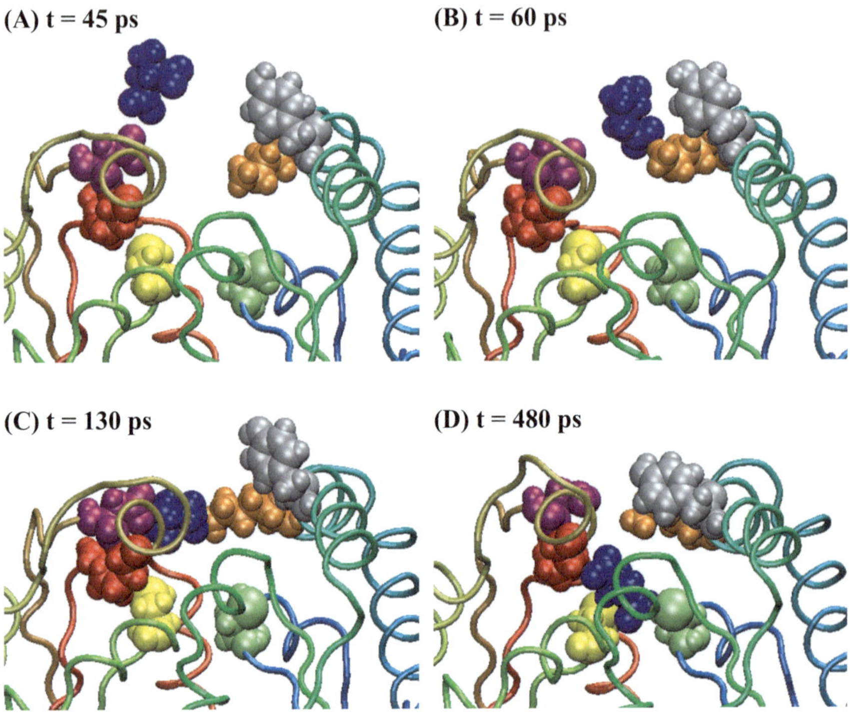

Fig. 21.3 The docking process of L-aspartic acid into the catalytic site of PhAspR. Schematic drawings for the position of the L-aspartic acid and the conformation of the catalytic site of PhAspR at 45 ps (**a**), 60 ps (**b**), 130 ps (**c**), and 480 ps (**d**) during the docking MD simulation are shown. The backbone is shown as tube model. L-Aspartic acid (*blue*) and the important residues around the catalytic site, Asp47 (*purple*), Arg48 (*red*), Lys164 (*orange*), Tyr160 (*gray*), Cys82 (*yellow*), and Cys194 (*green*), are drawn as *van der Waals* model (Reproduced from Fig. 9 of Yoshida et al. 2006)

SbAspR was expressed in *E. coli* and purified to homogeneity (Abe et al. 2006). The recombinant SbAspR displayed essentially identical properties with the native protein, including the possession of a bound PLP and its sensitivity to AMP and ATP. The recombinant SbAspR exhibited relatively low dehydratase activity toward L-*threo*-3-hydroxyaspartate to produce oxaloacetate. This activity may represent only a side reaction often observed in PLP-dependent enzymes due to the chemistry of PLP. However, it is also possible that the activity somehow reflects the structural homology with threonine dehydratases as well as serine racemase, which also shows dehydratase activity.

Significant levels of D-Asp are present in the cerebral ganglion of the F- and C-clusters of the invertebrate *A. californica*, and D-Asp appears to be involved in cell-cell communication in this system. Wang et al. identified a gene encoding an amino acid racemase from *A. californica* (DAR1) (Wang et al. 2011). DAR1 converts aspartate and serine to their other chiral forms in a PLP-dependent manner. DAR1 has a predicted length of 325 amino acids and is 55 % identical to SbAspR

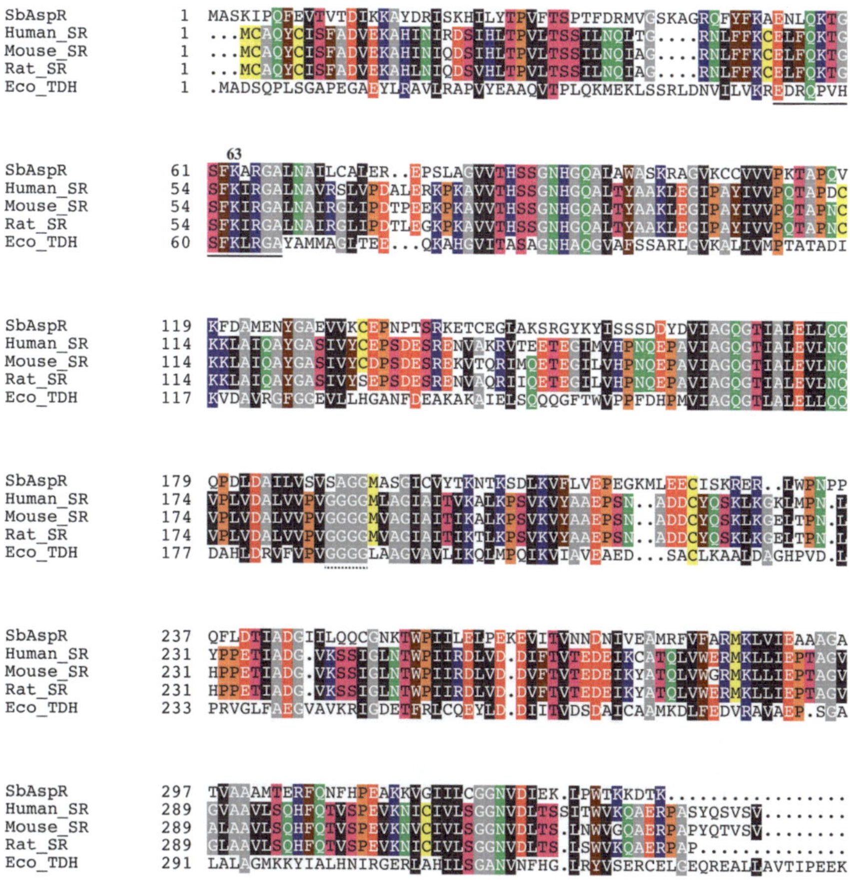

Fig. 21.4 A multiple alignment of the amino acid sequence of SbAspR with the human, mouse, and rat serine racemase (*SR*) homologues and *E. coli* biosynthetic threonine dehydratase (*Eco_TDH*). The putative PLP-binding lysine residue is Lys63 of SbAspR. A PLP-binding motif is indicated by the *underline*. The *dotted underline* indicates the tetraglycine loop (Modified from Figure 2 of Abe et al. 2006 JB)

and 41 % identical to mammalian serine racemase. However, it is only 14 % identical to the recently reported mammalian aspartate racemase. Using the recombinant DAR1, its activity toward serine and aspartate was confirmed. DAR1 is the first characterized eukaryotic racemase that can catalyze the racemization of two substrates. DAR1 exists as a homodimer, similar to mouse SerR and SbAspR. ATP and $MgCl_2$ promote the racemase activity of DAR1, and it has been proposed that ATP and $MgCl_2$ exert allosteric regulatory effects on mouse SerR by lowering the enzyme K_m.

In the acidothermophilic archaeon *Thermoplasma acidophilum*, the proportion of D-aspartate to total aspartate was as high as 39.7 % (Long et al. 2001). Crude extracts of *T. acidophilum* showed aspartate-specific racemase activity. The activity was very low in the absence of PLP and was insensitive to an SH-modifying reagent (Long et al. 2001). Although the genome of *T. acidophilum* has been completely sequenced, no aspartate racemase genes of either type were identified. Thus, the high levels of D-aspartate may be produced by a new type of PLP-dependent aspartate racemase in *T. acidophilum*.

21.3.2 Mammalian Aspartate Racemase

D-Aspartate exists in various mammalian organs and is especially abundant in the developing brain. However, the enzyme involved in the formation of D-aspartate was unknown until 2010. The search for the aspartate racemase gene in genome sequences by looking for homology to serine racemase or bacterial aspartate racemases was *in vain*.

Kim et al. were the first to identify and clone mammalian aspartate racemase, which co-localized with D-aspartate in the brain and neuroendocrine tissues (Kim et al. 2010). They were inspired by the previous findings that glutamate-oxaloacetate transaminase (GOT) from *E. coli* can generate small amounts of D-aspartate in the process of transaminating L-aspartate to L-glutamate, and its aspartate racemizing activity is enhanced by the double mutations of Trp140 to His and Arg292 to Lys. Kim et al. discovered a gene encoding a protein homologous to the mutant GOT. The gene was found to encode a protein that is more closely related to cytosolic and mitochondrial GOT than to serine racemase or serine dehydratase (Fig. 21.5).

When cloned and expressed, the putative aspartate racemase (DR), a 45.5-kDa protein, generates substantial D-aspartate but only one-fifth as much L-glutamate and very little D-glutamate. Like serine racemase and SbAspR, DR is PLP dependent. PLP binds to Lys249, and DR is inhibited by aminooxyacetic acid, an inhibitor of PLP-dependent enzymes. The K_m of the recombinant DR for L-aspartate is 3.1 mM, the V_{max} is 0.46 mmol/mg/min, the optimum pH is 7.5, and the optimum temperature is 37 °C. The importance of Lys136, the presumed proton donor, is highlighted by its mutation to tryptophan, which virtually abolishes the racemase activity. DR is expressed most abundantly in the brain, heart, and testes, with somewhat lower levels in the adrenal glands and negligible expression in the liver, lung, and kidney.

The depletion of DR by short hairpin RNA in the newborn neurons of the adult hippocampus elicits profound defects in the dendritic development and survival of the newborn neurons. Because D-aspartate is a potential endogenous ligand for NMDA receptors, the loss of which elicits a phenotype resembling DR depletion, D-aspartate may function as a modulator of adult neurogenesis.

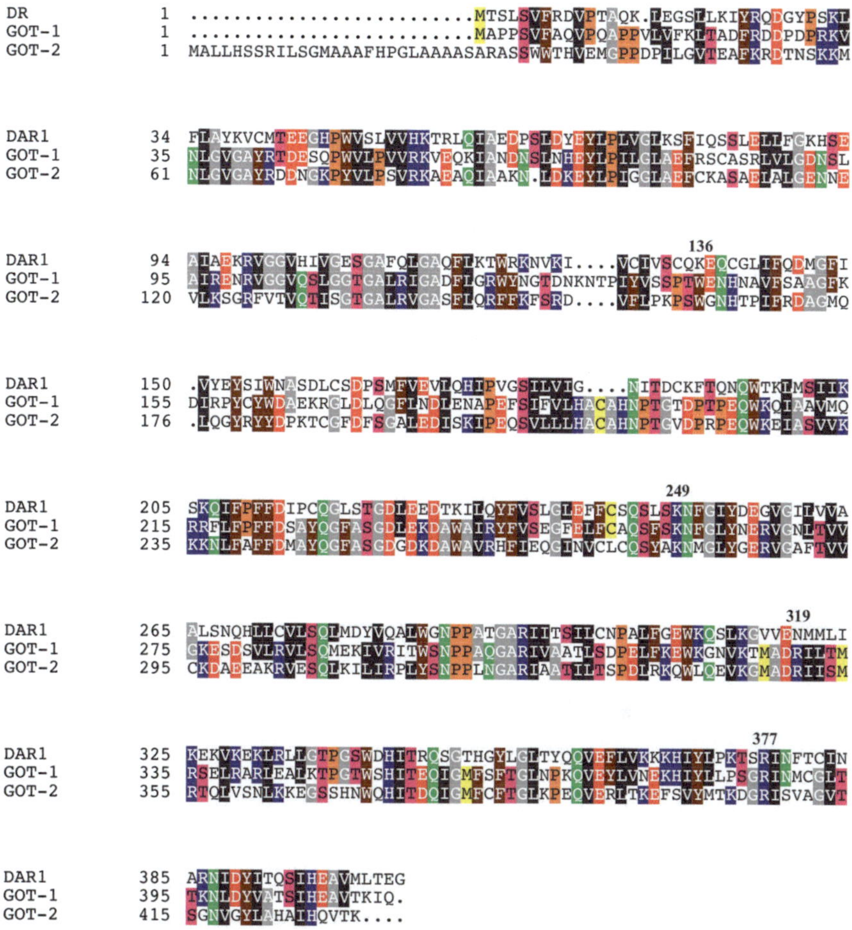

Fig. 21.5 An amino acid alignment of DAR1, mouse cytoplasmic glutamate-oxaloacetate transaminase (GOT-1), and mouse mitochondrial glutamate-oxaloacetate transaminase (GOT-2). Arg377 is a conserved amino acid residue that recognizes the carboxyl group of the substrate, amino acids, and α-keto acids. Both GOT-1 and GOT-2 have a tryptophan at position 167, whereas DR contains a lysine, which can donate protons for racemization. GOT-1 and GOT-2 also have an arginine at position 319, whereas DR has a glutamine, which allows the access of water molecules (Modified from Fig. S1 of Kim et al. 2010 PNAS S1)

21.4 Conclusion

Aspartate racemases are divided into PLP-independent and PLP-dependent aspartate racemases. The aspartate racemases of eubacteria and hyperthermophilic archaea are PLP independent. Their catalytic sites are constituted by two conserved cysteine residues. However, it is still unknown how the catalytic cysteine residues can abstract a proton from the α-carbon. The aspartate racemases of invertebrates and vertebrates are PLP dependent, and those of invertebrates, bivalves, and *A. californica* are homologous to serine racemases. It took many years to identify the aspartate racemase responsible for the synthesis of D-Asp in mammals due to the lack of homology with other amino acid racemases. The gene for the mammalian aspartate racemase was finally obtained *via* its homology with glutamate-oxaloacetate transaminase. Although the existence of an aspartate racemase in the acidothermophilic archaeon *T. acidophilus* was shown, it had not been identified until now.

More than 40 years have passed since the first discovery of aspartate racemase. The study of aspartate racemase has been relatively slow, reflecting the difficulty of the research on this enzyme. As we have identified almost all of the different types of aspartate racemases, I hope the study on the physiological function of aspartate racemases and D-aspartate will continue to progress.

References

Abe K, Takahashi S, Muroki Y, Kera Y, Yamada RH (2006) Cloning and expression of the pyridoxal 5′-phosphate-dependent aspartate racemase gene from the bivalve mollusk Scapharca broughtonii and characterization of the recombinant enzyme. J Biochem 139 (2):235–244. doi:10.1093/jb/mvj028

Johnston MM, Diven WF (1969) Studies on amino acid racemases. I. Partial purification and properties of the alanine racemase from Lactobacillus fermenti. J Biol Chem 244 (19):5414–5420

Kim PM, Duan X, Huang AS, Liu CY, Ming GL, Song H, Snyder SH (2010) Aspartate racemase, generating neuronal D-aspartate, regulates adult neurogenesis. Proc Natl Acad Sci U S A 107 (7):3175–3179. doi:10.1073/pnas.0914706107

Lamont HC, Staudenbauer WL, Strominger JL (1972) Partial purification and characterization of an aspartate racemase from Streptococcus faecalis. J Biol Chem 247(16):5103–5106

Liu L, Iwata K, Kita A, Kawarabayasi Y, Yohda M, Miki K (2002a) Crystal structure of aspartate racemase from Pyrococcus horikoshii OT3 and its implications for molecular mechanism of PLP-independent racemization. J Mol Biol 319(2):479–489. doi:10.1016/s0022-2836(02) 00296-6

Liu L, Iwata K, Yohda M, Miki K (2002b) Structural insight into gene duplication, gene fusion and domain swapping in the evolution of PLP-independent amino acid racemases. FEBS Lett 528 (1–3):114–118

Long Z, Lee JA, Okamoto T, Sekine M, Nimura N, Imai K, Yohda M, Maruyama T, Sumi M, Kamo N, Yamagishi A, Oshima T, Homma H (2001) Occurrence of D-amino acids and a pyridoxal 5′-phosphate-dependent aspartate racemase in the acidothermophilic archaeon, Thermoplasma acidophilum. Biochem Biophys Res Commun 281(2):317–321. doi:10.1006/ bbrc.2001.4353

Matsumoto M, Homma H, Long Z, Imai K, Iida T, Maruyama T, Aikawa Y, Endo I, Yohda M (1999) Occurrence of free D-amino acids and aspartate racemases in hyperthermophilic archaea. J Bacteriol 181(20):6560–6563

Nagata Y, Fujiwara T, Kawaguchi-Nagata K, Fukumori Y, Yamanaka T (1998) Occurrence of peptidyl D-amino acids in soluble fractions of several eubacteria, archaea and eukaryotes. Biochim Biophys Acta 1379(1):76–82

Ohtaki A, Nakano Y, Iizuka R, Arakawa T, Yamada K, Odaka M, Yohda M (2008) Structure of aspartate racemase complexed with a dual substrate analogue, citric acid, and implications for the reaction mechanism. Proteins 70(4):1167–1174. doi:10.1002/prot.21528

Okada H, Yohda M, Giga-Hama Y, Ueno Y, Ohdo S, Kumagai H (1991) Distribution and purification of aspartate racemase in lactic acid bacteria. Biochim Biophys Acta 1078 (3):377–382

Shibata K, Watanabe T, Yoshikawa H, Abe K, Takahashi S, Kera Y, Yamada R-h (2003a) Nucleotides modulate the activity of aspartate racemase of Scapharca broughtonii. Comp Biochem Physiol B Biochem Mol Biol 134(4):713–719. doi:10.1016/s1096-4959(03)00031-9

Shibata K, Watanabe T, Yoshikawa H, Abe K, Takahashi S, Kera Y, Yamada RH (2003b) Purification and characterization of aspartate racemase from the bivalve mollusk Scapharca broughtonii. Comp Biochem Physiol B Biochem Mol Biol 134(2):307–314

Staudenbauer W, Strominger JL (1972) Activation of D-aspartic acid for incorporation into peptidoglycan. J Biol Chem 247(16):5095–5102

Wang L, Ota N, Romanova EV, Sweedler JV (2011) A novel pyridoxal 5′-phosphate-dependent amino acid racemase in the Aplysia californica central nervous system. J Biol Chem 286 (15):13765–13774. doi:10.1074/jbc.M110.178228

Yamada RH, Kera Y, Takahashi S (2006) Occurrence and functions of free D-aspartate and its metabolizing enzymes. Chem Rec 6(5):259–266. doi:10.1002/tcr.20089

Yamauchi T, Choi SY, Okada H, Yohda M, Kumagai H, Esaki N, Soda K (1992) Properties of aspartate racemase, a pyridoxal 5′-phosphate-independent amino acid racemase. J Biol Chem 267(26):18361–18364

Yohda M, Okada H, Kumagai H (1991) Molecular cloning and nucleotide sequencing of the aspartate racemase gene from lactic acid bacteria Streptococcus thermophilus. Biochim Biophys Acta 1089(2):234–240

Yohda M, Endo I, Abe Y, Ohta T, Iida T, Maruyama T, Kagawa Y (1996) Gene for aspartate racemase from the sulfur-dependent hyperthermophilic archaeum, Desulfurococcus strain SY. J Biol Chem 271(36):22017–22021

Yoshida T, Seko T, Okada O, Iwata K, Liu L, Miki K, Yohda M (2006) Roles of conserved basic amino acid residues and activation mechanism of the hyperthermophilic aspartate racemase at high temperature. Proteins 64(2):502–512. doi:10.1002/prot.21010

Part VI
D-Amino Acids in Foods

Chapter 22
D-Amino Acids in Fermentative Foods

Yuta Mutaguchi, Jyunpei Kobayashi, Tadao Oikawa, and Toshihisa Ohshima

Abstract In recent years, D-amino acids have been shown to be present, in a free or bound state, in a wide variety of foods and beverages, and to exhibit properties that differ from those of L-isomers. It is anticipated that as more is learned about the properties of D-amino acids in foods, they will be applied for the improvement of food characteristics. In this section, we will summarize what is known about the distribution and metabolism of D-amino acids in foods and beverages, especially fermented types, as well as what is known about the enzymes involved in the metabolism of D-amino acids. In addition, applications in which D-amino acids are being used to improve the characteristics of food are described.

Keywords D-Amino acid • Functional foods • Fermented foods • Lactic acid bacteria • D-Amino acid metabolism • Racemase • Epimerase • Transaminase • Dehydrogenase • Oxidase

In recent years, D-amino acids have been shown to be present, in a free or bound state, in a wide variety of foods and beverages and to exhibit properties that differ from those of L-isomers. It is anticipated that as more is learned about the properties

---

Y. Mutaguchi
Faculty of Bioresource Sciences, Department of Biotechnology, Akita Prefectural University, 241-438 Shimoshinjo, Nakano, Akita 010-0195, Japan
e-mail: y-muta@akita-pu.ac.jp

J. Kobayashi
Department of Chemistry and Biotechnology, Graduate School of Engineering, Tottori University, Minami 4-101, Koyamacho, Tottori 680-8552, Japan
e-mail: jyumpei.kobayashi@gmail.com

T. Oikawa
Faculty of Chemistry, Materials and Biotechnology, Department of Life Science & Technology, Kansai University, 3-3-35 Yamate-cho, Suita-shi, Osaka-fu 564-8680, Japan
e-mail: oikawa@kansai-u.ac.jp

T. Ohshima (✉)
Faculty of Engineering, Department of Biomedical Engineering, Osaka Institute of Technology, 5-16-1 Ohmiya, Asahiku, Osaka-fu 535-8585, Japan
e-mail: ohshima@bme.oit.ac.jp

T. Yoshimura et al. (eds.), *D-Amino Acids*, DOI 10.1007/978-4-431-56077-7_22

of D-amino acids in foods, they will be applied for the improvement of food characteristics. In this section, we will summarize what is known about the distribution and metabolism of D-amino acids in foods and beverages, especially fermented types, as well as what is known about the enzymes involved in the metabolism of D-amino acids. In addition, applications in which D-amino acids are being used to improve the characteristics of food are described.

22.1 Characteristics of D-Amino Acids Related to Foods and Beverages

Among the amino acids used in foods, L-glutamate (monosodium salt) is the most widely used and has a kelp flavor component (umami). By contrast, D-glutamate, its stereoisomer (enantiomer), has no kelp flavor, but rather a bitter taste. Among the 20 proteinogenic amino acids, the L-isomers of 19 (all except glycine) are known to taste different than the D-isomers, as in the case of glutamate (Table 22.1) (Solms et al. 1965). For example, D-alanine, D-phenylalanine, D-tryptophan, D-serine, D-leucine, and D-valine all taste much sweeter than their L-isomers. Indeed, D-alanine, D-phenylalanine, and D-tryptophan are reportedly about 3, 5, and 35 times sweeter than sugar, respectively, and the sweet tastes of crab and shrimp are derived from the presence of D-alanine (Kawai et al. 2012). Up to now, however, much less attention has been paid to the D-amino acids in foods and beverages than to the L-amino acids. This is because nearly all proteins in foods are constituted exclusively of L-amino acids, and determination of D-amino acids in the mixture of D- and L-amino acids has not been easy. Consequently, details about the roles of D-amino acids as effectors of food and beverage on taste, flavor, and preservation have been remained unclear until quite recently. It was the fortunate recent development of analytical techniques for enantioselective determination of D- and L-amino acids using HPLC that enabled characterization of the distinct distributions of free and protein-bound forms of D-amino acids in foods and food ingredients. The results of those analyses suggest D-amino acids likely serve as nutrients (primary function) and effectors of food flavor and preservation (secondary function), as well as mediators of good health that may have antiaging properties (third function). In 2008, authors began to study the distribution and dynamic property behavior of D-amino acids in foods and beverages, focusing mainly on various fermented foods. In addition, authors investigated the metabolism of D-amino acids, the functions of related enzymes, and the mechanisms of their regulation. In the course of those studies, our analysis of the D-amino acid distributions in vinegars showed that lactic acid bacteria produce large amounts of D-amino acids.

Table 22.1 Taste of L- and D-amino acids

	Sweet	Not sweet
D-Amino acids	Ala, Val, Leu, Ile, Ser, Thr, Cys, Met, Phe, Trp, Tyr, Gln, Asn, Glu, Lys, Arg, His	Pro, Asp
L-Amino acids	Ala, Ser, Cys, Pro, Lys	Val, Leu, Ile, Thr, Met, Phe, Trp, Tyr, Gln, Asn, Glu, Asp, Arg, His

22.2 Distribution of D-Amino Acids in Vinegars and Involvement of Lactic Acid Bacteria in Their Production

The wide distribution of D-amino acids in foods containing milk (Rubio-Barroso et al. 2006), beer (Erbe and Brückner 2000), wine (Kato et al. 2011), vegetables, fruit, and rice (Brückner and Westhauser 2003; Gogami et al. 2009) is already well known. Fermented foods, in particular, are known to contain D-amino acids, including D-alanine, D-aspartate, and D-glutamate (Abe et al. 1999; Brückner and Hausch 1989), which could be derived from the starting materials used for the production of fermented foods or could be produced during microbial fermentation (Brückner and Hausch 1989). Authors first examined the source of D-amino acids in vinegars because such an analysis during production of vinegar is comparatively easy.

Vinegar is generally produced through the fermentation of grains, vegetables, or fruits. Yeasts and acetic acid bacteria are mainly responsible for the fermentation processes that produce vinegars. Yeasts first convert sugar to alcohol in a process known as the alcoholic fermentation, after which acetic acid bacteria, such as the *Acetobacter* species, convert the alcohol to acetic acid in acetic fermentation. When grains are used as the starting materials for the production of vinegars, starch is converted to sugars by *koji* fungi such as the *Aspergillus* species before alcoholic fermentation. In addition, the Central Research Institute of the Mizkan Group Corporation (Handa, Japan) has experimentally produced vinegar using a procedure that includes lactic fermentation. Authors determined the levels of 16 different D-amino acids in 11 vinegars produced from several different sources and through several different manufacturing processes (Mutaguchi et al. 2013a,). They found that high-brix nonglutinous brown rice black vinegar matured in barrels and lactic-fermented tomato vinegar contain much higher total D-amino acid concentrations than other vinegars. The highest total D-amino acid concentration was detected in lactic-fermented tomato vinegar (3773.2 μM). Notably, tomato vinegar produced from the same source material but without lactic fermentation showed a much lower total D-amino acid concentration (237.7 μM), which suggests lactic fermentation is mainly responsible for the production of D-amino acids.

Production of lactic-fermented tomato vinegar entails three fermentation steps: alcoholic, acetic, and lactic fermentation. To clarify which of those processes

Table 22.2 Analysis of D-amino acids in samples collected at different fermentation steps during production of lactic-fermented tomato vinegar

	Source (tomato juice)		After alcoholic fermentation		After acetic fermentation		Before lactic fermentation		After lactic fermentation	
Amino acid[a]	μM	% D[b]	μM	% D	μM	% D	μM	% D	μM	% D
D-Asp	68.2	0.3	68.8	0.4	69.0	0.4	99.4	0.4	599.0	2.7
D-Ser	–[c]	–	–	–	–	–	–	–	22.7	0.7
D-Arg	–	–	–	–	–	–	–	–	37.2	65.7
D-Ala	24.6	0.1	29.5	0.2	41.1	0.3	28.1	0.1	1962.4	10.7
D-Val	–	–	–	–	–	–	–	–	81.6	8.0
D-Met	–	–	–	–	–	–	–	–	32.3	21.6
D-Phe	–	–	–	–	–	–	–	–	89.7	3.2
D-*allo*-Ile	–	–	–	–	–	–	–	–	420.9	38.6
D-Leu	–	–	–	–	N.D.[d]	N.D.	–	–	285.0	37.3
D-Glu	67.6	0.2	66.8	0.2	64.7	0.2	81.7	0.2	209.1	0.7
D-Asn	78.0	0.5	71.4	0.6	62.9	0.5	71.7	0.4	71.0	0.5

[a]No other measured D-amino acids (D-glutamine, D-histidine, D-threonine, D-tyrosine, and D-tryptophan) were detected
[b]Relative percentages of D-amino acids were calculated as 100 × [D-amino acid]/[D-amino acid + L-amino acid]
[c]Not detected
[d]N.D., not determined

involves D-amino acid production, D-amino acids were analyzed in samples collected during the different fermentation steps in the production of lactic-fermented tomato vinegar. Using this approach, it was found that the increase of the D-amino acid levels was markedly greater during the lactic fermentation step than during the alcoholic or acetic fermentation steps (Table 22.2). The total concentration of D-amino acids in the sample after lactic fermentation step was 12.4 times higher than before that step, and the relative percentage of D-amino acids (100 × [D-amino acids]/[D-amino acids + L-amino acids]) in the sample after lactic fermentation was also much higher (14.3 times) than that before the fermentation. Furthermore, the lactic fermentation produced D-serine, D-arginine, D-valine, D-methionine, D-phenylalanine, D-*allo*-isoleucine, and D-leucine in addition to D-aspartate, D-alanine, D-glutamate, and D-asparagine, and it increased the concentrations of D-aspartate, D-alanine, and D-glutamate by 6.0, 69.8, and 2.6 times, respectively. It thus appears that most of the D-amino acids in lactic-fermented tomato vinegar are produced during lactic fermentation.

To further confirm the apparent contribution made by lactic acid bacteria to D-amino acid production in lactic-fermented tomato vinegar, the levels of D-amino acids in culture medium conditioned by each of eight strains of lactic acid bacteria were assessed (Table 22.3). In each case, the total D-amino acid concentration in the conditioned medium was higher than in medium without growth of lactic acid bacteria. In particular, there were marked increases in the D-alanine, D-glutamate, and D-aspartate concentrations. For example, the relative D-alanine percentages

Table 22.3 Concentrations (μM) of D-amino acids in culture medium conditioned by lactic acid bacteria

Amino acid[a]	No bacterium	*L. Lactis* subsp. *lactis*	*L. reuteri*	*L. gasseri*	*L. brevis*	*L. salivarius*	*L. otakiensis*	*L. kisoensis*	*L. rapi*
D-Asp	32.1	74.8	150.2	225.0	316.7	118.3	128.7	120.4	110.1
D-Ser	7.7	13.3	13.8	34.9	20.4	13.6	14.1	24.0	12.4
D-Arg	–[b]	–	–	–	–	35.9	—	—	—
D-Ala	33.3	402.5	286.0	1410.6	934.4	537.8	336.0	576.2	309.6
D-Val	–	–	–	–	–	–	42.9	–	–
D-Phe	N.D.[c]	6.3	N.D.	5.7	5.5	N.D.	10.7	N.D.	N.D.
D-*allo*-Ile	–	–	41.6	–	–	–	141.3	–	–
D-Leu	–	–	31.4	–	N.D.	N.D.	194.0	N.D.	N.D.
D-Glu	13.3	22.9	246.3	101.9	88.0	118.7	54.7	161.0	65.5
D-Asn	13.9	12.6	11.5	16.4	14.2	11.9	10.9	14.0	4.9
Total	100.3	532.4	780.8	1794.5	1379.2	836.2	933.3	895.6	502.5

[a]No other measured D-amino acids (D-glutamine, D-histidine, D-threonine, D-tyrosine and D-tryptophan) were detected
[b]Not detected
[c]N.D., not determined

(100 × [D-alanine]/[D-alanine + L-alanine]) within the cells of *L. lactis* subsp. *lactis* and *L. salivarius* were extremely high, 92.3 % and 89.6 %, respectively. In addition, increases in the concentrations of several other D-amino acids were also detected: D-serine in medium from all of the eight strains; D-arginine in media from *Lactobacillus salivarius*; D-valine in media from *Lactobacillus otakiensis*; D-phenylalanine in medium from *Lactococcus lactis* subsp. *lactis*, *Lactobacillus gasseri*, *Lactobacillus brevis*, and *L. otakiensis*; D-*allo*-isoleucine in medium from *Lactobacillus reuteri* and *L. otakiensis*; and D-leucine in medium from *L. reuteri* and *L. otakiensis*. These findings are all consistent with the idea that the high total D-amino acid concentrations in lactic-fermented tomato vinegar are mainly produced during lactic fermentation.

Through analysis of the differences in the D-amino acid contents of several kinds of vinegar, reflecting differences in both the sources and fermentation processes, it became clear that it is lactic fermentation that is mainly responsible for the production of many of the D-amino acids present in vinegars. Similar increases in total D-amino acid levels in fermented foods have been seen in beer, wine, and sake following inoculation and proliferation of lactic acid bacteria (Erbe and Brückner 2000; Kato et al. 2011; Gogami et al. 2011). Although it is known that yeast and acetic acid bacteria also produce D-amino acids during fermentation, comparison of D-amino acid production during alcoholic, acetic, and lactic fermentation using the same lots of source material confirmed the abundant contribution made by lactic fermentation to the production of D-amino acids.

In addition, the Oikawa group found that Japanese sake contains a number of D-amino acids, including D-alanine, D-asparagine, D-aspartic acid, D-arginine, D-glutamic acid, D-histidine, D-isoleucine, D-lysine, D-serine, D-tyrosine, D-phenylalanine, and D-proline, and sake produced using the kimoto method contains higher amounts of D-amino acids than sake produced using the sokujo-moto method (Oikawa 2014). The most important difference between the kimoto and sokujo-moto brewing processes is the origin of the lactic acid used in their pasteurization: lactic acid produced by lactic acid bacteria is used for pasteurization of raw materials in kimoto brewing, while commercially available food-grade lactic acid is used as an additive for sokujo-moto brewing. Accordingly, the lactic acid bacteria likely produce the D-amino acids in kimoto sake, and Oikawa et al. showed for the first time that the two lactic acid bacteria species isolated from kimoto sake, *Lactobacillus sakei* NBRC 15893 and *Leuconostoc mesenteroides* subsp. *sake* NBRC 102480, each produce several extracellular D-amino acids: *L. sakei* NBRC 15893 produces D-alanine, D-aspartic acid, and D-glutamate, while *L. mesenteroides* subsp. *sake* NBRC 102480 produces D-alanine, D-glutamate, and D-lysine.

It is well known that lactic acid bacteria strains are involved in the production of many fermented foods, including cheese, yogurt, soy sauce, pickles, and sauerkraut and, as a consequence, D-amino acids are present in many of these foods. Furthermore, many kinds of lactic acid bacteria are also deeply involved in human health as indigenous bacteria. Beyond our expectations, therefore, D-amino acids produced by lactic acid bacteria may not only be responsible for improving the taste and smell of fermented foods but for maintaining the health of the people who eat those foods.

For those reasons, elucidation of relationships between the different properties of foods and their D-amino acid contents is anticipated.

22.3 D-Amino Acid Metabolism and the Related Enzymes in Fermentation Microorganisms, Including Lactic Acid Bacteria

Understanding the mechanism(s) responsible for the high production of D-amino acids by lactic acid bacteria and other fermentative bacteria is an important objective that will shed light on the function of bacterial D-amino acids. In particular, information about D-amino acid metabolism and the mechanisms by which the related enzymes are regulated is useful for elucidating the properties of D-amino acids in foods, especially fermented foods. The main microbial enzymes functioning in D-amino acid metabolism are dehydrogenases, oxidases, aminotransferases, acetyltransferases, amidases, aldolases, dehydratase, racemases, aminomutase, and synthetases. In this section, recent research into D-amino acid metabolism and the related enzymes in fermentative microorganisms, including lactic acid bacteria, is described, with emphasis on analysis based on genome information and molecular biology.

22.3.1 Amino Acid Racemase

As mentioned above, lactic acid bacteria produce extracellular D-alanine, D-aspartate, and D-glutamate. This is a major reason why foods fermented using lactic acid bacteria contain large amounts of D-amino acids. Among the enzymes mentioned above, amino acid racemases (EC 5.1.1.-), which catalyze the conversion of L- or D-amino acids to their respective enantiomers, giving rise to a racemic mixture, appear pivotal for the production of D-amino acids. However, there was only fragmented information available on the racemases expressed in lactic acid bacteria before author's study. As a first step toward identifying the racemases in lactic acid bacteria, they screened for racemase genes using genomic information from lactic acid bacteria, and several genes encoding putative amino acid racemases were found.

As mentioned above, Oikawa's group found that in sake produced using the kimoto method, *L. sakei* NBRC 15893 produces D-alanine, D-aspartic acid, and D-glutamate, while *L. mesenteroides* subsp. *sake* NBRC 102480 produces D-alanine, D-glutamate, and D-lysine (Oikawa 2014). Bioinformatic analysis of the two strains revealed the presence of four putative alanine, aspartate, and glutamate racemase genes in the *L. sakei* NBRC 15893 genome and three putative alanine and glutamate racemase genes in the *L. mesenteroides* subsp. *sake* NBRC 102480 genome. These seven genes were expressed in *Escherichia coli*, and the enzymological properties of their products were studied in detail and are summarized in Table 22.4. The products of the putative alanine, aspartate, and glutamate racemase genes in *L. sakei* NBRC 15893 were identified as alanine, aspartate, and glutamate racemases, respectively. The enzymological properties of the enzymes are similar to those of other known bacterial racemases, with the exception of aspartate racemase, which had a lower optimal pH. Interestingly, the products of the three putative alanine and glutamate racemase genes in *L. mesenteroides* subsp. *sake* NBRC 102480 were identified as alanine, lysine, histidine, and glutamate racemases. To our knowledge, the histidine racemase from *L. mesenteroides* subsp. *sake* NBRC 102480 is the first example of an enzyme that specifically catalyzes the racemization of histidine. The substrate specificities of the amino acid racemases expressed by *L. sakei* NBRC 15893 and *L. mesenteroides* subsp. *sake* NBRC 102480 were consistent with the D-amino acids detected in sake, suggesting it is these enzymes that catalyze the production of D-amino acids in sake during the kimoto brewing process.

On the other hand, Yoshimura's group detected D-lysine in addition to D-alanine and D-glutamic acid in samples collected during the making of white wine (Kato et al. 2011). Genetic analysis of the microflora in the samples revealed that these D-amino acids are produced by another lactic acid bacterium, *Oenococcus oeni*, during the malolactic acid fermentation period. Mills et al. performed a genomic analysis of *O. oeni* PS-1 in 2005 (Mills et al. 2005) and found that the *O. oeni* PS-1 genome encodes two putative alanine racemase homolog genes. Based on genomic information from *O. oeni* PS-1, the two genes were amplified using polymerase chain reaction and a metagenomic DNA mixture prepared from a sample containing a high concentration of D-lysine and were then cloned into *E. coli* (Kato et al. 2012). The enzymological properties of the two gene products showed that one gene encodes an alanine racemase, while the other encodes a lysine racemase. The latter gene in *O. oeni* is composed of 1121 bp, encoding 371 amino acid residues. The primary structure of the enzyme (Oo-LysR) is similar to that of the alanine racemase from *Geobacillus stearothermophilus* (similarity: approximately 60 %). This enzyme (a homodimer with a molecular mass of 41 kDa) is stable at up to 50 °C (60 min incubation), and the optimum pH for reaction is pH 9.0. Oo-LysR showed the greatest activity toward lysine, but also showed activity toward arginine and ornithine residues. No other amino acids, including alanine, aspartic acid, and glutamic acid, could serve as substrates. Accordingly, this enzyme was named lysine racemase, as distinct from the arginine racemase in *Pseudomonas taetrolens* NBRC 3460 (Matsui et al. 2009). The crystal structure of Oo-LysR was previously solved as an alanine racemase (Palani et al. 2013).

Like other amino acid racemase genes, the proline racemase, histidine racemase, and other racemase genes have already been annotated in the genome database of lactic acid bacteria. The growing genomic information on amino acid racemases is expected to shed light on the production mechanisms and functions of D-amino acids in lactic acid bacteria. On the other hand, identification and characterization of racemases on the basis of their enzyme activity within cells are important for

Table 22.4 Basic enzymological properties of the products of seven amino acid racemase homolog genes from *Lactobacillus sakei* NRRC 15893 and *Leuconostoc mesenteroides* subsp. *sake* NBRC 15893

Origin	*Lactobacillus sakei* NRRC 15893			*Leuconostoc mesenteroides* subsp. *sake* NBRC 15893			
Enzyme name	Alanine racemase	Aspartate racemase	Glutamate racemase	Alanine racemase	Glutamate racemase	Lysine racemase	Histidine racemase
Substrate specificity	Ala, Ser	Asp	Glu	Ala, Ser	Glu	Lys, Arg, Orn	His
Molecular mass (kDa)[a]	91	52	35	82	42	89	89
Subunit molecular mass (kDa)[b]	42	26	30	42	30	42	42
Quaternary structure	α_2	α_2	α	α_2	α	α_2	α_2
Cofactor	PLP	none	none	PLP	none	PLP	PLP
Optimum temperature (°C)	35	45	37	15	25	25	50
Thermal stability (°C)[c]	45	40	30	40	55	35	35
Optimum pH	10.5	6.5	10.5	6.5	7.5	8.0	9.5
pH stability[d]	10.0	7.0–7.5	10.0	7.5	7.5	7.5–8.0	7.0

[a]Determined using gel filtration
[b]Determined using SDS-polyacrylamide gel electrophoresis
[c]For 1 h
[d]0 °C for 1 h

investigation of the production mechanisms and special functions of bacterial D-amino acids.

22.3.2 Finding a Novel Racemase, Isoleucine 2-Epimerase in Lactic Acid Bacteria

Alanine, glutamate, aspartate, and lysine racemases have all been identified as D-amino acid-metabolizing enzymes and characterized as described above. In addition, it is now known that many bacteria produce free D-branched amino acids (D-BCAAs) such as D-isoleucine, D-leucine, and D-valine (Lam et al. 2009), though the mechanism of their production and their physiological functions in bacteria remain unknown.

When authors analyzed D-amino acid production by *Lactobacillus* strains, they found that the growth of *Lactobacillus otakiensis* JCM 15040 cultures is accompanied by secretion of D-leucine, D-*allo*-isoleucine, and D-valine into the culture medium (Fig. 22.1) (Mutaguchi et al. 2013a, b). In addition, cell extract from this bacterium contained epimerase and racemase activities that may be responsible for production of these D-BCAAs. This epimerase catalyzed epimerization at the C-2 position between L-isoleucine and D-*allo*-isoleucine, and the specific activity of the isoleucine 2-epimerase was higher than the specific activities of the leucine and valine racemases in cell-free extract. The protein exhibiting the isoleucine 2-epimerase activity was purified from the cell extract of *L. otakiensis* JCM 15040, and the N-terminal amino acid sequence of the purified enzyme was determined. A homology search based on the N-terminal amino acid sequence revealed it to be completely identical to those deduced from the putative γ-aminobutyrate aminotransferase (GABA-AT) gene of *Lactobacillus buchneri* NRRL B-30929. *L. buchneri* JCM 1115 is a similar strain to *L. buchneri* NRRL B-30929 and exhibits a pattern of time-dependent accumulation of D-leucine, D-*allo*-isoleucine, and D-valine in the culture medium that is similar to the pattern observed with *L. otakiensis* JCM 15040. Then, after the putative GABA-AT gene from *L. buchneri* JCM 1115 was expressed in *E. coli*, the purified recombinant enzyme exhibited isoleucine 2-epimerase activity, but not GABA-AT activity.

Detailed characterization of the isoleucine 2-epimerase from *L. buchneri* JCM 1115 revealed that the enzyme preferentially catalyzes epimerization between L-isoleucine and D-*allo*-isoleucine, but also catalyzes racemization of nonpolar amino acids, including leucine and valine (Table 22.5). The arginine racemase (EC 5.1.1.9) from *Pseudomonas graveolens* and amino acid racemase from *Pseudomonas putida* (EC 5.1.1.10) have been identified as amino acid racemases with broad substrate specificity (Yorifuji et al. 1971; Lim et al. 1993). However, these two racemases show much less activity toward leucine than the main substrate lysine and none toward isoleucine or valine. In addition, alanine racemase from *P. putida* (Liu et al. 2012) acts on L-isoleucine, but that enzyme shows much less

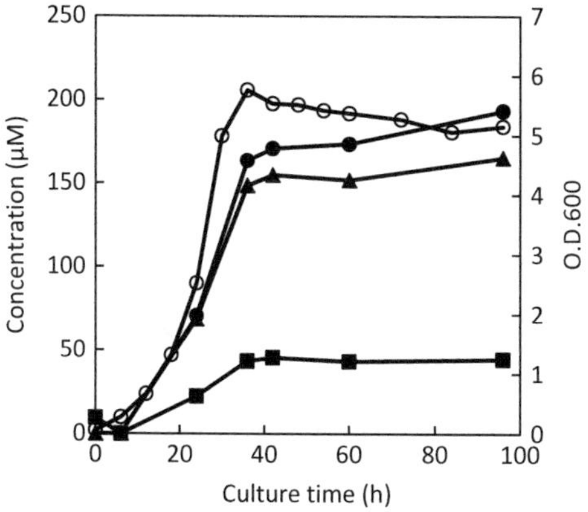

Fig. 22.1 Time course of D-BCAA accumulation in culture medium from *L. otakiensis* JCM 15040. *Solid circles*, *solid triangles*, and *solid squares* show the concentrations of D-leucine, D-*allo*-isoleucine, and D-valine, respectively. The cell concentration is shown as the O.D.600 (*open circles*)

Table 22.5 Substrate specificity of the racemase and epimerase activities of isoleucine 2-epimerase from *L. buchneri* JCM 1115

L→D			D→L		
L-Form substrate	Specific activity (μmol mg^{-1} min^{-1})	Relative activity (%)	D-Form substrate	Specific activity (μmol mg^{-1} min^{-1})	Relative activity (%)
Isoleucine	149	100	*allo*-Isoleucine	221	100
Norvaline	83.4	56	Norvaline	183	83
Norleucine	74.1	50	Valine	131	59
Valine	71.7	48	Norleucine	116	52
2-Aminobutanoic acid	47.7	32	Methionine	93.7	42
Leucine	44.2	30	2-Aminobutanoic acid	66.3	30
Phenylalanine	35.7	24	Leucine	54.8	25
Methionine	32.0	21	Isoleucine	44.5	20
allo-Isoleucine	27.8	19	Phenylalanine	13.6	6
Serine	9.58	6	Serine	4.91	2
Alanine	3.95	3	Alanine	4.43	2

activity toward L-isoleucine than L-alanine (relative activity: 38.7%). These findings suggest that isoleucine 2-epimerase from *L. buchneri* is the first reported enzyme catalyzing racemization of BCAAs as its main substrate in all organisms. In many racemase reactions, the k_{cat}/K_m for the L-amino acid equals that for the

D-amino acid, because the racemization $K_{eq} = 1$ in the Haldane equation. However, in this isoleucine 2-epimerase reaction, the value of K_{eq} with L-isoleucine as a substrate is 1.31, and the value of [k_{cat}/K_m for L-isoleucine]/[k_{cat}/K_m for D-*allo*-isoleucine] is 1.41. That this value is >1 probably reflects the fact that the reaction is an epimerization, not a racemization. Both optimum pHs for epimerization by this enzyme are fairly acidic (pH 5.0, 6.0), while other amino acid racemases have pH optima around 8.0. Another difference between this enzyme and other racemases is its homotetrameric structure (four identical subunits of 49 kDa); most reported amino acid racemases exist as monomers or dimers. Furthermore, the isoleucine 2-epimerase is completely inhibited by hydroxylamine and other inhibitors of PLP, indicating that it requires PLP as a coenzyme. Racemases are generally divided into two groups, PLP dependent and PLP independent (Yoshimura and Esaki 2003), and the isoleucine 2-epimerase falls in the PLP-dependent group. All PLP-dependent enzymes whose structures have been solved so far are classified as one of five fold types (I–V) based on similarities in their primary and secondary structures (Eliot and Kirsch 2004). However, there is no correlation between fold type and reaction specificity. The PLP-dependent racemases identified so far have been classified as fold type I, II, or III, and the bacterial PLP-dependent racemases are included in fold types I and III. An amino acid sequence homology search of the Protein Data Bank for *L. buchneri* isoleucine 2-epimerase using BLAST showed that the enzyme is classified into the fold type I group of PLP-dependent enzymes. The first bacterial racemase classified into the fold type I group was α-amino-ε-caprolactam racemase from *Achromobacter obae* (Okazaki et al. 2009). The isoleucine 2-epimerase from *L. buchneri* is the second example of a bacterial racemase belonging to the fold type I group. Collectively, the characteristics summarized above indicate that the isoleucine 2-epimerase from *L. buchneri* is a novel type of amino acid racemase. It is expected that analysis of the physiological functions of this enzyme will reveal special functions of D-amino acids, including D-BCAAs, in lactic acid bacteria and other strains.

22.3.3 D-Amino Acid Transaminase

As mentioned above, racemases play pivotal roles in D-amino acid production from corresponding enantiomers. However, to further investigate D-amino acid metabolism in lactic acid bacteria, more information about D-amino acid-related enzymes other than racemases is necessary. One such D-amino acid-metabolizing enzyme is D-amino acid transaminase (D-AAT) that catalyzes transamination between D-amino acids and 2-oxo acids (Yonaha et al. 1975; Tanizawa et al. 1989; Lee et al. 2006). In the genome database for lactic acid bacteria, authors found a putative D-AAT gene in *Lactobacillus salivarius*. After expressing the *L. salivarius* D-AAT gene in *E. coli*, the product was purified as a His-tagged form using metal affinity chromatography. The purified *L. salivarius* D-AAT was PLP dependent and contained 0.91 molecules of PLP per subunit molecule and was a homodimer with a molecular mass of 31.5 kDa. Maximum activity was observed at a temperature of 60 °C, which is similar to enzymes from the *Bacillus* and *Geobacillus* species [25–27]. On the other hand, *L. salivarius* D-AAT retained less than 20 % residual activity after incubation at 60 °C. It is thus a little less stable than the enzymes from *Bacillus* and *Geobacillus* species, which do not lose activity below 60 °C. The *L. salivarius* D-AAT exhibited maximum activity at pH 6.0. There was no loss of activity after incubation for 30 min at pHs between 5.5 and 9.5 (30 °C), and it retained half its activity at pH 4.5 and 11.0 under the same conditions. These characteristics differ from those of *Bacillus* and *Geobacillus* D-AATs, which show maximum activities around pH 8.0 and are less stable under acidic conditions.

L. salivarius D-AAT has broad amino donor specificity, with high activity toward several D-amino acids, including D-*allo*-isoleucine, D-α-aminobutyrate, and D-methionine, in addition to D-alanine (Table 22.6). D-*allo*-Isoleucine was the most preferable amino donor, but interestingly D-isoleucine was totally inert. In

Table 22.6 Amino donor specificity of D-AATs

	Relative activity (%)			
Substrate	*L. salivarius*	*Bacillus* sp. YM[c]	*B. sphaericus*[d]	*G. toebii* SK1[e]
D-Alanine	100	100	100	100
D-*allo*-Isoleucine	104 ± 15	–[b]	0	–
D-α-Amino butyrate	90 ± 2	98	97	89
D-Methionine	89 ± 3	19	61	0.7
D-Leucine	84 ± 4	2.4	8	–
D-Norvaline	83 ± 9	34	83	8.2
D-Valine	83 ± 2	3.2	0	1.9
D-Histidine	75 ± 6	1.9	6	1.9
D-Norleucine	71 ± 2	5	46	–
D-Pyroglutamate	41 ± 2	–	–	–
D-Threonine	13 ± 1	0	0	0.9
D-Arginine	5 ± 1	1	6	-
D-Serine	N.D.[a]	60	30	70
D-Cysteine	N.D.	25	31	11
D-Proline	N.D.	1.2	0	12
D-Asparagine	N.D.	0.9	32	–
D-Isoleucine	N.D.	0.2	0	–
D-*allo*-Threonine	N.D.	0.1	0	–
D-Phenylalanine	N.D.	0	0	–
β-Alanine	N.D.	0	0	–
D-Aspartate	N.D.	0	0	–
Glycine	N.D.	–	–	–

[a]N.D., not detected
[b]No data
[c]Reference (Tanizawa et al. 1989)
[d]Reference (Yonaha et al. 1975)
[e]Reference (Lee et al. 2006)

addition, very broad amino acceptor specificity for 2-oxo acids was observed. Alpha-oxobutyrate, glyoxylate, indole-3-pyruvate, 2-oxovalerate, 3-methyl-2-oxobutyrate, and 4-hydroxyphenylpyruvate all showed greater activity than 2-oxoglutarate or pyruvate as amino acceptors, and 2-oxobutyrate exhibited much greater activity than other 2-oxo acids. This means that the amino donor specificity of *L. salivarius* D-AAT differs substantially from those of other D-AATs. In the case of D-AATs from *Bacillus* sp., BCAAs are known to be inert or to have extremely low activity as amino donors (Yonaha et al. 1975; Tanizawa et al. 1989; Lee et al. 2006). In addition, *L. salivarius* D-AAT showed much higher activities toward 2-oxobutyrate and glyoxylate than 2-oxoglutarate as an amino acceptor. This differs markedly from observations made with the *Bacillus* sp. strain YM1, *B. sphaericus*, and *G. toebii* SK1 enzymes. It thus appears that the *L. salivarius* D-AAT is a novel enzyme and may have a specific function in the D-amino acid metabolism of the bacterium. However, it is still unclear whether this unique enzyme, along with various racemases, is responsible for the high production of many D-amino acids. To elucidate the functions of the D-amino acids produced by lactic acid bacteria, further enzymatic and metabolic studies will be needed.

22.3.4 Other D-Amino Acid-Related Enzymes (Dehydrogenases and Oxidases)

In addition to racemases and transaminases, oxidoreductases, including both dehydrogenases and oxidases, are also considered to be D-amino acid-related enzymes. There has been no reported direct evidence of D-amino acid dehydrogenases or oxidases in microorganisms related to fermented foods and any other foods. In general, D-amino acid dehydrogenases and oxidases are flavin (FAD or FMN)-dependent enzymes and catalyze irreversible oxidative deamination of D-amino acids to their oxo analogs. The electron of $FADH_2$, a product of an enzyme reaction, is transferred to the electron transport system to yield ATP. These enzymes are thus known to function in the FAD-dependent degradation of D-amino acids to produce cell energy. For a long time, it was thought that D-amino acid oxidase did not exist in prokaryotic organisms. However, the enzyme was recently identified in *Arthrobacter protophormiae*, *Streptomyces coelicolor*, and *Rubrobacter xylanophilus* (Takahashi et al. 2014). In addition to these bacterial species, a search of the prokaryotic genome database revealed a wide distribution of proteins homologous to D-amino acid oxidase in bacteria, particularly in *Actinobacteria*. Although actinobacterial strains are soil bacteria and not directly related to food, their D-amino acid oxidases (EC 1.4. -) may have some function in D-amino acid degradation in food. FAD-dependent D-amino acid dehydrogenases (EC 1.4.99.-) have also been identified in *E. coli* and other bacteria, as well as archaea, and a few microbial enzymes have been characterized so far (Satomura et al. 2002, 2015).

This dehydrogenase may play a role in D-amino acid degradation similar to D-amino acid oxidase. Finally, D-amino acid dehydratases such as D-serine dehydratase (EC 4.3.1.18) and D-threonine aldolase are present in some bacteria. The functions of those enzymes and their role in D-amino acid metabolism are as yet unknown.

22.4 Physiological Function and Application of D-Amino Acids in Functional Food Production

As described in other chapters in this book, studies of the physiological functions of D-amino acids have mainly focused on D-aspartate and D-serine in mammalian cells, and the physiological functions of D-amino acids in food have remained unclear up to now. In 2013, Oikawa's group performed a sensory evaluation of 141 bottles of Japanese sake and used principal component analysis to assess the relationship between the D-amino acid concentration and sake taste (Oikawa 2014). They found that D-alanine, D-aspartate, and D-glutamate in sake increase the taste and the overall balance of sake tastes. Interestingly, no other D-amino acids, including D-serine, D-valine, D-asparagine, D-glutamine, D-lysine, D-tyrosine, D-histidine, D-phenylalanine, and D-proline, had any effect on the taste or the overall balance of sake tastes. This is the first example clearly showing a function of D-amino acids in food. In general, kimoto sake, which is produced using a traditional method, is known to have a better taste and to be of higher quality than other Japanese sakes, and it is thought that the higher levels of D-alanine, D-aspartate, and D-glutamate in the sake produced using the kimoto method may be responsible for the improvement in taste.

Recently, much attention has focused on producing D-amino acid-enriched functional foods that are effective for improving human health. In 2010, Shiseido Company (Tokyo, Japan) found that various D-amino acids are present in the human stratum corneum and that the concentration of D-aspartate declines as individuals age. D-Aspartate is more effective than L-aspartate at stimulating collagen production and contributes to the activation of fibroblasts in thick (cornified) skin to increase the collagen concentration. The company has demonstrated that after 2 months of D-aspartate ingestion, the concentration of D-aspartate and water in thick skin is clearly higher, and the moisture content of the skin is elevated (http://wol.nikkeibp.co.jp/article/nhpro/20100625/107625/). Based on this finding, a functional food called "kireinosusume" (http://www.shiseido.co.jp/kiresusu/) was produced. To our knowledge, this is first example of a functional food enriched in D-amino acids as an ingredient. In addition, D-alanine reportedly improves production of laminin-5 in the basement membrane of skin and contributes to a restoration of laminin-5 in aged people, who often exhibit a laminin-5 deficiency (http://news.mynavi.jp/news/2013/05/22/260/). Based on this finding, Shiseido intends to sell a cosmetic supplemented with D-alanine as an antiaging ingredient in China (private note). In 2010, Oikawa et al. detected several D-amino acids in "Nigolin," a sake liqueur produced by Kiku-Masamune (Kobe, Japan). Nigolin has an approximately 8 times higher concentration of D-amino acids than Japanese sake and an approximately 44 times higher concentration of D-amino acids than makkoli, a Korean sake produced using a similar brewing process. Particularly enriched is D-aspartate, which is known to contribute to skin health, as described above. Nigolin is now available as "Nigolisparkling" produced by Kiku-Masamune (http://www.kikumasamune.co.jp/products/shochu_liqueur/866_nigorispa_250.html). D-Amino acid-enriched health food is currently under development and will be produced in the near future. It is expected that a variety of D-amino acid-based health foods will be produced and sold worldwide.

References

Abe H, Park JN, Fukumoto Y, Fujita E, Tanaka T, Washio T, Otsuka S, Shimizu T, Watanabe K (1999) Occurrence of D-amino acids in fish sauces and other fermented fish products. Fish Sci 65:637–641

Brückner H, Hausch M (1989) Gas chromatographic detection of D-amino acids as common constituents of fermented foods. Chromatographia 28:487–492

Brückner H, Westhauser T (2003) Chromatographic determination of L- and D-amino acids in plants. Amino Acids 24:43–55

Eliot AC, Kirsch JF (2004) Pyridoxal phosphate enzymes: mechanistic, structural, and evolutionary considerations. Annu Rev Biochem 73:383–415

Erbe T, Brückner H (2000) Chromatographic determination of amino acid enantiomers in beers and raw materials used for their manufacture. J Chromatogr A 881:81–91

Gogami Y, Ito K, Kamitani Y, Matsushima Y, Oikawa T (2009) Occurrence of D-serine in rice and characterization of rice serine racemase. Phytochemistry 70:380–387

Gogami Y, Okada K, Oikawa T (2011) High-performance liquid chromatography analysis of naturally occurring D-amino acids in sake. J Chromatogr B 879:3259–3267

Kato S, Ishihara T, Hemmi H, Kobayashi H, Yoshimura T (2011) Alternations in D-amino acid concentrations and microbial community structures during the fermentation of red and white wines. J Biosci Bioeng 111:104–108

Kato S, Hemmi H, Yoshimura T (2012) Lysine racemase from a lactic acid bacterium, *Oenococcus oeni*: structural basis of substrate specificity. J Biochem 152:505–508

Kawai M, Sekine-Hayakawa Y, Okiyama A, Ninomiya Y (2012) Gustatory sensation of (L)- and (D)-amino acids in humans. Amino Acids 43:2349–2358

Lam H, Oh DC, Cava F, Takacs CN, Clardy J, de Pedro MA, Wandor MK (2009) D-Amino acids govern stationary phase cell wall remodeling in bacteria. Science 325:1552–1555

Lee SG, Hong SP, Song JJ, Kim SJ, Kwak MS, Sung MH (2006) Functional and structural characterization of thermostable D-amino acid aminotransferases from *Geobacillus* spp. Appl Environ Microbiol 72:1588–1594

Lim YH, Yokoigawa K, Esaki N, Soda K (1993) A new amino acid racemase with threonine α-epimerase activity from *Pseudomonas putida*: purification and characterization. J Bacteriol 175:4213–4217

Liu JL, Liu XQ, Shi YW (2012) Expression, purification, and characterization of alanine racemase from *Pseudomonas putida* YZ-26. World J Microbiol Biotechnol 28:267–274

Matsui D, Oikawa T, Arakawa N, Osumi S, Lausberg F, Stäbler N, Freudl R, Eggeling L (2009) A periplasmic, pyridoxal-5′-phosphate-dependent amino acid racemase in *Pseudomonas taetrolens*. Appl Microbiol Biotechnol 83:1045–1054

Mills DA, Rawsthorne H, Parker C, Tamir D, Makarova K (2005) Genomic analysis of *Oenococcus oeni* PSU-1 and its relevance to winemaking. FEMS Microbiol Rev 29:465–475

Mutaguchi Y, Ohmori T, Akano H, Doi K, Ohshima T (2013a) Distribution of D-amino acids in vinegars and involvement of lactic acid bacteria in the production of D-amino acids. Springer Plus 2:691

Mutaguchi Y, Ohmori T, Wakamatsu T, Doi K, Ohshima T (2013b) Identification, purification, and characterization of a novel amino acid racemase, isoleucine 2-epimerase, from *Lactobacillus* species. J Bacteriol 195:5207–5215

Oikawa T (2014) Bioindustry 31:33–40 (in Japanese)

Okazaki S, Suzuki A, Mizushima T, Kawano T, Komeda H, Asano Y, Yamane T (2009) The novel structure of a pyridoxal 5′-phosphate-dependent fold-type I racemase, α-amino-ε-caprolactam racemase from *Achromobacter obae*. Biochemistry 48:941–950

Palani K, Burley SK, Swaminathan S (2013) Structure of alanine racemase from *Oenococcus oeni* with bound pyridoxal 5′-phosphate. Acta Crystallogr Sect F: Struct Biol Cryst Commun 69:15–19

Rubio-Barroso S, Santos-Delgadoa MJ, Martín-Olivara C, Polo-Díeza LM (2006) Indirect chiral HPLC determination and fluorimetric detection of D-amino acids in milk and oyster samples. J Dairy Sci 89:82–89

Satomura T, Kawakami R, Sakuraba H, Ohshima T (2002) Dye-linked D-proline dehydrogenase from hyperthermophilic archaeon *Pyrobaculum islandicum* is a novel FAD-dependent amino acid dehydrogenase. J Biol Chem 277:12861–12867

Satomura T, Ishikura M, Koyanagi T, Sakuraba H, Ohshima T, Suye S (2015) Dye-linked D-amino acid dehydrogenase from the thermophilic bacterium *Rhodothermus marinus* JCM9785: characteristics and role in trans-4-hydroxy-L-proline catabolism. Appl Microbiol Biotechnol 99:4265–4275

Solms J, Vuataz L, Egli RH (1965) The taste of L- and D-amino acids. Experientia 21:692–694

Takahashi S, Furukawara M, Omae K, Tadokoro N, Saito Y, Abe K, Kera Y (2014) A highly stable D-amino acid oxidase of the thermophilic bacterium *Rubrobacter xylanophilus*. Appl Environ Microbiol 80:7219–7229

Tanizawa K, Asano S, Masu Y, Kuramitsu S, Kagamiyama H, Tanaka H, Soda K (1989) The primary structure of thermostable D-amino acid aminotransferase from a thermophilic *Bacillus* species and its correlation with L-amino acid aminotransferases. J Biol Chem 264:2450–2454

Yonaha K, Misono H, Yamamoto T, Soda K (1975) D-Amino acid aminotransferase of *Bacillus sphaericus*. Enzymologic and spectrometric properties. J Biol Chem 250:6983–6989

Yorifuji T, Ogata K, Soda K (1971) Arginine racemase of *Pseudomonas graveolens*. I. Purification, crystallization, and properties. J Biol Chem 246:5085–5092

Yoshimura T, Esaki N (2003) Amino acid racemases: functions and mechanisms. J Biosci Bioeng 96:103–109

MIX
Papier aus verantwortungsvollen Quellen
Paper from responsible sources
FSC® C105338

If you have any concerns about our products,
you can contact us on
ProductSafety@springernature.com

In case Publisher is established outside the EU,
the EU authorized representative is:
Springer Nature Customer Service Center GmbH
Europaplatz 3, 69115 Heidelberg, Germany

Printed by Libri Plureos GmbH
in Hamburg, Germany